# AMERICAN MATHEMATICAL SOCIETY TRANSLATIONS

## Series 2

## Volume 80

### THIRTEEN PAPERS ON FUNCTIONS OF REAL AND COMPLEX VARIABLES

by


V. S. Azarin
N. N. Čaus
Ju. F. Korobeĭnik
B. M. Levitan
M. S. Mel′nikov
S. M. Nikol′skiĭ
G. H. Sindalovskiĭ
G. P. Tolstov
A. G. Vituškin
V. A. Zmorovič



Published by the
AMERICAN MATHEMATICAL SOCIETY
Providence, Rhode Island
1969

Standard Book Number 821-81780-1

Library of Congress Catalog Number A51-5559

Printed in the United States of America

## TABLE OF CONTENTS

# INEQUALITIES FOR ENTIRE FUNCTIONS OF FINITE DEGREE AND THEIR APPLICATION TO THE THEORY OF DIFFERENTIABLE FUNCTIONS OF SEVERAL VARIABLES

S. M. NIKOL'SKIĬ

In the first and second sections of the article, we examine entire functions $g = g_{\nu_1 \cdots \nu_n}(z_1, \cdots, z_n)$ of several variables $z_1, \cdots, z_n$ of finite degree $\nu_1, \cdots, \nu_n$ with respect to each corresponding variable. 1) Given any $\epsilon > 0$, any such function satisfies the inequality

$$|g_{\nu_1 \ldots \nu_n}| < Ae^{\sum_1^n (\nu_k + \varepsilon) | z_k |}, \qquad A = A(\varepsilon).$$

We shall limit ourselves to examining only functions $g$ which are $p$th power integrable over the whole space $R_n$ of the real variables $x_1, \cdots, x_n$. For them the norm

$$\|g\|_p = \|g_{\nu_1 \ldots \nu_n}\|_p = \Big(\int_{-\infty}^{\infty} \cdots \int_{-\infty}^{\infty} |g_{\nu_1 \ldots \nu_n}(x_1, \ldots, x_n)|^p \, dx_1 \ldots dx_n\Big)^{\frac{1}{p}} < \infty$$

is finite. Here $p$ can be any number satisfying the inequality $1 \leq p \leq \infty$. The case $p = \infty$, which expresses that $g$ is bounded on $R_n$, also enters into consideration as a special case.

If $p$ is finite and $\|g\|_p$ is finite, then $g \to 0$ when all the $z_i \to \infty$; but then, obviously, the norm $\|g\|_p$ is also finite for $p' > p$. In §1 this property is given a quantitative characterization in the form of the inequality

$$\|g_{\nu_1 \ldots \nu_n}\|_{p'} \leqslant 2^n \Big(\prod_1^n \nu_i\Big)^{\frac{1}{p} - \frac{1}{p'}} \|g_{\nu_1 \ldots \nu_n}\|_p \, (1 \leqslant p \leqslant \infty),$$

which is valid for all $g$ with finite norm $\|g\|_p$. In a certain sense it is inverse to Hölder's inequality.

The other inequality (1.12) obtained in §1 makes it possible to estimate the norm $\|g\|_p^{(m)}$ of a function $g$ of finite degree as defined only on the space $R_m$ using the norm $\|g\|_p^{(n)}$, which is defined on the whole space $R_n$ $(m < n)$.

---

1) They are also called functions of exponential type, but we shall adhere to S. N. Bernšteĭn's terminology.

Both of these inequalities serve as an essential solving apparatus in the investigations carried out in §4. They, in turn, are obtained using other inequalities which tie together integrals (norms) of the function $g$ with their corresponding integral sums. My inequalities here are closely related to inequalities of a similar type obtained by M. Plancherel and G. Pólya [1] [9, 10]. To obtain them these authors used the Fourier series representation of the function $g$ as their point of departure. The tool I used was an inequality of S. N. Bernšteĭn, which I had to generalize in a suitable way. Here for the subsequent applications it was very important for me to attain precision of the constants in these inequalities relative to the $\nu_i$ and $p (1 \leq p \leq \infty)$, at least in the sense of order.

In §3 a theory of best approximations is constructed for differentiable functions of several variables defined in the infinite space $L_p^{(n)}$ $(1 \leq p \leq \infty;$ $n$ is the dimension of the space). The main problem consisted in successfully defining classes of differentiable functions so that the usual theory of best approximations turns out to be closed for them. In the remainder I followed procedures already familiar in the theory of approximations, altering them in line with my purpose.

To prove converse theorems, I applied S. N. Bernšteĭn's method, which is, incidentally, still the only method known for proving propositions of this type. I also used procedures employed by Bernšteĭn in Theorems 6 [3] and 7 [2]. We should keep in mind that Bernšteĭn first introduced into the field the very notion of the best approximation of a function defined in an infinite space by using functions of finite degree, and he created on its base a theory of approximations on the infinite axis or in space (especially in the metric of functions bounded in space; the case $p = \infty$ was fully examined by him).

In §4 we can see how all these special inequalities work for general function theory. Theorem 12 generalizes to the case of infinite space well-known propositions of S. L. Sobolev [6] and V. I. Kondrašov [4] (proved by these authors in a completely different way) about the imbedding and compactness of classes of differentiable functions of several variables.

The generalization proceeds in the following directions:

1) The method used here makes it possible to investigate functions having differential properties not only of the same, but also of different orders in each of its variables.

2) Differential properties of functions are defined not only if they have

---

1) Inequalities of this type for trigonometric polynomials in one variable were obtained still earlier by J. Marcinkiewicz and A. Zygmund [8].

partial derivatives of specified orders $r$, but also if they satisfy a Lipschitz condition or a generalized Lipschitz condition of order $\alpha$, where $0 < \alpha \leq 1$. This allows us to examine fractional, as well as integral, values of the parameters $r$.

3) Finally, in an example in the fundamental case it is shown that these propositions, in some sense, bear a final, completed character.

Inequalities for (generally multiple) integrals are written throughout the article without any explanation. They are all based on Hölder's inequalities and on the more general

$$\left(\int \left| \int K(u, v)\, du \right|^p dv\right)^{1/p} \leqslant \int \left(\int \left| K(u, v) \right|^p dv\right)^{1/p} du.$$

In particular, this inequality will very often be used in the following circumstances:

$$\left(\int_{-\infty}^{\infty} \left| \int_{-\infty}^{\infty} K(t-x) f(t)\, dt \right|^p dx\right)^{1/p} = \left(\int_{-\infty}^{\infty} \left| \int_{-\infty}^{\infty} K(t) f(t+x)\, dt \right|^p dx\right)^{1/p}$$

$$\leqslant \int_{-\infty}^{\infty} |K(t)| \left(\int_{-\infty}^{\infty} |f(t+x)|^p dx\right)^{1/p} dt = \int_{-\infty}^{\infty} |K(t)|\, dt \left(\int_{-\infty}^{\infty} |f(t)|^p dt\right)^{1/p},$$

where $K \in L$ and $f \in L_p$.

## §1. Inequalities for entire functions of finite degree

Let $R_n$ be the $n$-dimensional Euclidean space of points $(x_1, \cdots, x_n)$ with real coordinates. We let $L_p^{(n)}$ designate the space of functions $f = f(x_1, \cdots, x_n)$ defined and measurable in $R_n$ and (if $1 \leq p < \infty$) $p$th power integrable over $R_n$. Hence for every function $f \in L_p^{(n)}$ we have

$$\|f\|_p = \left(\int_{-\infty}^{\infty} \cdots \int_{-\infty}^{\infty} |f(x_1, \ldots, x_n)|^p dx_1 \ldots dx_n\right)^{1/p} < \infty.$$

As usual, it is understood that a function $f \in L_\infty^{(n)}$ is essentially bounded with finite essential maximum

$$\|f\|_\infty = \sup \text{vrai} \, |f(x_1, \cdots, x_n)| < \infty.$$

We shall also consider entire functions

$$g_{\nu_1 \ldots \nu_n} = g_{\nu_1 \ldots \nu_n}(z_1, \cdots, z_n)$$

of $n$ complex variables $z_1, \cdots, z_n$ which belong to $L_p^{(n)}$ and are of exponential type with degree $\nu_1, \cdots, \nu_n$ with respect to these variables.

Following S. N. Bernšteĭn, we shall call these functions simply functions of degree $\nu_1, \cdots, \nu_n$. Any such function $g_{\nu_1 \cdots \nu_n}(z_1, \cdots, z_n)$ has the following properties:

($\alpha$) It expands into a power series

$$g_{\nu_1 \cdots \nu_n}(z_1, \ldots, z_n) = \sum_{i_1=0}^{\infty} \cdots \sum_{i_n=0}^{\infty} a_{i_1 \cdots i_n} z_1^{i_1}, \ldots, z_n^{i_n},$$

which converges absolutely for any complex $z_1, \cdots, z_n$.

($\beta$) For any $\epsilon > 0$ there exists an $A > 0$ such that for all complex $z_1, \cdots, z_n$

$$|g_{\nu_1 \cdots \nu_n}(z_1, \ldots, z_n)| < Ae^{\sum\limits_1^n (\nu_k + \varepsilon)|z_k|}$$

($\gamma$) As a function of the real numbers $x_1, \cdots, x_n$

$$g_{\nu_1 \cdots \nu_n} \in L_p^{(n)}.$$

We have the following lemma, which somewhat broadens Bernšteĭn's well-known inequality:

**Lemma 1.** *If for almost every point* $(x_2, \cdots, x_n)$ *of* $(n-1)$*-dimensional space the function* $\phi(x_1, \cdots, x_n) \in L_p^{(n)}$ *is an entire function of degree* $\nu$ *with respect to* $x_1$, *then the partial derivative* $\partial\phi/\partial x_1$ *belongs to* $L_p^{(n)}$ *and satisfies the inequality*

$$\left\| \frac{\partial \varphi}{\partial x_1} \right\|_p \leqslant \nu \| \varphi \|_p. \tag{1.1}$$

**Proof.** Obviously $\partial\phi/\partial x_1$ exists almost everywhere and is a function measurable in $R_n$. For $p = \infty$, (1.1) is Bernšteĭn's inequality, so it remains for us to prove (1.1) for finite $p$. We note that if the function $g_\nu(z)$ of degree $\nu$ in one variable $z$ belongs to $L_p$, where $1 \le p < \infty$, then it tends to zero as the real number $x$ increases without bound ([9], part II, p. 148). Consequently $g_\nu(x)$ is bounded on the real axis, and we know (see for example [1], 1947 ed., p. 154) that its derivative is represented in the form of a series converging on the real axis

$$g'_\nu(x) = \frac{\nu}{\pi^2} \sum_{-\infty}^{\infty} (-1)^{k+1} \frac{g_\nu\left(\dfrac{\left(k - \frac{1}{2}\right)\pi}{\nu} + x\right)}{\left(k - \frac{1}{2}\right)^2},$$

where

$$\frac{1}{\pi^2}\sum_{-\infty}^{\infty}\frac{1}{\left(k-\frac{1}{2}\right)^2}=1. \tag{1.2}$$

Now let a function $\phi$ be given which satisfies the conditions of the lemma for $1 \leq p < \infty$. Then, for points $(x_2, \cdots, x_n)$ forming a set $\mathfrak{E}$ which differs from the $(n-1)$-dimensional subspace of points $(x_2, \cdots, x_n)$ by a set of measure zero, with respect to $x_1$ the function $\phi$ is of degree $\nu$ and is $p$th power integrable on the real axis. Consequently for these points and any $x_1$ (i.e. almost everywhere in $R_n$) the partial derivative $\partial\phi/\partial x_1$ exists and equals

$$\frac{\partial\varphi}{\partial x_1}=\frac{\nu}{\pi^2}\sum_{-\infty}^{\infty}(-1)^{k+1}\frac{\varphi\left(\frac{\left(k-\frac{1}{2}\right)\pi}{\nu}+x_1, x_2, \ldots, x_n\right)}{\left(k-\frac{1}{2}\right)^2}.$$

Because $\phi \in L_p^{(n)}$ and because of (1.2), the series on the right converges in the $L_p^{(n)}$ sense to some function, and since it converges to $\partial\phi/\partial x_1$ almost everywhere in $R_n$, this function is $\partial\phi/\partial x_1$. We have here the inequality

$$\left\|\frac{\partial\varphi}{\partial x_1}\right\|_p \leqslant \frac{\nu}{\pi^2}\sum_{-\infty}^{\infty}\frac{\|\varphi\|_p}{\left(k-\frac{1}{2}\right)^2} \leqslant \nu\|\varphi\|_p, \tag{1.3}$$

as was to be proved.

In particular, it follows from the lemma that the following inequality holds for an entire function $g_{\nu_1\cdots\nu_n}$ of degree $\nu_1, \cdots, \nu_n$ which belongs to $L_p^{(n)}$:

$$\left\|\frac{\partial g_{\nu_1\cdots\nu_n}}{\partial x_k}\right\|_p \leqslant \nu_k \|g_{\nu_1\cdots\nu_n}\|_p.$$

Hence, setting $z_k = x_k + iy_k$ $(k = 1, \cdots, n)$ and expanding the function $g = g_{\nu_1\cdots\nu_n}$ in powers of $iy_k$ in a neighborhood of the real point $(x_1, \cdots, x_n)$, we obtain

$$g_{\nu_1\cdots\nu_n}(x_1+iy_1, \ldots, x_n+iy_n)$$

$$=\sum_0^{\infty}\cdots\sum_0^{\infty}\frac{1}{k_1!\ldots k_n!}\frac{\partial^{k_1+\ldots+k_n}}{\partial x_1^{k_1}\ldots\partial x_n^{k_n}}(iy_1)^{k_1}\ldots(iy_n)^{k_n}.$$

After integrating over $R_n$, we easily obtain from (1.3) the inequality

$$\|g_{\nu_1 \ldots \nu_n}\|_p \leqslant \sum_0^\infty \ldots \sum_0^\infty \frac{|\nu_1 y_1|^{k_1} \ldots |\nu_n y_n|^{k_n}}{k_1! \ldots k_n!} \|g_{\nu_1 \ldots \nu_n}\|_p$$

$$= e^{\sum_1^n \nu_k |y_k|} \|g_{\nu_1 \ldots \nu_n}\|_p.$$

If we set $p = \infty$ in it, we then have the inequality

$$|g_{\nu_1 \ldots \nu_n}(x_1 + iy_1, \ldots, x_n + iy_n)| \leqslant \|g_{\nu_1 \ldots \nu_n}\|_\infty e^{\sum_1^n \nu_k |y_k|},$$

which is therefore equivalent to inequality ($\beta$) (provided ($\alpha$) and ($\gamma$) are already satisfied).

**Theorem 1.** *Let $g_\nu(z)$ be an entire function of degree $\nu$ belonging to $L_p^{(1)}$ $(1 \le p \le \infty)$, let $\alpha > 0$, $h = \alpha/\nu$, $x_k = kh$ $(k = 0, \pm 1, \cdots)$ and $x_k \le x'_k \le x_k + rh$, where $r$ in an integer. Then*

$$\left(h \sum_{-\infty}^{\infty} |g_\nu(x'_k)|^p\right)^{1/p} \leqslant (1 + r\alpha)\left(\int_{-\infty}^{\infty} |g_\nu(x)|^p \, dx\right)^{1/p}. \tag{1.3}$$

**Proof.** The assertion of the theorem is trivial for $p = \infty$. Let $1 \le p < \infty$. Obviously,

$$\int_{-\infty}^{\infty} |g_\nu(x)|^p \, dx = \sum_{-\infty}^{\infty} \int_{x_k}^{x_{k+1}} |g_\nu(x)|^p \, dx = h \sum_{-\infty}^{\infty} |g_\nu(\xi_k)|^p,$$

where $x_k < \xi_k < x_{k+1}$. Further, using Bernšteĭn's inequality and Hölder's inequality and taking $1/p + 1/q = 1$, we obtain

$$h^{1/p} \left| \left(\sum_{-m}^{m} |g_\nu(x'_k)|^p\right)^{1/p} - \left(\sum_{-m}^{m} |g_\nu(\xi_k)|^p\right)^{1/p} \right|$$

$$\leqslant h^{1/p} \left(\sum_{-m}^{m} |g_\nu(x'_k) - g_\nu(\xi_k)|^p\right)^{1/p} \leqslant h^{1/p} \left[\sum_{-m}^{m} \left(\int_{x_k}^{x_k + rh} |g'_\nu(t)| \, dt\right)^p\right]^{1/p}$$

$$\leqslant h^{1/p} \left[\sum_{-m}^{m} \int_{x_k}^{x_k + rh} |g'_\nu|^p dt \, (rh)^{p/q}\right]^{1/p} = h \left(r^{1 + p/q} \int_{-\infty}^{\infty} |g'_\nu|^p \, dt\right)^{1/p}$$

$$= rh \|g'_\nu\|_p \leqslant r\alpha \|g_\nu\|_p.$$

Therefore

$$h^{1/p}\left(\sum_{-m}^{m}|g_\nu(x_k')|^p\right)^{1/p}=h^{1/p}\left[\left(\sum_{-m}^{m}|g_\nu(x_k')|^p\right)^{1/p}-\left(\sum_{-m}^{m}|g(\xi_k)|^p\right)^{1/p}\right]$$

$$+h^{1/p}\left(\sum_{-m}^{m}|g_\nu(\xi_k)|^p\right)^{1/p}\leqslant(1+r\alpha)\,\|g_\nu\|_p$$

and, setting $m=\infty$, we obtain (1.3).

Theorem 2. *Let $g_{\nu_1\cdots\nu_n}(z_1,\cdots,z_n)$ be an entire function which belongs to $L_p^{(n)}$ of degree $\nu_1,\cdots,\nu_n$ with respect to $z_1,\cdots,z_n$, $\alpha_k>0$, $h_k=\alpha_k/\nu_k$ $(k=0,1,\cdots,n)$, $x_l^{(k)}=lh_k$ $(l=0,\pm1,\pm2,\cdots)$ and $x_l^{(k)}\le \overline{x}_l^{(k)}\le x_l^{(k)}+rh_k$, where $r$ is an integer. Then*

$$\sup_{u_i}\left[\prod_{k-1}^{n}h_k\sum_{-\infty}^{\infty}\cdots\sum_{-\infty}^{\infty}|g_{\nu_1\cdots\nu_n}(\overline{x}_{l_1}^{(1)}-u_1,\cdots,\overline{x}_{l_n}^{(n)}-u_n)|^p\right]^{1/p}$$

$$\le\prod_{1}^{n}(1+r\alpha_k)\,\|g_{\nu_1\cdots\nu_n}\|_p \tag{1.4}$$

*where the upper limit is taken over all real $u_1,\cdots,u_n$, and the summation is taken over all possible systems of integers $l_1,\cdots,l_n$.*

Proof. For any fixed real numbers $u_i$, the function $g_{\nu_1\cdots\nu_n}(z_1-u_1,\cdots,z_n-u_n)$ is an entire function of degree $\nu_1,\cdots,\nu_n$ for which

$$\|g_{\nu_1\cdots\nu_n}(x_1-u_1,\cdots,x_n-u_n)\|_p=\|g_{\nu_1\cdots\nu_n}(x_1,\cdots,x_n)\|_p.$$

It therefore suffices to prove the inequality

$$\prod_{1}^{n}h_k\sum_{-\infty}^{\infty}\cdots\sum_{-\infty}^{\infty}|g_{\nu_1\cdots\nu_n}(\overline{x}_{l_1}^{(1)},\ldots,\overline{x}_{l_n}^{(n)})|^p\leqslant\prod_{1}^{n}(1+r\alpha_k)\,\|g_{\nu_1\cdots\nu_n}\|_p. \tag{1.5}$$

But it is already proved for $n=1$ in the preceding theorem. Suppose it is proved for $m=n-1$ and let $g=g_{\nu_1\cdots\nu_n}\in L_p^{(n)}$.

There exists a set $\mathfrak{S}$ of points $x_1$ which differs from the real axis by a set of measure zero such that for all $x_1\in\mathfrak{S}$ the function $g$ belongs to $L_p^{(n-1)}$ when considered as a function of $x_2,\cdots,x_n$. Therefore

$$\left(\prod_{k=2}^{n}(1+r\alpha_k)\right)^{p}\int_{-\infty}^{\infty}\cdots\int_{-\infty}^{\infty}|g(x_1,x_2,\ldots,x_n)|^p\,dx_2,\ldots,dx_n\geqslant$$

$$\geqslant \prod_{k=2}^{n} h_k \sum_{-\infty}^{\infty} \cdots \sum_{-\infty}^{\infty} |g(x_1, \overline{x}_{l_2}^{(2)}, \ldots, \overline{x}_{l_n}^{(n)})|^p,$$

from which, after integrating with respect to $x_1$ and raising to the power $p^{-1}$, we obtain

$$\prod_{k=2}^{n} (1 + r\alpha_k) \| g \|_p \geqslant \left( \prod_{k=2}^{n} h_k \right)^{1/p} \sum_{l_2=-\infty}^{\infty} \cdots \sum_{l_n=-\infty}^{\infty} \int_{-\infty}^{\infty} |g(x_1, \overline{x}_{l_2}^{(2)}, \ldots, \overline{x}_{l_n}^{(n)})|^p dx_1$$

$$\geqslant \frac{1}{1 + r\alpha_1} \left( \prod_{1}^{n} h_k \right)^{1/p} \left( \sum_{-\infty}^{\infty} \cdots \sum_{-\infty}^{\infty} |g(\overline{x}_{l_1}^{(1)}, \ldots, \overline{x}_{l_k}^{(n)})|^p \right)^{1/p}$$

and the theorem is proved.

**Theorem 3.** *Let $g_{\nu_1 \cdots \nu_n}(z_1, \cdots, z_n)$ be an entire function of degree $\nu_1, \cdots, \nu_n$ for which the left side of (1.4) is finite. Then $g_{\nu_1 \cdots \nu_n} \in L_p^{(n)}$ and, in the notation of Theorem 2,*

$$\| g_{\nu_1 \ldots \nu_n} \|_p \leqslant \left[ (2r)^n \prod_{1}^{n} h_k \right]^{1/p}$$

$$\times \sup_{u_i} \left( \sum_{-\infty}^{\infty} \cdots \sum_{-\infty}^{\infty} |g_{\nu_1 \ldots \nu_n}(\overline{x}_{l_1}^{(1)} - u_1, \ldots, \overline{x}_{l_n}^{(n)} - u_n)|^p \right)^{1/p}. \quad (1.6)$$

*In this inequality the coefficient of $(2r)^{n/p}$ on the right can be replaced by one in the case of equally spaced points of interpolation $\overline{x}_{l_k}^{(k)} = x_{l_k}^{(k)}$ ($l_k = 0, \pm 1, \cdots$).*

**Proof.** We have

$$\int_{-\infty}^{\infty} \cdots \int_{-\infty}^{\infty} |g_{\nu_1 \ldots \nu_n}(x_1, \ldots, x_n)|^p dx_1 \ldots dx_n$$

$$= \sum_{-\infty}^{\infty} \cdots \sum_{-\infty}^{\infty} \int_{\overline{x}_{l_1-1}^{(1)}}^{\overline{x}_{l_1}^{(1)}} \cdots \int_{\overline{x}_{l_n-1}^{(n)}}^{\overline{x}_{l_n}^{(n)}} |g|^p dx_1 \ldots dx_n$$

$$= \sum_{-\infty}^{\infty} \cdots \sum_{-\infty}^{\infty} \int_{0}^{2rh_1} \cdots \int_{0}^{2rh_n} |g(\overline{x}_{l_1}^{(1)} - u_1, \ldots, \overline{x}_{l_n}^{(n)} - u_n)|^p du_1 \ldots du_n$$

$$\leqslant \int_{0}^{2rh_1} \cdots \int_{0}^{2rh_n} \sum_{-\infty}^{\infty} \cdots \sum_{-\infty}^{\infty} |g(\overline{x}_{l_1}^{(1)} - u_1, \ldots, \overline{x}_{l_n}^{(n)} - u_n)|^p du_1 \ldots du_n$$

$$\leqslant (2r)^n \prod_{1}^{n} h_k \sup_{u_i} \sum_{-\infty}^{\infty} \cdots \sum_{-\infty}^{\infty} |g(\overline{x}_{l_1}^{(1)} - u_1, \ldots, \overline{x}_{l_n}^{(n)} - u_n)|^p.$$

Thus the left side of this chain of inequalities is finite. After raising to the power $p^{-1}$ we obtain inequality (1.6). If all points of interpolation were equally spaced, then obviously it would be sufficient to take the integrals $\int_0^{h_i}$ instead of $\int_0^{2rh_i}$ in the appropriate place in this chain, which would eliminate the coefficient of $(2r)^{n/p}$.

Application. Inequalities (1.4) and (1.6) are, in some sense, converse to each other.

If we introduce the notation

$$((g_{\nu_1\ldots\nu_n}))_p = \sup_{u_i}\left(\prod_1^n h_k \sum_{-\infty}^{\infty}\cdots\sum_{-\infty}^{\infty} |g_{\nu_1\ldots\nu_n}(x_{l_1}^{(1)}-u_1,\ldots,x_{l_n}^{(n)}-u_n)\right)^{1/p},$$

$$x_l^{(i)} = lh_i \quad (l=0,\pm 1,\ldots), \tag{1.7}$$

where the sum in the right side is taken over equally spaced points of interpolation, we obtain the inequalities

$$\|g_{\nu_1\ldots\nu_n}\|_p \leqslant ((g_{\nu_1\ldots\nu_n}))_p \leqslant \prod_1^n (1+\alpha_k)\,\|g_{\nu_1\ldots\nu_n}\|_p, \tag{1.8}$$

showing that for fixed $n$ they are precise in the sense of order.

It should be said that in [9] inequalities of the form

$$\left(\sum_{-\infty}^{\infty}\cdots\sum_{-\infty}^{\infty}|g_{\nu_1\ldots\nu_n}(x_{k_1}^{(1)},\ldots,x_{k_n}^{(n)})|^p\right)^{1/p} \leqslant C_{\nu_1\ldots\nu_n}^{(n)}\|g_{\nu_1\ldots\nu_n}\|_p \tag{1.9}$$

are obtained. However, the constants $C^{(n)}_{\nu_1\cdots\nu_n}$ obtained by these authors in them are far from precise in the sense of order, and this will be important for us in what follows. Inequalities of the form (1.9) are invertible only if the points of interpolation $x^{(i)}_{k_i}$ are placed sufficiently thickly in comparison with the quantities $\nu_i$ (see [9], part II, p. 149).

Lemma 2. *For any nonnegative numbers* $a_1, \cdots, a_n$

$$\left(\sum_1^n a_k^{p'}\right)^{1/p'} \leqslant \left(\sum_1^n a_k^p\right)^{1/p} \qquad (1\leqslant p\leqslant p'<\infty). \tag{1.9'}$$

Proof. For $n=2$ the inequality follows from the easily proved inequality

$$\frac{(1+x^{p'})^p}{(1+x^p)^{p'}} \leqslant 1 \qquad (0\leqslant x<\infty).$$

If now inequality (1.9′) is true for $n-1$, then

$$\left(\sum_1^n a_k^{p'}\right)^{1/p'} = \left[\left(\sum_1^{n-1} a_k^{p'}\right)^{\frac{1}{p'}p'} + a_n^{p'}\right]^{1/p'} \leqslant \left[\left(\sum_1^{n-1} a_k^{p'}\right)^{p/p'} + a_n^p\right]^{1/p}$$

$$\leqslant \left(\sum_1^{n-1} a_k^p + a_n^p\right)^{1/p} = \left(\sum_1^{n-1} a_k^p\right)^{1/p}$$

**Theorem 4.** *If the entire function $g_{\nu_1\cdots\nu_n}(z_1, \cdots, z_n)$ of degree $\nu_1, \cdots, \nu_n$ belongs to $L_p^{(n)}$ and $1 \le p < p' \le \infty$, then it also belongs to $L_{p'}^{(n)}$, and*

$$\| g_{\nu_1\ldots\nu_n} \|_{p'} \leqslant 2^n \left(\sum_1^n \nu_k\right)^{1/p-1/p'} \| g_{\nu_1\ldots\nu_n} \|_p. \tag{1.10}$$

*This inequality is precise in the sense of order for fixed $n$ and arbitrary $\nu_i$.*

**Proof.** From inequalities (1.7) and (1.8) and Lemma (2), we have

$$\|g_{\nu_1\ldots\nu_n}\|_{p'} \leqslant \left(\prod_1^n h_k\right)^{1/p'} \sup_{u_i}\left(\sum_{-\infty}^{\infty}\cdots\sum_{-\infty}^{\infty} | g_{\nu_1\cdots\nu_n}(x_{l_1}^{(1)}-u_1, \cdots, x_{l_n}^{(n)}-u_n)|\right)^{1/p}$$

$$\leqslant \left(\prod_1^n h_k\right)^{1/p'} \sup_{u_i}\left(\sum_{-\infty}^{\infty}\cdots\sum_{-\infty}^{\infty} | g_{\nu_1\ldots\nu_n}(x_{l_1}^{(1)}-u_1, \ldots, x_{l_n}^{(n)}-u_n)|^p\right)^{1/p}$$

$$\leqslant \left(\prod_1^n h_k\right)^{1/p'-1/p} \prod_1^n (1+\alpha_k) \| g_{\nu_1\ldots\nu_n} \|_p$$

$$= \prod_1^n (1+\alpha_k) \left(\frac{\nu_k}{\alpha_k}\right)^{1/p-1/p'} \| g_{\nu_1\ldots\nu_n} \|_p,$$

for any positive numbers $\alpha_k$ $(k = 1, 2, \cdots, n)$. We set

$$\frac{1}{p} - \frac{1}{p'} = w.$$

The function

$$\psi(\alpha) = \frac{1+\alpha}{\alpha^w}$$

attains a minimum on the semiaxis $0 < \alpha < \infty$ equal to

$$\lambda_w = \frac{1}{w^w (1-w)^{1-w}} \leqslant 2. \tag{1.11}$$

We can thereby write

$$\| g_{\nu_1\cdots\nu_n} \|_{p'} \leqslant (\lambda_w)^n \left( \prod_1^n \nu_k \right)^{1/p-1/p'} \| g_{\nu_1\cdots\nu_n} \|_{p},$$

from which (1.10) follows by (1.11).

To prove the second assertion of the theorem, we consider the function

$$g^*_{\nu_1\cdots\nu_n} = \prod_1^n \frac{\sin^2 \frac{\nu_k z_k}{2}}{z_k^2},$$

which obviously belongs to $L_p^{(n)}$ for any $p$ satisfying the inequalities $1 \leq p \leq \infty$ and is an entire function of degree $\nu_1, \cdots, \nu_n$. Its norm equals

$$\| g^*_{\nu_1\cdots\nu_n} \|_p = \left( 2^n \prod_1^n \int_0^\infty \left| \frac{\sin^2 \frac{\nu_k t}{2}}{t^2} \right|^p dt \right)^{1/p}$$

$$\approx c_p \left( \prod_1^n \nu_k \right)^{2-1/p} \quad (1 \leqslant p \leqslant \infty) \ (\nu_k \to \infty)$$

and consequently there exists a positive constant $\gamma$ such that as $\nu_k \longrightarrow \infty$ we have

$$\| g_{\nu_1\cdots\nu_n} \|_{p'} > \gamma \left( \prod_1^n \nu_k \right)^{1/p-1/p'} \| g_{\nu_1\cdots\nu_n} \|_{p}.$$

By applying Lemma 2 we can also prove the following theorem, which shows the connection between the norm of a function $g_{\nu_1\cdots\nu_n}$ defined on $R_m$ and its norm defined on the linear subspace $R_n$.

Theorem 5. *If an entire function* $g = g_{\nu_1\cdots\nu_n}(z_1, \cdots, z_n)$ *of degree* $\nu_1, \cdots, \nu_n$ *belongs to* $L_p^{(n)}$, $1 \leq p \leq \infty$ *and* $1 \leq m < n$, *then for any* $x_{m+1}, \cdots, x_n$, *the function* $g$ *is pth power integrable with respect to* $x_1, \cdots, x_m$, *and the following inequality holds precisely for fixed* $n$ *and arbitrary* $\nu_i$ *in the sense of order:*

$$\left( \int_{-\infty}^{\infty} \cdots \int_{-\infty}^{\infty} | g(u_1, \ldots, u_m, x_{m+1}, \ldots, x_n) |^p du_1 \ldots du_m \right)^{1/p}$$

$$\leqslant 2^n \left( \prod_{m+1}^n \nu_k \right)^{1/p} \| g \|_p.$$

Proof. By (1.7) and (1.8)

$$\prod_1^n (1+\alpha_k) \| g \|_p \geqslant \left(\prod_1^n h_k\right)^{1/p}$$

$$\times \sup_{u_i} \left\{ \sum_{l_{m+1},\ldots,l_n} \sum_{l_1,\ldots,l_m} | g(x_{l_1}^{(1)} - u_1, \ldots, x_{l_n}^{(n)} - u_n)|^p \right\}^{1/p}$$

$$= \left(\prod_1^n h_k\right)^{1/p} \sup_{u_i} \sup_{l_{m+1},\ldots,l_n} \left\{ \sum_{l_1,\ldots,l_m} | g(x_{l_1}^{(1)} - u_1, \ldots, x_{l_n}^{(n)} - u_n)|^p \right\}^{1/p}$$

$$\geqslant \left(\prod_{m+1}^n h_k\right)^{1/p} \left(\int_{-\infty}^{\infty} \cdots \int_{-\infty}^{\infty} | g(u_1, \ldots, u_m, x_{m+1}, \ldots, x_n)|^p du_1 \cdots du_m\right)^{1/p},$$

from which, setting $\alpha_k = 1$, we obtain (1.12).

The precision of inequality (1.12) can be verified by the function $g^*_{\nu_1 \cdots \nu_n}$, which has already served as an example in a similar case in the proof of Theorem 4.

## §2. Inequalities for trigonometric polynomials

Any trigonometric polynomial of order $\nu_1, \cdots, \nu_n$ in the variables $x_1, \cdots, x_m$ can be written in the form

$$T_{\nu_1 \ldots \nu_n}(x_1, \ldots, x_n) = \sum_{k_1=-\nu_1}^{\nu_1} \cdots \sum_{k_n=-\nu_n}^{\nu_n} c_{k_1 \ldots k_n} e^{i \sum_{s=1}^n k_s x_s}.$$

We let $L_p^{*(n)}$ designate the space of functions $f(x_1, \cdots, x_n)$ which are measurable in $R_n$, periodic with period $2\pi$ in each of the variables $x_i$ and $p$th power integrable $(1 \leq p < \infty)$ over the period. Thus for every function $f \in L_p^{*(n)}$ we have

$$\| f \|_p^* = \left(\int_0^{2\pi} \cdots \int_0^{2\pi} | f(x_1, \ldots, x_n)|^p dx_1 \ldots dx_n\right)^{1/p} < \infty; \quad (2.1)$$

and for $p = \infty$,

$$\| f \|_\infty^* = \sup_{x_i} \operatorname{vrai} | f(x_1, \ldots, x_n)|. \quad (2.2)$$

For trigonometric polynomials whose norms are defined by (2.1) and (2.2) we have properties completely analogous to the properties obtained in §1 for entire functions of degree $\nu_1, \cdots, \nu_n$. For example, the following inequalities hold:

$$\|T_{\nu_1\ldots\nu_n}\|_p \leqslant \left(\prod_1^n h_i \max_{u_i} \sum_1^{N_1} \cdots \sum_1^{N_n} |T_n(x_{l_1}^{(1)} - u_1, \ldots, x_{l_n}^{(n)} - u_n)|^p\right)^{1/p}$$

$$\leqslant \prod_1^n (1 + h_i \nu_i) \|T_{\nu_1\ldots\nu_n}\|_p, \tag{2.3}$$

where

$$1 \leqslant p \leqslant \infty, \qquad N_i \text{ is integral}, \qquad h_i = \frac{2\pi}{N_i}, \qquad x_k^{(i)} = kh_i$$

$$(i = 1, 2, \ldots, n; \quad k = 0, \pm 1, \pm 2, \ldots).$$

Their proof is completely analogous to the proof of (1.8), except that we must integrate everywhere with respect to the variables $x_1, \cdots, x_n$ and the summation over $x_{l_1}^{(1)}, \cdots, x_{l_n}^{(n)}$ ranges over the cube $0 \leq x_k < 2\pi$ of the space $R_n$ determining a period, rather than over the whole space.

Marcinkiewicz and Zygmund [8] obtained inequalities of the form

$$A\left(\int_0^{2\pi} |T_n(x)|^p\, dx\right)^{1/p} \leqslant \left(\frac{2\pi}{N}\sum_1^N |T_n(x_k)|^p\right)^{1/p} \leqslant B\left(\int_0^{2\pi} |T_n(x)|^p\, dx\right)^{1/p}$$

where $A$ and $B$ are constants not depending on the arbitrary trigonometric polynomials $T_n(x)$ of one variable. However, we should keep in mind that the first of these inequalities is fulfilled only if $N$ is sufficiently large compared to $n$. For example, even if $N = 2n + 1$ and $p = 1$ it is not true, as we can see by setting

$$T_n(x) = \frac{\sin \frac{2n+1}{2} x}{\sin \frac{x}{2}}.$$

Inequality (2.3) with $p = 1$, $n = 1$ and $N = N_1 = 2\nu_1 + 1$ is essentially contained in the article [12] of S. M. Lozinskiĭ, p. 215.

Using inequalities (2.3), with $N_i = \nu_i$, and arguments analogous to those of Theorem 4, we obtain the following inequality for any trigonometric polynomial:

$$\|T_{\nu_1\ldots\nu_n}\|_{p'} \leqslant 2^n \left(\prod_1^n \nu_k\right)^{1/p - 1/p'} \|T_{\nu_1\ldots\nu_n}\|_p \qquad (1 \leqslant p < p' \leqslant \infty) \tag{2.4}$$

or, as in Theorem 5, we have the inequality

$$\left(\int_0^{2\pi}\cdots\int_0^{2\pi}|T_{\nu_1\ldots\nu_n}(u_1,\ldots,u_m,x_{m+1},\ldots,x_n)|^p du_1\ldots du_n\right)^{1/p}$$
$$\leqslant 2^n\left(\prod_{m+1}^{n}\nu_k\right)^{1/p}\|T_{\nu_1\ldots\nu_n}\|_p \tag{2.5}$$
$$(1\leqslant p<\infty).$$

## §3. Classes of functions. Best approximations using entire functions of finite degree

We agree to say that the function $f(x)$ has order $r$ on the real axis if the derivative $f^{(r-1)}(x)$ of order $r-1$ exists and is absolutely continuous on any finite interval of the axis. Thus the derivative $f^{(r)}(x)$ actually exists only almost everywhere, but $f^{(r-1)}(x)$ is its indefinite integral. If $r=1$, then $f(x)$ is simply absolutely continuous on any finite interval of the $x$-axis. It is in this sense that we must interpret the statement that a function $f(x_1,\cdots,x_n)$ of several variables has a partial derivative of order $r$ on the $x_1$-axis for fixed $x_2,\cdots,x_n$.

We shall say that a measurable function $f(x_1,\cdots,x_n)$ belongs to the class $H^{(r)}_{px_1}(M)$ $(r>0,\ 1\le p\le\infty)$ if it satisfies the following condition.

We represent $r$ in the form $r=\rho+\alpha$, where $\rho$ is an integer and $0<\alpha\le 1$. For almost all $x_2,\cdots,x_n$ the partial derivative $\partial^\rho f/\partial x_1^\rho$ exists, so that it is defined almost everywhere in $R_n$, it belongs to $L^{(n)}_p$, and the following inequality holds for any $h$:[1])

$$\left\|\frac{\partial^\rho f(x_1+h,x_1,\ldots,x_n)}{\partial x_1^\rho}-2\frac{\partial^\rho f(x_1,\ldots,x_n)}{\partial x_1^\rho}\right.$$
$$\left.+\frac{\partial^\rho f(x_1-h,x_2,\ldots,x_n)}{\partial x_1^\rho}\right\|_p\leqslant M|h|^\alpha. \tag{3.1}$$

S. N. Bernšteĭn [3] examined similar classes for $p=\infty$ and arbitrary $n$, and Zygmund [11] treated the periodic case with $n=1$, $1\le p\le\infty$.

We shall further say that a function $f(x_1,\cdots,x_n)$ belongs to the class $H_p^{(r_1,\cdots,r_n)}(M_1,\cdots,M_n)$ $(1\le p\le\infty,\ r_i>0)$ if it belongs to all of the classes $H^{(r_1)}_{px_1}(M_1),\cdots,H^{(r_n)}_{px_n}(M_n)$.

---

1) In the case $0<\alpha<1$ it is possible, as will be evident later on, to replace this inequality by the simpler

$$\left\|\frac{\partial^\rho f(x_1+h,x_2,\ldots,x_n)}{\partial x_1^\rho}-\frac{\partial^\rho f(x_1,\ldots,x_n)}{\partial x_1^\rho}\right\|_p\leqslant M|h|^\alpha,$$

but it would then be necessary to consider the cases $\alpha<1$ and $\alpha=1$ separately.

In those cases when we will not be interested in the values of the constants $M, M_1, \cdots, M_n$, we shall omit them and write $H^{(r_1)}_{px_1}, \cdots, H^{(r_n)}_{px_n}$, $H^{(r_1,\cdots,r_n)}_p$.

Let a function $f \in L^{(n)}_p$ be given. Along with it we shall consider all possible functions $g_\nu(x_1, \cdots, x_n)$ also belonging to $L^{(n)}_p$ and having the property that for almost all $(x_2, \cdots, x_n)$ they are entire functions of degree $\nu$ in the variable $x_1$.

By the best approximation of the function $f$ using the functions $g_\nu(x_1, \cdots, x_n)$, we mean the greatest lower bound

$$A_{\nu x_1}(f)_p = \inf_{g_\nu} \|f - g_\nu\|_p,$$

ranging over all possible functions $g_\nu$.

We further consider the difference of the function $f$ with respect to the variable $x_1$ with given increment $h$; that is, we set

$$\Delta_{x_1}(f, h) = \Delta_{x_1}(f) = \Delta'_{x_1} f = f(x_1 + h, x_2, \ldots, x_n) - f(x_1, \ldots, x_n),$$

$$\Delta^\rho_{x_1}(f, h) = \Delta^\rho_{x_1} f = \Delta_{x_1}\Delta^{\rho-1}_{x_1} f \ (\rho = 2, 3, \ldots), \ \Delta^0_{x_1} f = f(x_1, \ldots, x_n).$$

**Lemma 3.** *If the function $f \in L^{(n)}_p$ has a partial derivative of order $\rho$ with respect to $x_1$ which belongs to $L^{(n)}_p$, then for any integer $s \geq 0$ we have*

$$\|\Delta^{(\rho+s)}_{x_1} f\|_p \leqslant h^\rho \|\Delta^s_{x_1} f^{(\rho)}_{x_1}\|_p. \tag{3.2}$$

**Proof.** We prove (3.2) for $\rho = 1$ and $s = 0$:

$$\|\Delta_{x_1} f\|_p = \left(\int_{-\infty}^{\infty} \cdots \int_{-\infty}^{\infty} |\Delta_{x_1} f|^p dx_1 \ldots dx_n\right)^{1/p}$$

$$= \left(\int_{-\infty}^{\infty} \cdots \int_{-\infty}^{\infty} \left| \int_{x_1}^{x_1+h} f'_{x_1}(t, x_2, \ldots, x_n)\, dt \right|^p dx_1 \ldots dx_n\right)^{1/p}$$

$$\leqslant \left(\int_{-\infty}^{\infty} \cdots \int_{-\infty}^{\infty} |f'_{x_1}|^p dt\, h^{p/q} dx_1 \ldots dx_n\right)^{1/p} = h^{1/p+1/q}\left(\int_{-\infty}^{\infty} \cdots \int_{-\infty}^{\infty} |f'_{x_1}|^p dx_1, \ldots, dx_n\right)^{1/p}$$

$$= h\|f'_{x_1}\|_p \quad \left(\frac{1}{p} + \frac{1}{q} = 1\right).$$

Hence

$$\|\Delta^{\rho+s}_{x_1} f\|_p \leqslant h \|\Delta^{\rho+s-1}_{x_1} f'_{x_1}\|_p \leqslant \ldots \leqslant h^\rho \|\Delta^s_{x_1} f^{(\rho)}_{x_1}\|_p.$$

**Theorem 6.** *For any positive* $r$ *and* $\nu$ *there exists an entire function* $K_\nu^{(r)}(u)$ *of degree* $\nu$ *which is integrable on the real axis and has the properties*

$$1)\quad \int_{-\infty}^{\infty} K_\nu^{(r)}(u)\,du = 1, \tag{3.3}$$

$$2)\quad \int_{-\infty}^{\infty} |K_\nu^{(r)}(u)|\,du < a_r. \tag{3.4}$$

3) *For all functions* $f$ *belonging to the class* $H_{px_1}^{(r)}$ *we have*

$$A_{\nu x_1}(f)_p \leqslant \frac{b_r}{\nu^r} \tag{3.5}$$

*where* $a_r$ *and* $b_r$ *depend on* $r$ *but not on* $\nu$.

*It follows from* (3.5) *that*

$$\left\| f - \int_{-\infty}^{\infty} K_\nu^{(r)}(u - x_1)\, f(u, x_2, \ldots, x_n)\,du \right\|_p \leqslant \frac{b_r M}{\nu^r}. \tag{3.6}$$

S. N. Bernšteĭn [3] proved this theorem (and Theorems 7 and 10) for $p = \infty$, and N. I. Ahiezer [1] for $n = 1$, $p \geq 1$.

**Proof.** In the proof we shall follow Bernšteĭn's reasoning, which quickly leads to the goal if we do not care about the magnitude of the constant. Let $g(x)$ be an entire function of degree $q$ such that

$$1)\quad \int_{-\infty}^{\infty} g(t)\,dt = 1, \tag{3.7}$$

$$2)\quad \int_{-\infty}^{\infty} |g(t)|\,dt < \infty, \tag{3.8}$$

$$3)\quad b_r = \int_{-\infty}^{\infty} |g(t)|\,|t|^r dt < \infty.^{1)} \tag{3.9}$$

We note that

$$(-1)^{\rho+1}\Delta^{\rho+2}(\varphi, h) = \sum_{i=0}^{\rho+2} (-1)^{i-1} C_{\rho+2}^i \varphi(x + ih) = \sum_{i=1}^{\rho+2} p_i \varphi(x + ih) - \varphi(x),$$

where

---

1) For example, we can set $g(x) = K_\nu (\sin(x/\lambda)/x)^\lambda$ where $\lambda \geq r + 2$ is even and $K_\lambda$ is chosen so that (3.7) is fulfilled.

$$\sum_{i+1}^{\rho+2} p_i = 1. \tag{3.10}$$

Now let $f \in H^{(r)}_{px_1}$. We consider the entire function of degree $\nu$ in $x_1$ defined as follows:

$$g_\nu = g_\nu(x_1, \ldots, x_n)$$

$$= \int_{-\infty}^{\infty} g(t) \left\{ (-1)^{\rho+1} \Delta_{x_1}^{\rho+2} \left[ f, \frac{t}{\nu};\ x_1, \ldots, x_n \right] + f(x_1, \ldots, x_n) \right\} dt$$

$$= \int_{-\infty}^{\infty} g(t) \sum_{i=1}^{\rho+2} p_i f\left( x_1 + \frac{t_i}{\nu}, x_2, \ldots, x_n \right) dt = \int_{-\infty}^{\infty} K_\nu(t - x_1) f(t_1, x_1, \cdots, x_n) dt,$$

where

$$K_\nu^{(r)}(u) = \sum_{1}^{\rho+2} \frac{p_i \nu}{i} g\left( \frac{\nu u}{i} \right) \tag{3.11}$$

is an entire function of degree $\nu$. Then

$$g_\nu - f = (-1)^{\rho+1} \int_{-\infty}^{\infty} g(t) \Delta_{x_1}^{\rho+2} \left[ f, \frac{t}{\nu};\ x_1, \ldots, x_n \right] dt,$$

$$\|g_\nu - f\|_p \leqslant \int_{-\infty}^{\infty} g(t) \left( \int_{-\infty}^{\infty} \cdots \int_{-\infty}^{\infty} \left| \Delta_{x_1}^{\rho+2} \left[ f, \frac{t}{\nu};\ x_1, \ldots, x_n \right] \right|^p dx_1 \ldots dx_n \right)^{1/p} dt,$$

from which, in view of (3.2), letting $r = \rho + \alpha$ ($\rho$ integral, $0 < \alpha \leq 1$), we have

$$\|f - g_\nu\|_p = \frac{1}{\nu^{\rho+\alpha}} \int_{-\infty}^{\infty} |g(t)|\, |t|^{\rho+\alpha} dt = \frac{b_r}{\nu^r}. \tag{3.12}$$

(3.3) follows from (3.7), (3.10) and (3.11); further,

$$\int_{-\infty}^{\infty} |k_\nu^{(r)}(u)|\, du \leqslant \sum_{1}^{\rho+2} |p_i| \int_{-\infty}^{\infty} \frac{\nu}{i} \left| g\left( \frac{\nu u}{i} \right) \right| du = \sum_{1}^{\rho+2} |p_i| \int_{-\infty}^{\infty} |g(u)|\, du = a_r. \tag{3.13}$$

Theorem 7. *If $f \in H_p^{(r_1, \cdots, r_n)}(M_1, \cdots, M_n)$, then its best approximation using entire functions $g_{\nu_1 \cdots \nu_n}$ of degree $\nu_1, \cdots, \nu_n$ in $x_1, \cdots, x_n$ satisfies the inequality*

$$A_{\nu_1\ldots\nu_n}(f)_p \leqslant d\sum_1^n \frac{M_k}{\nu_k^{r_k}}, \tag{3.14}$$

*where $d$ depends on $r_1, \cdots, r_n$.*

**Proof.** We shall only consider the case of a function $f(x_1, x_2, x_3)$ of three variables.

We consider the entire function of degree $\nu_1, \nu_2, \nu_3$ in $x_1, x_2, x_3$ defined by the equation

$$g_{\nu_1\nu_2\nu_3}(x_1, x_2, x_3) = \int_{-\infty}^{\infty}\int_{-\infty}^{\infty}\int_{-\infty}^{\infty} K_{\nu_1}^{(r_1)}(u_1)K_{\nu_2}^{(r_2)}(u_2)K_{\nu_3}^{(r_3)}(u_3) f(x_1+u_1,$$
$$x_2+u_2, x_3+u_3)\,du_1\,du_2\,du_3,$$

where $K_\nu^{(r)}$ is the kernel of (3.11). Equation (3.3) is valid for it, so that

$$f - g_{\nu_1\nu_2\nu_3} = \int_{-\infty}^{\infty}\int_{-\infty}^{\infty}\int_{-\infty}^{\infty} K_{\nu_1}^{(r_1)}(u_1)K_{\nu_2}^{(r_2)}(u_2)K_{\nu_3}^{(r_3)}(u_3)\,[f(x_1, x_2, x_3)$$
$$- f(x_1+u_1, x_2+u_2, x_3+u_3)]\,du_1\,du_2\,du_3 = \psi_1+\psi_2+\psi_3,$$

where

$$\psi_1 = \int\int\int K_{\nu_1}^{(r_1)}K_{\nu_2}^{(r_2)}K_{\nu_3}^{(r_3)}[f(x_1, x_2, x_3) - f(x_1+u_1, x_2, x_3)]\,du_1\,du_2\,du_3$$
$$= \int K_{\nu_1}^{(r_1)}(u_1)\,[f(x_1, x_2, x_3) - f(x_1+u_1, x_2, x_3)]\,du_1,$$

$$\psi_2 = \int\int\int K_{\nu_1}^{(r_1)}K_{\nu_2}^{(r_2)}K_{\nu_3}^{(r_3)}\,[f(x_1+u_1, x_2, x_3) - f(x_1+u_1, x_2+u_2, x_3)]du_1du_2du_3$$
$$= \int\int K_{\nu_1}^{(r_1)}(u_1)K_{\nu_2}^{(r_2)}(u_2)[f(x_1+u_1, x_2, x_3) - f(x_1+u_1, x_2+u_2, x_3)]\,du_1\,du_2$$
$$= \int K_{\nu_1}^{(r_1)}(u_1)\,h_2(x_1+u_1, x_2, x_3)\,du.$$

$$\psi_3 = \int\int\int K_{\nu_1}^{(r_1)}K_{\nu_2}^{(r_2)}K_{\nu_3}^{(r_3)}\,[f(x_1+u_1, x_2+u_2, x_3) - f(x_1+u_1, x_2+u_2, x_3+u_3)]$$
$$\times\,du_1\,du_2\,du_3 = \int\int K_{\nu_1}^{(r_1)}(u_1)K_{\nu_2}^{(r_2)}(u_2)\,h_3(x_1+u_1, x_2+u_2, x_3)\,du_1\,du_2.$$

From this and (3.5) we get

$$\|\psi_1\|_p \leqslant \frac{b_{r_1}M_1}{\nu_1^{r_1}},$$

$$\|\psi_2\|_p \leqslant \int |K_{\nu_1}^{(r_1)}(u_1)|\,du_1\,\left(\int\int\int |h_2(x_1+u_1, x_2, x_3)|^p dx_1dx_2dx_3\right)^{1/p}$$
$$\leqslant a_{r_1}\|h_2\|_p < \frac{a_{r_1}b_{r_2}M_2}{\nu_2^{r_2}};$$

$$\|\psi_3\|_p \leqslant \iint |K_{\nu_1}^{(r_1)}(u_1)K_{\nu_2}^{(r_2)}(u_2)|\, du_1\, du_2 \left(\iiint |h_3(x_1+u_1, x_2+u_2, x_3)|^p dx_1 dx_2 dx_3\right)^{1/p}$$

$$< a_{r_1} a_{r_2} \|h_3\|_p < \frac{a_{r_1} a_{r_2} b_{r_3} M_3}{\nu_3^{r_3}};$$

and consequently

$$\|f - g_{\nu_1\nu_2\nu_3}\|_p \leqslant \sum_1^3 \|\psi_k\|_p < \frac{b_{r_1} M_1}{\nu_1^{r_1}} + \frac{a_{r_1} b_{r_2} M_2}{\nu_2^{r_2}} + \frac{a_{r_1} a_{r_2} b_{r_3} M_3}{\nu_3^{r_3}}$$

$$< d \sum_1^3 \frac{M_k}{\nu_k^{r_k}}.$$

**Lemma 4.** *Let there be given a sequence of functions* $f_k(x_1, \cdots, x_n)$ $(k = 1, 2, \cdots)$ *belonging to* $L_p^{(n)}$ *and having partial derivative* $\partial^r f_k / \partial x_1^r$ *of order* $r$ *with respect to* $x_1$ *which also belongs to* $L_p^{(n)}$. *If here*

$$\| f - f_k \|_p \to 0 \qquad (k \to \infty),$$

$$\left\| \varphi - \frac{\partial^r f_k}{\partial x^r} \right\|_p \to 0 \qquad (k \to \infty), \tag{3.15}$$

*then* $f$ *can be altered on a set of measure zero of the space* $R_n$ *so as to have a partial derivative* [1] *with respect to* $x$ *of order* $r$ *which equals*

$$\frac{\partial^r f}{\partial x_1^r} = \phi.$$

Proof. For simplicity we shall limit ourselves to two variables $x = x_1$ and $y = x_2$.

We note that if $f \in L_p^{(2)}$, then $f$ is integrable over every finite rectangle and consequently the integral

$$\int_c^b f(x, v)\, dv$$

exists for almost all $x$, where $(c, b)$ is any finite interval.

Since $\|f - f_k\|_p \longrightarrow 0$ $(k \longrightarrow \infty)$, it follows that the $f_k$ converge in measure to $f$ on the rectangle $|x| < a$, $|y| < b$, which we designate by $\Delta$, and consequently there is a sequence of natural numbers $n_1 < n_2 < \cdots$ such that

1) This must be interpreted in the sense that for any $x_1$ and almost all $x_2, \cdots, x_n$ the partial derivative $\partial^{r-1} f / \partial x_1^{r-1} f$ of order $r - 1$ with respect to $x_1$ exists and is absolutely continuous on any finite interval of the $x_1$-axis.

$$\lim f_{n_k}(x, y) = f(x, y) \tag{3.16}$$

on a set $E_1$ of measure $\operatorname{mes} E_1 = \operatorname{mes} \Delta$.

We represent $f_k(x, y)$ in the form

$$f_k(x, y) = \sum_{s=0}^{r-1} f_k^{(s)}(0, y)\, x^s + \frac{1}{\Gamma(r)} \int_0^x (x - t)^{r-1} \frac{\partial^r f_k}{\partial x^r}(u, y)\, du, \tag{3.17}$$

which is possible for almost all $s$ by assumption.

We take any value of $x$ satisfying the inequality $|x| < a$. Because $\| \phi - \partial^r f_k / \partial x^r \|_p \longrightarrow 0$ $(k \longrightarrow \infty)$, it follows that

$$\int_{-b}^{b} \left| \int_0^x (x-u)^{r-1} \frac{\partial^r f_k(u, y)}{\partial x^r}\, du - \int_0^x (x - u)^{r-1} \varphi(u, y)\, du \right| dy$$

$$\leqslant M \left( \int_{-b}^{b} \int_0^x \left| \frac{\partial^r f_k(u, y)}{\partial x^r} - \varphi(u, y) \right|^p du\, dy \right)^{1/p} (2ba)^{1/p} \to 0 \ (k \to \infty),$$

where

$$M = \max_\Delta |x - u|^{r-1}, \qquad \frac{1}{p} + \frac{1}{q} = 1,$$

and a subsequence of $n'_k$ can be chosen from the sequence of $n_k$ so that the limit

$$\lim_{k \to \infty} \int_0^x (x - u)^{r-1} \frac{\partial^r f_k(u, y)}{\partial x^r}\, du = \int_0^x (x - u)^{r-1} \varphi(u, y)\, du \tag{3.18}$$

exists almost everywhere when $k$ runs through the $n'_k$.

We now choose $r$ values of $x$

$$-a < x_1 < x_2 < \cdots < x_r < a$$

such that for every $l = 1, 2, \cdots, r$ and almost all $y$, the points $(x_l, y)$ belong to $E_1$. This is possible because $\operatorname{mes} E_1 = \operatorname{mes} \Delta$. We can obviously also choose a subsequence $\{n'_k\}$ from the sequence $\{n_k\}$ such that (3.18) will be fulfilled for $x = x_l$ $(l = 1, \cdots, r)$ and for $k$ running through this subsequence. It will then follow from (3.16) and (3.17) that

$$\lim \sum_{s=0}^{r-1} f_{n'_k}^{(s)}(0, y)\, x_l^s \qquad (l = 1, 2, \ldots, r)$$

exists for almost all $y$, and consequently, for almost all $y$, so do the limits

$$\lim_{n'_k \to \infty} f^{(s)}_{n'_k}(0, y) = A_s(y) \qquad (s = 0, 1, \ldots, r-1). \tag{3.19}$$

Now let $e$ be the set of values of $x$ in the interval $|x| < a$ such that $(x, y) \in E_1$ for almost all $y$. Obviously mes $e = 2a$. For each $x \in e$ a subsequence $\{n_k^{(x)}\}$ can be chosen from the sequence $\{n'_k\}$ such that equations (3.16), (3.18) and (3.19) will be simultaneously fulfilled for almost all $y$ when $k$ runs through the subsequence $\{n_k^{(x)}\}$. But then (3.17) will lead to the equation

$$f(x, y) = \sum_{s=0}^{r-1} A_s(y)\, x^s + \frac{1}{\Gamma(x)} \int_0^x (x-u)^{r-1} \varphi(u, y)\, du$$

for every $x \in e$ and for almost all $y$. This equation will thereby be valid for almost all $(x, y)$ in $\Delta$. Its right side is a function which is the $r$-fold indefinite integral of the integrable function $\phi(u, y)$ for almost all $y$. With this the lemma is proved.

Theorem 8. *Let $r > 0$, $f \in L_p^{(u)}$ and*

$$A_{\nu x_1}(f)_p < \frac{K}{\nu^r} \tag{3.20}$$

*for all $\nu$ running through the geometric progression $\nu = a^k$ $(k = 0, 1, \cdots)$ with ratio $a > 1$. Further, let $g$ designate an entire function of degree one in $x_1$ for which*

$$\|f - g\|_p \leq K. \tag{3.21}$$

*Then the function $f_1 = f - g$ belongs to the class $H^{(r)}_{px_1} M$. Here*

$$M = c_r K, \tag{3.22}$$

*where $c_r$ depends on $r$ but not on $K$.*

Proof. We let $g_\nu = g_\nu(x_1, \cdots, x_n)$ denote entire functions of degree $\nu$ in $x_1$ for which

$$\|f - g_\nu\|_p < \frac{K}{\nu^r},$$

where $\nu = a^k = m_k$ $(k = 0, 1, \cdots)$.

Then the function $f$ satisfying the conditions of the theorem is represented in the form

$$f = g + \phi,$$

where $g = g_{m_0}$ and

$$\varphi = \sum_1^\infty (g_{m_k} - g_{m_{k-1}}).$$

We differentiate this series termwise $\rho$ times with respect to $x_1$. We then obtain

$$\varphi_{x_1}^{(\rho)} = \sum_1^\infty Q_k, \tag{3.23}$$

where

$$Q_k = \frac{\partial^\rho (g_{m_k} - g_{m_{k-1}})}{\partial x_1^\rho}.$$

By (1.1) and (3.20),

$$\| Q_k \|_p \leqslant m_k^\rho \frac{2K}{m_{k-1}^r} = a^{k\rho} \frac{2K}{a^{(k-1)(\rho+\alpha)}} = \frac{a^{r+1} K}{a^{k\alpha}}.$$

Since the series with terms of the form $a^{-k\alpha}$ converges, the series (3.23) converges in the $L_p^{(n)}$ sense and the termwise differentiation is legitimate from the previous lemma. We have thereby proved that the function $\phi$ has a derivative of order $\rho$ belonging to $L_p^{(n)}$. The series (3.23) converges to it in the $L_p^{(n)}$ sense.

We now take a positive number $h$ and for it select a natural number $N$ such that

$$\frac{1}{a^{N+1}} < h \leqslant \frac{1}{a^N}.$$

For the second difference of the function $\phi_{x_1}^{(\rho)}$ with respect to $x_1$ we have

$$\| \Delta_{x_1}^2 (\varphi_{x_1}^{(\rho)}, h) \|_p \leqslant \sum_1^N \| \Delta_{x_1}^2 (Q_k, h) \|_p + 4 \sum_{k=N+1}^\infty \| Q_k \|_p.$$

From Lemmas 1 and 3 (inequalities (1.1) and (3.2)) we have

$$\sum_{1}^{N} \| \Delta^2_{x_1} (Q_{k_1}, h) \|_p$$

$$\leqslant h^2 \sum_{1}^{n} \| Q''_{kx_1} \|_p \leqslant h^2 \sum_{1}^{N} a^{2k} \frac{a^{r+1} K}{a^{k\alpha}} = a^{r+1} K h^2 \frac{a^{(2-\alpha)(N+1)} - a^{2-\alpha}}{a^{2-\alpha} - 1}$$

$$\leqslant a^{r+1} K h^2 a^{(2-\alpha)(N+1)} = a^{r+1} K h^2 a^{2-\alpha} h^{\alpha-2} \leqslant a^{r+3} K h^\alpha,$$

$$4 \sum_{N+1}^{\infty} \| Q_k \|_p \leqslant a^{r+3} c \sum_{N+1}^{\infty} \frac{1}{a^{k\alpha}} \leqslant a^{r+3} K \frac{a h^\alpha}{a^\alpha - 1} = \frac{a^{r+4} K}{a^\alpha - 1} h^\alpha.$$

This implies the theorem if we set

$$c_r = a^{r+s} + \frac{a^{r+4}}{a^\alpha - 1}. \tag{3.24}$$

We note that the function $g$ of the first degree in $x_1$ that figures in Theorem 6 also belongs to the class $H^{(r)}_{px_1}$, but only with another constant depending on $\|f\|_p$. In fact, it follows from (3.21) that

$$\|g\|_p \leq \|f\|_p + K \tag{3.25}$$

and by Lemma 1

$$\|g^{(\rho)}_{x_1}\|_p \leq \|f\|_p + K,$$

from which we get

$$\|\Delta^2_{x_1} g^{(\rho)}_{x_1}\|_p \leq 4(\|f\|_p + K)$$

and by Lemma 3

$$\|\Delta^2_{x_1} g^{(\rho)}_{x_1}\|_p \leqslant h \| g^{(\rho+1)}_{x_1} \|_p \leqslant h (\| f \|_p + K).$$

The last two inequalities lead to the inequality

$$\| \Delta^2_{x_1} g^{(\rho)}_{x_1} \|_p \leqslant 4 (\| f \|_p + K) h^\alpha, \tag{3.26}$$

which is valid for any $h$.

By virtue of this remark, Theorems 5 and 6 imply the following assertion.

Theorem 9. *For the function $f \in L^{(n)}_p$ to belong to the class $H^{(r)}_{px_1}$ (with some constant $M$), it is necessary and sufficient that there exist a constant $K$ for which*

$$A_{\nu x_1}(f)_p < \frac{K}{\nu^r}$$

*for all* $\nu \geq 1$ *or else for all* $\nu$ *running through the increasing geometric progression* $\nu = a^k$ $(a > 1,\ k = 0, 1, \cdots)$.

**Theorem 10.** *Let the best approximation of the function* $f \in L_p^{(n)}$ *using the entire functions* $g_{\nu_1 \cdots \nu_n}$ *of degree* $\nu_1, \cdots, \nu_n$ *in* $x_1, \cdots, x_n$ *satisfy the inequality*

$$A_{\nu_1 \ldots \nu_n}(f)_p < \sum_1^n \frac{K_i}{\nu_i^{r_i}}, \tag{3.27}$$

*where* $K_i$ *is a constant,* $r_i > 0$ *and the* $\nu_i$ *run through the geometric progressions* $\nu_i = a_i^k$ $(k = 0, 1, \cdots)$ *with ratios* $a_i > 1$. *Further, let* $g$ *be an entire function of degree one in each of the variables* $x_1, \cdots, x_n$, *and such that*

$$\| f - g \|_p < \sum_1^n K_i \tag{3.28}$$

*(it exists by* (3.27) *and belongs to the class* $H_p^{(r_1, \cdots, r_n)}$ *with constants depending not only on* $K$ *but also on* $\|f\|_p$*). Then the function* $\phi = f - g$ *belongs to the class* $H_p^{(r_1, \cdots, r_n)}(M_1, \cdots, M_n)$, *where*

$$M_i = C_{r_i} \sum_{i=1}^n K_i.$$

**Proof.** For the best approximation of order $\nu_i$ in the variable $x_i$ of the function $f$ satisfying the conditions of the theorem, we have the obvious inequality

$$A_{\nu_i x_i}(f)_p \leqslant A_{\nu_1 \ldots \nu_n}(f)_p < \sum_1^n \frac{K_i}{\nu_i^{r_i}},$$

for any $\nu_2, \cdots, \nu_n$. Hence

$$A_{\nu_1 x_1}(f)_p \leqslant \frac{K_1}{\nu_1^{r_1}} \leqslant \frac{\sum_1^n K_i}{\nu_1^{r_1}}. \tag{3.29}$$

It then follows from (3.28), (3.29) and Theorem 8 that the function $\phi$ as a function of $x_1$ belongs to the class $H_{px_1}^{(r_1)}(M_1)$, where

$$M_1 = C_{r_1} \sum_1^n K_k.$$

We reason analogously in the case of the remaining variables $x_2, \cdots, x_n$.

We consider the class $E$ of entire functions $g = g_{\nu_1 \cdots \nu_n}(x_1, \cdots, x_n)$ with fixed degrees $\nu_1, \cdots, \nu_n$ which belong to $L_p^{(n)}$ and have bounded norm

$$\|g_{\nu_1 \cdots \nu_n}\|_p \leq M. \tag{3.30}$$

These functions are bounded in $R_n$ by (1.10) and their norm in the $L_\infty^{(n)}$ sense satisfies the inequality

$$\| g_{\nu_1 \ldots \nu_n} \|_\infty \leqslant 2^n \left( \prod_1^n \nu_k \right)^{1/p} M = M_1.$$

But then, as we know from S. N. Bernšteĭn's theory, [1] the class $E$ has the property of compactness on any bounded set $\Omega \subset R_n$. In other words, from any sequence of functions of class $E$ we can extract a sequence converging uniformly on $\Omega$ to some function $g^*_{\nu_1 \cdots \nu_n}$. We can always assume here that the limit function $g^*$ belongs to $L_p^{(n)}$, since by the diagonal process we can obtain a subsequence $g^{(k)}$ $(k = 1, 2, \cdots)$ converging uniformly, and hence in the $L_p$ sense, on each of the concentric spheres $\zeta_1, \zeta_2, \cdots$ with radius increasing without bound; but then

$$\left( \int_{\zeta_n} \cdots \int | g^* |^p \, dx_1 \ldots dx_n \right)^{1/p} = \lim_{k \to \infty} \left( \int_{\zeta_n} \cdots \int | g^{(k)} |^p \, dx_1 \ldots dx_n \right)^{1/p} \leqslant M$$

for any sphere $\zeta_n$, and hence

$$\|g^*\|_p^{(n)} \leq M.$$

We note that in the sense of the norm defined on the whole space $R_n$ the collection of functions (3.30) does not form a compact set. This certainly follows from the infinite dimensionality of this collection. For example, the functions

$$\varphi_k(x) = \sum_{l=1}^{k} \sin \frac{x}{l} \qquad (k = 1, 2, \ldots),$$

each of the first degree, form an infinite linearly independent sequence.

Theorem 11. *Let $\mathfrak{S}$ be the set of functions $f \in L_p^{(n)}$ with the following properties:*

1) $\|f\|_p \leq M$ *for all* $f \in \mathfrak{S}$,

2) $\sup_{f \in \mathfrak{S}} A_k(f)_p = \eta_k \longrightarrow 0 \ (k \longrightarrow \infty)$,

---

1) This can be obtained by expanding $g_{\nu_1 \cdots \nu_n}$ in a Fourier series and noting that its coefficients are bounded by Bernšteĭn's inequality (1.1).

*where* $A_k(f)_p$ *is the best approximation of the function* $f$ *using entire functions* $g_{\nu_1^{(k)}\cdots\nu_n^{(k)}}$ *with degrees* $\nu_1^{(k)}, \cdots, \nu_n^{(k)}$ $(k = 1, 2, \cdots)$, *where* $\nu_i^{(k)} \to \infty$ *as* $k \to \infty$.

*Then the set* $\mathfrak{S}$ *is compact in the* $L_p$ *sense on any bounded measurable set* $\Omega$ *belonging to* $R_n$.

**Proof.** We consider the sequence of functions $f_1, f_2, \cdots$ of $\mathfrak{S}$ and let $g_k(f) = g_{\nu_1^{(k)}\cdots\nu_n^{(k)}}(x_1, \cdots, x_n)$ $(k = 1, 2, \cdots)$ be the best approximating function of degrees $\nu_1^{(k)}, \cdots, \nu_n^{(k)}$ for $f$, i.e.

$$A_k(f)_p = \|f - g_k(f)\|_p.$$

By the diagonal process we can extract a subsequence

$$f_{n_1} = \phi_1, \ f_{n_2} = \phi_2, \cdots$$

such that $g_k(\phi_l)$ converges uniformly on $\Omega$ as $l \to \infty$ for any $k$. This is due to the fact that

$$\| g_k(f) \|_p \leqslant \| f \|_p + A_k(f)_p \leqslant M + \eta_k < M_1.$$

Further,

$$\Big(\int_\Omega \cdots \int |\varphi_l - \varphi_m|^p \, dx_1 \dots dx_n\Big)^{1/p} \leqslant A_k(\varphi_l)_p$$

$$+ \Big(\int_\Omega \cdots \int |g_k(\varphi_l) - g_k(\varphi_m)|^p \, dx, \dots, dx_n\Big)^{1/p} + A_k(\varphi_m)_p.$$

By the second condition of the theorem we can find $k$ so large that the first and third terms of the right side will be less than any $\epsilon/3$ independently of $m$ and $l$. We can further find an $N$ such that for all $k$, $l > N$ the second term will be less than $\epsilon/3$. This proves the theorem.

**Remark.** All the theorems proved in this section remain valid if $H^{(r)}_{px_i}$ and $H_p^{(r_1, \cdots, r_n)}$ are understood as periodic functions with period $2\pi$ and with norms defined on the cube determining the period, and if $A_{\nu x_i}(f)_p$ and $A_{\nu_1 \cdots \nu_n}(f)_p$ are understood as the corresponding best approximations using the trigonometric polynomials $T_{\nu x_i}$ and $T_{\nu_1 \cdots \nu_n}$. The proof is completely analogous. The same remark applies to §4.

## §4. Theorems on imbedding the classes $H_p^{(r_1, \cdots, r_n)}$

**Theorem 12.** *Let* $1 \le p \le p' \le \infty$, $r_i > 0$ $(i = 1, 2, \cdots, n)$, $1 \le m \le n$ *and*

$$\varkappa_m = 1 - \Big(\frac{1}{p} - \frac{1}{p'}\Big) \sum_1^m \frac{1}{r_i} - \frac{1}{p} \sum_{m+1}^n \frac{1}{r_i} > 0. \tag{4.1}$$

*Then, if the function f belongs to the class* $H_p^{(r_1,\cdots,r_n)}(M_1,\cdots,M_n)$, *then it also belongs to the class* $H_p^{(\rho_1^{(m)},\cdots,\rho_m^{(m)})}(M_i^*,\cdots,M_m^*)$ *as a function of the variables* $x_1,\cdots,x_m$ *for any fixed* $x_{m+1},\cdots,x_n$, *where*

$$\rho_i^{(m)} = r_i\kappa_m;\quad M_i^* \leq \alpha\,\|f\|_p^{(n)} + \beta\sum_1^n M_k$$

*and* $\alpha$ *depends only on n and* $\beta$ *on* $n, m, r_1, r_2, \cdots, r_n$. *In particular, for* $m = n$,

$$H_p^{(r_1,\ldots,r_n)}(M_1,\ldots,M_n) \subset H_{p'}^{(\rho_1,\ldots,\rho_n)}(M_1^*,\ldots,M_n^*),$$
$$\rho_i = \varkappa r_i,\quad \varkappa_n = 1-\left(\frac{1}{p}-\frac{1}{p'}\right)\sum_1^n \frac{1}{r_i}. \tag{4.2}$$

*If*

$$1-\frac{1}{p}\sum_1^n\frac{1}{r_i} > 0,\qquad 1<p,$$

*then for any* $p'$ *and* $m$, $1<p\leq p'\leq\infty$, $1\leq m\leq n$, *we can find a function f belonging to the class* $H_p^{(r_1,\cdots,r_n)}$, *but not (for any M) to the class* $H_{p'}^{(\rho_1^{(m)},\cdots,\rho_{i-1}^{(m)},\rho_{i-1}^{(m)}+z,\rho_{i+1}^{(m)},\cdots,\rho_n^{(m)})}$.

**Proof.** We first prove the first part of the theorem. Let

$$f\in H_p^{(r_1,\cdots,r_n)}(M_1,\cdots,M_n).$$

We set

$$\nu_i^{r_i} = 2^s\qquad (s=0,1,2,\cdots).$$

Thus $\nu_i = \nu_i(s)$ is a function of $s$. Further, let

$$g_{\nu_1\ldots\nu_n} = g_s\ (s=0,1,2,\ldots)$$

be the best entire function of degree $\nu_1,\cdots,\nu_n$ in $x_1,\cdots,x_n$ for $f$, with

$$A_{\nu_1\ldots\nu_n}(f)_p = \|f-g_s\|_p.$$

By Theorem 7,

$$A_{\nu_1\ldots\nu_n}(f)_p \leqslant \frac{d\sum_1^n M_k}{2^s}, \tag{4.3}$$

where $\alpha$ depends on $r_1,\cdots,r_n$.

Therefore, the function $f$ can be represented in the form of a series converging to it in the $L_p^{(n)}$ sense:

$$f = g_0 + \sum_1^\infty (g_s - g_{s-1}) = Q_0 + \sum_1^\infty Q_s. \tag{4.4}$$

Here

$$\| Q_s \|_p = \| g_s - f \|_p + \| f - g_{s-1} \|_p \leqslant \frac{2d \sum_1^n M_k}{2^{s-1}} \qquad (s = 1, 2, \ldots)$$

and $Q_s$ has degree $2^{s/r_i}$ in $x_i$ $(i = 1, 2, \cdots, n)$.

Consequently, by inequalities (1.10) and (1.12), on the subspace $R_m$ of points $u_1, \cdots, u_m, x_{m+1}, \cdots, x_n$, where the $u_i$ are arbitrary and the $x_{m+1}, \cdots, x_n$ are fixed, the norm $\| Q_s \|_{p'}^{(m)}$ satisfies the inequalities

$$\| Q_s \|_{p'}^{(m)} = \Big( \int_{-\infty}^{\infty} \cdots \int_{-\infty}^{\infty} | Q_s (u_1, \ldots, u_m, x_{m+1}, \ldots, x_n) |^{p'} du_1 \ldots du_m \Big)^{1/p'}$$

$$\leqslant 2^n \Big( \prod_{m+1}^n \nu_k \Big)^{1/p'} \| Q_s \|_{p'}^{(n)} \leqslant 2^{2n} \Big( \prod_1^m \nu_k \Big)^{1/p - 1/p'} \Big( \prod_{m+1}^n \nu_k \Big)^{1/p} \| Q_s \|_p^{(n)}$$

$$\leqslant 2^{2n+2} \frac{d \sum_1^n M_k}{2^{s\varkappa_m}}. \tag{4.5}$$

Therefore

$$\Big\| f - Q_0 - \sum_{s=1}^{\mu-1} Q_s \Big\|_{p'}^{(m)} = \sum_{s=\mu}^{\infty} \| Q_s \|_{p'} \leqslant 2^{2n+2} d \sum_1^n M_k \sum_{s=\mu}^{\infty} \frac{1}{2^{s\varkappa_m}} < \frac{C \sum_1^n M_k}{2^{\mu\varkappa_m}},$$

where $C$ is a constant depending on $r_i$ and $\kappa_m$, and since $Q_0 + \Sigma_1^{\mu-1} Q_s$ has degree no greater than $2^{\mu/r_i}$ in $x_i$, it follows that

$$\Big\| f - Q_0 - \sum_{s=1}^{\mu-1} Q_s \Big\|_{p'}^{(m)} \leqslant \frac{C \sum_1^n M_k}{\nu_i^{\rho_i^{(m)}}}$$

for any $x_{m+1}, \cdots, x_n$.

By Theorem 8 this means that the function $f - g_0 (g_0 = Q_0)$ as a function of $x_i (i = 1, 2, \cdots, n)$ belongs to the function class $H_{p' x_i}^{(\rho_i^{(m)})} (M_i)$, defined for the subspace $R_m$, where

$$\overline{M}_i = C_{\rho_i^{(m)}} C \sum_1^n M_k$$

and the $C_{\rho_i^{(m)}}$ are defined by equation (3.24) ($a = 2$ for $r = \rho_i^{(m)}$).

Consequently

$$f - g_0 \in H_{p'}^{(\rho_1^{(m)}, \ldots, \rho_m^{(m)})}(\overline{M}_1, \ldots, \overline{M}_m),$$

for any $x_{m+1}, \cdots, x_n$.

As for the function $g_0$ of degree one in all the $x_i$, from the inequality

$$\| g_0 \|_p \leqslant \| f \|_p^{(n)} + d \sum_1^n M_k$$

(see (4.3)) by (1.10) and (1.12) it follows that

$$\| g_0 \|_{p'}^{(n)} < 2^n \| g_0 \|_{p'}^{(m)} \leqslant 2^{2n} \| g_0 \|_p^{(n)} = 2^{2n} \left( \| f \|_p^{(n)} + d \sum_1^n M_k \right) \tag{4.6}$$

and by (3.30) and (3.31) it follows that $g_0$ also belongs to the class $H_{p'}^{(\rho_1^{(n)}, \ldots, \rho_n^{(m)})}(K, \cdots, K)$ with constant $K = 4 \cdot 2^{2n} (\|f\|_p^{(n)} + d \Sigma_1^n M_k)$. The sum of the constants $M_i$ and $K$ can obviously not exceed $\alpha \|f\|_p^{(n)} + \beta \Sigma_1^n M_k$ where $\alpha$ depends on $n$ and $\beta$ on $n, m, r_1, \cdots, r_n$. This proves that $f \in H_{p'}^{(\rho_1^{(m)}, \ldots, \rho_m^{(m)})}(M_1^*, \cdots, M_m^*)$ for any $x_{m+1}, \cdots, x_n$, where $M_i^* \leq \alpha \|f\|_p^{(n)} + \beta \Sigma_1^n M_k$.

**Remark.** It is remarkable that the transformation of the numbers $r_i$ into $\rho_i^{(m)}$ using the factor $\kappa_m$ is transitive. If we are given numbers $r_1, \cdots, r_n$ and $p$ with $1 \leq p < \infty$, and are then given $p'$ and $p''$, $p \leq p' \leq p'' \leq \infty$, and subspaces $R_m$ and $R_l$ corresponding to them, where $R_l \subseteq R_m \subseteq R_n$, then going from $(p, R_n)$ to $(p', R_m)$ and then from $(p', R_m)$ to $(p'', R_l)$, we arrive at numbers $\rho_1^{(l)}, \cdots, \rho_l^{(l)}$ which can be obtained by going directly from $(p, R_n)$ to $(p'', R_l)$. This follows from the equation

$$\kappa_m \left\{ 1 - \left( \frac{1}{p'} - \frac{1}{p''} \right) \sum_1^l \frac{1}{\kappa_m r_i} - \frac{1}{p'} \sum_{l+1}^m \frac{1}{\kappa_m r_i} \right\}$$

$$= 1 - \left( \frac{1}{p} - \frac{1}{p''} \right) \sum_1^l \frac{1}{r_i} - \frac{1}{p} \sum_{l+1}^n \frac{1}{r_i},$$

where $\kappa_m$ is (4.1).

We now bring in an example of a function satisfying the second assertion of the theorem. By the remark just made, it suffices to construct the required function for the case $p' = \infty$. This same function can serve as an example for the case $p' < \infty$ as well, since if it belonged to a class characterized by numbers $\rho_i$ larger than specified by the first part of the theorem, then by the transitivity of the operation of going from $r_i$ to $\rho_i$ it also could not be an unbounded function in the case $p' = \infty$.

For the same reasons, it is easily seen that it suffices to give an example of a function for the case of going from $n$ to $m = 1$, i. e., to examine the passage from $(p, R_n)$ to $(\infty, 1)$. For simplicity, we shall limit ourselves to three variables. We can then give this function in the form of the series

$$f(x, y, z) = \sum_{\nu=1}^{\infty} \varphi_\nu (x, y, z),$$

$$\varphi_\nu (x, y, z) = \frac{1}{2^{\left(r_1 r_2 r_3 + \frac{r_2 r_3 + r_1 r_3 + r_1 r_2}{2}\right)}} \frac{\sin 2^{r_2 r_3 \nu} x}{x} \frac{\sin 2^{r_1 r_3 \nu} y}{y} \frac{\sin 2^{r_1 r_2 \nu} z}{z}.$$

The function $\phi_\nu$ is an entire function of degree $2^{r_2 r_3 \nu}$, $2^{r_1 r_3 \nu}$, $2^{r_1 r_2 \nu}$ in $x$, $y$ and $z$. Hence the sum

$$\sum_{\nu=1}^{\mu-1} \varphi_\nu$$

is also an entire function of degree $l = 2^{r_2 r_3 \mu}$, $m = 2^{r_1 r_3 \mu}$, $n = 2^{r_1 r_2 \mu}$ in $x$, $y$ and $z$. Further, from the equation

$$\left( \int_{-\infty}^{\infty} \left| \frac{\sin nx}{x} \right|^p dx \right)^{1/p} = n^{1/q} \left( \int_{-\infty}^{\infty} \left| \frac{\sin x}{x} \right|^p dx \right)^{1/p} = n^{1/q} A, \quad \frac{1}{p} + \frac{1}{q} = 1$$

we have

$$\left\| f - \sum_{1}^{\mu-1} \varphi_\nu \right\|_p \leq \sum_{\mu}^{\infty} \| \varphi_\nu \|_p = \sum_{\nu=\mu}^{\infty} \frac{A}{2^{r_1 r_2 r_3 \nu}} = \frac{C}{l^{\rho_1}} = \frac{C}{m^{\rho_2}} = \frac{C}{n^{\rho_3}}.$$

By Theorem 10, in this case $f$ belongs to the class $H_p^{(r_1, r_2, r_3)}$. We prove that for any $\epsilon > 0$ it does not belong to the class $H_{\infty x_1}^{(\rho_1 + \epsilon)}$ as a function of the variable $x$, where $\rho_i = \kappa r_i$, $\kappa = \kappa_1 = \kappa_2 = 1 - (1/p) \sum_1^3 1/r_i$ and it hence also does not belong to the class $H_\infty^{(\rho_1 + \epsilon, \rho_2, \rho_3)}$. Setting $y = z = 0$, we obtain

$$\varphi_\nu(x, 0, 0) = \frac{1}{2^{r_1 r_2 r_3 \left[1 - \frac{1}{p} \sum_1^3 \frac{1}{r_i}\right]}} \frac{\sin 2^{r_2} x^{r_3}}{2^{r_2 r_3 \nu} x} = \frac{1}{x} \int_0^x \frac{\cos a^\nu x \, dx}{a^{r_1 \varkappa \nu}},$$

where $a = 2^{r_2 r_3} > 1$.

The question is thereby reduced to proving that the function

$$F_{\rho_1}(x) = f(x, 0, 0) = \frac{1}{x} \int_0^x \sum_1^\infty \frac{\cos a^\nu x}{a^{r_1 \varkappa \nu}} dx = \frac{1}{x} \int_0^x \sum_1^\infty \frac{\cos a^\nu x}{a^{\rho_1 \nu}} dx \qquad (\rho_1 = r_1 \varkappa)$$

does not belong to $H_\infty^{(\rho_1 + \epsilon)}$.

We set $\rho_1 = k + \alpha$, where $k$ is an integer and $0 < \alpha \leq 1$, and we consider the function

$$\varphi_{k+\alpha}(t) = \sum_{\nu=1}^\infty \frac{\cos a^\nu t}{a^{(k+\alpha)\nu}},$$

which we know belongs to $H_\infty^{(k+\alpha)}$ but not to $H_\infty^{(k+\alpha+\epsilon)}$.

Integrating its expansion by Taylor's formula

$$\varphi_{k+\alpha}(t) = a_0 + a_1 t + \cdots + a_{k-1} t^{k-1} + \frac{1}{(k-1)!} \int_0^t (t-u)^{k-1} \varphi_{k+\alpha}^{(k)}(u) \, du$$

and dividing by $x$, we obtain

$$F_{k+\alpha}(x) = P_{k-1}(x) + \frac{1}{x} \frac{1}{(k-1)!} \int_0^x \int_0^t (t-u)^{k-1} \varphi_{k+\alpha}^{(k)}(u) \, du \, dt$$

$$= P_{k-1}(x) + \frac{1}{x} \frac{1}{k!} \int_0^x (x-u)^k \varphi_{k+\alpha}^{(k)}(u) \, du,$$

where $P_{k-1}(u)$ are certain polynomials of degree $k-1$. Hence

$$\frac{d^k F_{k+\alpha}(x)}{dx^k} = \sum_{l=0}^k C_k^l (-1)^l x^{-l-1} \int_0^x (x-u)^l \varphi_{k+\alpha}^{(k)}(u) \, du$$

$$= \frac{1}{x} \int_0^x \sum_0^k C_k^l \left(\frac{u}{x} - 1\right)^l \varphi_{k+\alpha}^{(k)}(u) \, du = \frac{1}{x^{k+1}} \int_0^x u^k \varphi_{k+\alpha}^{(k)}(u) \, du, \qquad (4.7)$$

$$F_{k+\alpha}^{(k)}(0) = \lim_{x \to 0} \frac{1}{x^{k+1}} \int_0^x u^k \varphi_{k+\alpha}^{(k)}(u) \, du = \frac{\varphi_{k+\alpha}^{(k)}(0)}{k+1}. \qquad (4.8)$$

Case 1. Let $k$ be even. Then

$$\varphi_{k+\alpha}^{(k)}(t) = \pm \sum_{\nu=1}^{\infty} \frac{\cos a^{\nu} t}{a^{\alpha\nu}},$$

$$\varphi_{k+\alpha}^{(k)}(0) - \varphi_{k+\alpha}^{(k)}(t) = \sum_{\nu=1}^{\infty} \frac{1-\cos a^{\nu} t}{a^{\alpha\nu}} = \sum_{\nu=1}^{\infty} \frac{2\sin^2 \frac{a^{\nu} t}{2}}{a^{\alpha\nu}}$$

$$\geqslant \sum_{1}^{n} \frac{2\sin^2 a^{\nu}\frac{t}{2}}{a^{\alpha\nu}} \geqslant \quad (\text{for} \quad a^{-n-1} < t \leqslant a^{-n}) \geqslant \frac{2}{\pi^2} t^2 \sum_{\nu=1}^{n} a^{(2-\alpha)\nu} \tag{4.9}$$

$$= \frac{2}{\pi^2} t^2 \frac{a^{(2-\alpha)(n+1)} - a^{2-\alpha}}{a^{2-\alpha} - 1} > Ct^{\alpha} \qquad (C > 0).$$

By (4.7) and (4.8)

$$\left| F_{k+\alpha}^{(k)}(x) - 2F_{k+\alpha}^{(k)}(0) + F_{k+\alpha}^{(k)}(-x) \right|$$

$$= \left| \frac{1}{x^{k+1}} \int_0^x u^k \left[\varphi_{k+\alpha}^{(k)}(u) - 2\varphi_{k+\alpha}^{(k)}(0) + \varphi_{k+\alpha}^{(k)}(-u)\right] du \right|$$

$$=^{1)} \frac{2}{x^{k+1}} \int_0^x u^k \left| \varphi_{k+\alpha}^{(k)}(u) - \varphi_{k+\alpha}^{(k)}(0) \right| du$$

$$>^{2)} \frac{2C}{x^{k+1}} \int_0^x u^{k+\alpha} du > C_1 x^{\alpha} \tag{4.10}$$

for small positive $x$. But this says that the function $f_{k+\alpha}$ does not belong to the class $H_{\infty}^{(k+\alpha+\epsilon)}$. For $\alpha < 1$ this is obvious. If $\alpha = 1$, then, assuming that $F_{k+\alpha} \in H_{\infty}^{(k+1-\epsilon)}$, we must conclude that it has a continuous derivative of order $k+1$ satisfying the Lipschitz condition of degree $\epsilon$. Hence for $x > 0$

$$\left| F_{k+1}^{(k)}(x) - 2F_{k+1}^{(k)}(0) + F_{k+1}^{(k)}(-x) \right|$$

$$= x \left| F_{k+1}^{(k+1)}(\theta x) - F_{k+1}^{(k+1)}(-\theta_1 x) \right| \leqslant Cx^{1+\varepsilon},$$

which contradicts inequality (4.10) for $\alpha = 1$.

Case 2. Let $k$ be odd. Then, to within a sign,

---

1) By the evenness of $\phi_{k+\alpha}^{(k)}(u)$.

2) By (4.9).

$$\varphi_{k+\alpha}^{(k)}(x) = \sum_{\nu=1}^{\infty} \frac{\cos a^{\nu} x}{a^{\nu(1+\alpha)}},$$

and since formula (4.9) remains true if we replace $\alpha$ by $1+\alpha$ in it, it follows that

$$|\varphi_{k+\alpha}^{(k)}(0) - \varphi_{k+\alpha}^{(k)}(t)| > C_1 x^{1+\alpha} \qquad (C_1 > 0).$$

Therefore, by (4.7),

$$F_{k+\alpha}^{(k-1)}(0) - F_{k+\alpha}^{(k-1)}(x) = \frac{1}{x^k}\int_0^x u^{k-1}[\varphi_{k+\alpha}^{(k-1)}(0) - \varphi_{k+\alpha}^{(k-1)}(u)]\,du$$

$$> \frac{C_1}{x^k}\int_0^x u^{k+\alpha}\,du > C_2 x^{1+\alpha}.$$

But in this case our function $F_{k+\alpha}$ cannot belong to the class $H_{\infty}^{(k+\alpha+\epsilon)}$ for $\alpha < 1$, since, taking into consideration that $\phi_{k+\alpha}^{(k)}(0) = 0$, and hence by (4.8)

$$F_{k+\alpha}^{(k)}(0) = 0,$$

we obtain for a function of this class, applying Taylor's formula for $\alpha + \epsilon < 1$:

$$|F_{k+\alpha}^{(k-1)}(x) - F_{k+\alpha}^{(k-1)}(0)| = |x F_{k+\alpha}^{(k)}(\theta x)|$$

$$= |x(F_{k+\alpha}^{(k)}(\theta x) - F_{k+\alpha}^{(k)}(0))| \leqslant Cx^{1+\alpha+\epsilon}.$$

If now $\alpha = 1$, then to within a sign

$$\varphi_{k+1}^{(k)}(u) = \sum_{1}^{\infty} \frac{\sin a^{\nu} u}{a^{\nu}}.$$

Consequently, if $n$ is chosen so that $a^{-n-1} \le x < a^{-n}$, we have

$$\frac{F_{k+1}^{(k)}(x) - F_{k+1}^{(k)}(0)}{x} = \frac{F_{k+1}^{(k)}(x)}{x} \geqslant \frac{1}{x^{k+2}}\left\{\int_0^x u^k \sum_{\nu=1}^{n} \frac{\sin a^{\nu} u}{a^{\nu}}\,du\right.$$

$$\left. - \int_0^x u^k \sum_{n+1}^{\infty} \frac{1}{a^{\nu}}\,du\right\} \geqslant C_2 n - C_3 \frac{1}{xa_n} > C_2 n - C_4 \underset{x\to 0}{\to} \infty$$

and the function $F_{k+1}^{(k)}(x)$ does not have a derivative at the point $x = 0$ and so cannot belong to the class $H_{\infty}^{(k+1+\epsilon)}$.

Remark. As noted in the Introduction, Theorem 12 is a generalization of propositions of S. L. Sobolev and V. I. Kondrašov. Since our results are formulated in terms different from those used by these authors, we shall dwell on this matter.

We first note that the generalized unmixed derivative in the sense of S. L. Sobolev is nothing other than the usual partial derivative of the given function as defined precisely by us above, if we alter the function when necessary on a set of measure zero in the space $R_n$.

In fact, let the function $f$ have unmixed partial derivative $\partial^\rho f/\partial x_1^\rho$ of order $\rho$ with respect to $x_1$ belonging to $L_p^{(n)}$ for almost all $x_2, \cdots, x_n$.[1) ] Then, if $\Psi$ is any function continuous together with its partial derivatives up to order $\rho$ which vanishes on the boundary of the cube $a_i \le x_i \le b_i$ $(i = 1, \cdots, n)$ of the space $R_n$ designated by $\Omega$, then after $\rho$-fold integration by parts we obtain the equation

$$\int_{a_1}^{b_1} f \frac{\partial^\rho \Psi}{\partial x_1{}^\rho}\, dx_1 = (-1)^\rho \int_{a_1}^{b_1} \Psi \frac{\partial^\rho f}{\partial x_1{}^\rho}\, dx_1$$

for almost all $x_2, \cdots, x_n$; after integrating with respect to $(x_2, \cdots, x_n)$, this leads to the equation

$$\int_\Omega \left[ f \frac{\partial^\rho \Psi}{\partial x_1{}^\rho} + (-1)^{\rho+1} \Psi \varphi \right] dv = 0,$$

which says that $\phi = \partial^\rho f/\partial x_1^\rho$ is the generalized derivative in the sense of Sobolev.

Conversely, let $\phi \in L_p^{(n)}$ be the generalized derivative of order $\rho$ with respect to $x_1$ of the function $f \in L_p^{(n)}$. As Sobolev showed (see [7], §5), to the function $f$ there corresponds an averaging function $f_h \in L_p^{(n)}$ which depends on the parameter $h$ and has continuous partial derivatives of any order, such that

$$\|f - f_h\|_p \longrightarrow 0 \quad (h \longrightarrow 0) \tag{4.11}$$

and also

$$\frac{\partial^\rho f_h}{\partial x_1{}^\rho} = \varphi_h \in L_p^{(n)}; \qquad \left\| \varphi - \frac{\partial^\rho f_h}{\partial x_1{}^\rho} \right\|_p \to 0 \qquad (h \to 0). \tag{4.12}$$

---

1) More precisely, $f$ has a derivative $\partial^{\rho-1} f/\partial x_1^{\rho-1}$ which is absolutely continuous on any finite interval relative to $x_1$ for almost all $x_2, \cdots, x_n$.

But we proved in Lemma 4 that (4.11) and (4.12) imply that the function $f$ can be altered on a set of measure zero so that then

$$\varphi = \frac{\partial^\rho f}{\partial x_1^\rho}.$$

It is now easily seen that for $r_1 = \cdots = r_n = r$ Theorem 12 is a generalization (in the case of infinite space) of theorems of Sobolev and Kondrašov.

In fact, let the function $f \in L_p^{(n)}$ have all generalized derivatives of order $r$ in the sense of Sobolev which belong to $L_p^{(n)}$. The unmixed generalized derivatives are the usual derivatives. Hence, by Lemma 3,

$$\| \Delta_{x_i}^2 f^{(r-1)} \|_p \leqslant h \| \Delta_{x_i} f^{(r)} \|_p = 2 \| f_{x_i}^{(r)} \|_p$$

and $f \in H_p^{(r_1, \cdots, r_n)} (M_1, \cdots, M_n)$, where $M_i = 2 \| f_{x_i}^{(n)} \|_p$, and consequently by Theorem 12 the function $f$, considered on the linear subspace $x_1, \cdots, x_m$, belongs to the class $H_{p'}^{(\rho, \cdots, \rho)} (M^*, \cdots, M^*)$, for any $x_{m+1}, \cdots x_n$, where

$$\rho = r \varkappa_m = r - \frac{m}{p'} - \frac{n}{p}; \qquad M^* \leqslant \alpha \| f \|_p^{(n)} + \beta n M,$$

if only $\rho > 0$. Thus, if we set $\rho = \bar{\rho} + \alpha$, where $\bar{\rho}$ is integral and $0 < \alpha \leq 1$, then the function $f$ will have partial derivatives of order $\bar{\rho}$ with respect to the variables $x_1, \cdots, x_m$ which belong to $L_{p'}^{(m)}$. Estimates for the norms of these derivatives can be calculated without trouble by estimating series (4.4) termwise using inequalities (4.5).

We further note that if we take $r_1 = \cdots = r_n = r$ in Theorem 12, we only start from the single condition that

$$\varkappa_m = 1 - \frac{n}{pr} + \frac{m}{p'r} > 0,$$

or, which is the same thing, $p' < mp/(n - rp)$.

It follows directly from our theory that the unmixed partial derivatives of $f$ exist and belong to $L_{p'}^{(m)}$. This also holds for mixed partial derivatives, as we see from the following theorem.

**Theorem 13.** *If the function $f$ belongs to the class $H_p^{(r_1, \cdots, r_n)}$ and the nonnegative integers $\rho_k$, together with the number $\alpha$, $0 < \alpha \leq 1$, satisfy the equation*

$$\sum_1^{i-1} \frac{\rho_k}{r_k} + \frac{\rho_i + \alpha}{r_i} + \sum_{i+1}^{n} \frac{\rho_k}{r_k} = 1,$$

*then the mixed partial derivative*

$$\varphi = \frac{\partial^{\rho_1+\cdots+\rho_n} f}{\partial x_1^{\rho_1}\cdots\partial x_n^{\rho_n}}$$

*exists, belongs to* $L_p^{(n)}$ *and satisfies the condition*

$$\| \Delta^2_{x_i h}\varphi \|_p \leqslant c\,|h|^{\alpha},$$

*where* $c$ *is a constant not depending on* $h$.

For $p = \infty$ this theorem was proved in my article [5], and for $1 \le p < \infty$ the proof is analogous.

We prove a final theorem on the compactness of classes, which corresponds to theorems of V. I. Kondrašov.

**Theorem 14.** *From any sequence bounded in the* $L_p^{(n)}$ *sense of functions belonging to the class* $H_p^{(r_1,\cdots,r_n)}(M_1, \cdots, M_n)$ *it is possible to extract a subsequence of functions* $f_1, f_2, \cdots$ *converging in the* $L_{p'}^{(m)}$ *sense on any bounded subset* $\Omega$ *of the subspace* $R_m$ *along with its unmixed partial derivatives up to orders* $\rho_1^*, \cdots, \rho_m^*$ *inclusive, respectively, where* $\rho_i^*$ *are positive integers for which* $\rho_i = \rho_i^* + \alpha_i$ $(0 < \alpha_i \le 1)$ *and* $\rho_i$ *and* $p'$ *are defined as in Theorem* 12.

**Proof.** In the process of proving Theorem 12 we established that if $f \in H_p^{(r_1,\cdots,r_n)}(M_1, \cdots, M_n)$, then it can be represented in the form of a series of entire functions which converges to it in the $L_{p'}^{(m)}$ sense on $R_m$:

$$f = \sum_0^\infty Q_s,$$

where (see (4.5))

$$\| Q_s \|_{p'}^{(m)} < \frac{c}{2^{s\kappa_m}} \qquad (s = 1,\ 2,\ \ldots),$$

where $c$ is a constant and $Q_s$ are entire functions of degree $2^{s/r_i}$ in $x_1, \cdots, x_m$.

By Lemma 4 we can differentiate our series termwise with respect to the variable $x_i$ $(i = 1, \cdots, m)$ $\tau$ times, if only $\tau \le \rho_i^*$, where $\rho_i = r_i\kappa_m$ and $\rho_i = \rho_i^* + \alpha_i$. This follows from the fact that, because of Bernšteĭn's inequality (Lemma 1), we have

$$\left\| \frac{\partial^\tau Q_s}{\partial x_i^\tau} \right\|_{p'}^{(m)} < \frac{c}{2^{s\kappa_m}}\, 2^{\frac{s\tau}{r_i}} = \frac{c}{2^{s\frac{\rho_i-\tau}{r_i}}},$$

where $\rho_i - \tau > 0$ and the series obtained by differentiating converges in the $L_{p'}^{(m)}$ sense.

Hence if we let $A_k(f)_{p'}^{(m)}$ designate the best approximation of the function $f$ on $R_m$ using entire functions of degrees $2\mu/r_i$ $(i = 1, 2, \cdots, m)$, then for every function $f \in H_p^{(r_1,\cdots,r_n)}(M_1, \cdots, M_n)$ we have

$$A_k\left(\frac{\partial^\tau f}{\partial x_i^\tau}\right)_{p'}^{(m)} \leqslant \left\| \frac{\partial^\tau f}{\partial x_i^\tau} - \sum_0^{\mu-1} \frac{\partial^\tau Q_s}{\partial x_i^\tau} \right\|_{p'}^{(m)} \leqslant \sum_{\mu}^{\infty} \frac{c}{2^{s\frac{\rho_i-\tau}{r_i}}} \to 0 \qquad (\mu \to \infty).$$

In addition, by the uniform boundedness of the norms of our functions assumed in the theorem $(\|f\|_p^{(n)} \leq M)$, the norms $\|\partial^\tau f/\partial x_i^\tau\|_{p'}^{(m)}$ will also be bounded. In fact,

$$\left\| \frac{\partial^\tau f}{\partial x_i^\tau} \right\|_{p'}^{(m)} \leqslant \sum_0^\infty \left\| \frac{\partial^\tau Q_s}{\partial x_i^\tau} \right\|_{p'}^{(m)},$$

where by (4.6), keeping in mind that $Q_0$ is of the first degree, we have

$$\left\| \frac{\partial^\tau Q_0}{\partial x_i^\tau} \right\|_{p'} \leqslant 2^{2n}\left(\|f\|_p^{(n)} + c\sum_1^n M_k\right) \leqslant A,$$

$$\sum_1^\infty \left\| \frac{\partial^\tau Q_s}{\partial x_i^\tau} \right\|_{p'} \leqslant \sum_1^\infty \frac{c}{2^{s\frac{(\rho_i-\tau)}{r_i}}} \leqslant B.$$

But then by Theorem 11 we can extract a subsequence $n_1, n_2, \cdots$ from any sequence of functions $f_1, f_2, \cdots$ belonging to $H_p^{(r_1,\cdots,r_n)}(M_1, \cdots, M_n)$ such that $\partial^\tau f_n^k/\partial r_i^\tau$ converges in the $L_p^{(m)}$ sense on any bounded measurable set of the subspace $R_m$. By extracting new subsequences when necessary, it is obviously possible to find a subsequence $f_{n_1}, f_{n_2}, \cdots$ which converges in the $L_p^{(m)}$ sense together with its partial derivatives of the orders indicated in the theorem to the specified function $f \in L_p^{(m)}$ and to its corresponding partial derivatives on any bounded measurable set of the subspace $R_m$. The limit function $f$ exists because the space $L_p$ is complete, and because of Lemma 4.

It is easy to see that the subsequence of functions which was just proved to exist converges weakly (on $R_m$) to the limit function $f$.

We further note that the article [13] by I. G. Petrovskiĭ and K. N. Smirnov is related to this question.

## BIBLIOGRAPHY

[1] N. I. Ahiezer, *Lectures on the theory of approximation*, OGIZ, Moscow, 1947; English transl., Ungar, New York, 1956. MR 10, 33; MR 20 #1872.

[2] S. N. Bernšteĭn, *On best approximation of continuous functions by polynomials of given degree*, Soobšč. Har'kov. Mat. Obšč. (2) 13 (1912), 49–194; in *Collected works*. Vol. I: *The constructive theory of functions*, Izdat. Akad. Nauk SSSR, Moscow, 1952, pp. 11–104. MR 14, 2.

[3] ———, *On properties of homogeneous functional classes*, Dokl. Akad. Nauk SSSR 57 (1947), 111–114. (Russian) MR 9, 235.

[4] V. I. Kondrašov, *On some properties of functions in spaces*, Dokl. Akad. Nauk SSSR 48 (1945), 563–566. (Russian)

[5] S. M. Nikol'skiĭ, *A generalization of a theorem of S. N. Bernšteĭn to differentiable functions of several variables*, Dokl. Akad. Nauk SSSR 59 (1948), 1533–1536. (Russian) MR 9, 427.

[6] S. L. Sobolev, *On a theorem of functional analysis*, Mat. Sb. 4 (46) (1938), 471–497; English transl., Amer. Math. Soc. Transl. (2) 34 (1963), 39–68.

[7] ———, *Applications of functional analysis in mathematical physics*, Izdat. Leningrad. Gos. Univ., Leningrad, 1950; English transl., Transl. Math. Monographs, vol. 7, Amer. Math. Soc., Providence, R.I., 1963. MR 14, 565; MR 29 #2624.

[8] J. Marcinkiewicz and A. Zygmund, *Mean values of trigonometrical polynomials*, Fund. Math. 28 (1937), 9–166.

[9] M. Plancherel and G. Pólya, a) *Fonctions entières et intégrales de Fourier multiples*. I, Comment. Math. Helv. (1937), 224–248. b) II, Comment. Math. Helv. (1938), 110–163.

[10] M. Plancherel, *Intégrales de Fourier et fonctions entières. Analyse harmonique*, Colloq. Internat. Centre National de la Recherche Scientifique, no. 15, Centre National Recherche Sci., Paris, 1949, pp. 31–43. MR 11, 340.

[11] A. Zygmund, *Smooth functions*, Duke Math. J. 12 (1945), 47–76. MR 7, 60.

[12] S. M. Lozinskiĭ, *On convergence and summability of Fourier series and interpolation processes*, Mat. Sb. 14 (56) (1944), 175–268. MR 6, 264.

[13] I. G. Petrovskiĭ and K. N. Smirnov, *On equicontinuity conditions for a family of functions*, Bjull. Moskov. Gos. Univ. Sect. A 1938, no. 10, 1–15. (Russian)

Translated by:

N. Koblitz

# PROPERTIES OF CERTAIN CLASSES OF FUNCTIONS OF SEVERAL VARIABLES ON DIFFERENTIABLE MANIFOLDS [1]

S. M. NIKOL'SKIĬ

CONTENTS



## Introduction

This paper is a further development in a certain direction of some results published in our paper [1]. We have in mind the imbedding theorem (see [1], Theorem 12), which in its turn is a development of the corresponding results of S. L. Sobolev [5], [6] and V. I. Kondrašov [7]. Here, properly speaking, we shall take up only a special case of this theorem, when the function of $n$ variables, having the given differentiable properties, is considered in one and the same metric $L_p (1 \le p \le \infty)$ from the point of view of its behavior on a certain $m$-dimensional manifold, $m < n$.

It is convenient for our purposes to formulate that special case of the theorem as follows.

*Suppose that a function defined on a real n-dimensional space $R_n$ belongs to the class* [2] $H_p^{(r_1, \cdots, r_n)}$ *and that for certain nonnegative integers* $\lambda_{m+1}, \cdots, \lambda_n$, *forming a system* $(\lambda)$, *the inequality*

---

1) A short exposition of the basic results of this paper was published in the notes [2] and [4]. Among other results they were reported on at the conference on differential equations in Moscow in May 1952.

2) For the definition of this class see §2.

$$\rho_i^{(\lambda)} = r_i\left(1 - \sum_{j=m+1}^{n} \frac{\lambda_j}{r_j} - \frac{1}{p} \sum_{j=m+1}^{n} \frac{1}{r_j}\right) > 0 \tag{1}$$

*is satisfied. Then the partial derivative*

$$\frac{\partial^{\lambda_{m+1}+\cdots+\lambda_n} f}{\partial x_{m+1}^{\lambda_{m+1}} \cdots \partial x_n^{\lambda_n}} = \varphi_{(\lambda)}(x_1, \ldots, x_m), \tag{2}$$

*as a function of* $x_1, \cdots, x_m$, *for any fixed* $x_{m+1}, \cdots, x_n$ *lies in the class* $H_p^{(\rho_1^{(\lambda)}, \ldots, \rho_m^{(\lambda)})}$.

Examples show (see also [8]) that this assertion, at least in the terms of the classes in question, cannot be strengthened.

In §3 of this paper we show that the converse to the proposition just stated holds as well. In fact, *if we give in advance arbitrary functions* $\phi_{(\lambda)}(x_1, \cdots, x_m)$ *of the variables* $x_1, \cdots, x_m$, *corresponding to distinct (admissible) systems* $(\lambda)$ *of nonnegative numbers for which condition* (1) *holds, each of the functions* $\phi_{(\lambda)}$ *belonging to the corresponding class* $H_p^{(\rho_1^{(\lambda)}, \ldots, \rho_m^{(\lambda)})}$, *then it is possible to construct in* $R_n$ *a function* $f(x_1, \cdots, x_n) \in H_p^{(r_1, \cdots, r_n)}$ *such that for* $x_{m+1} = \cdots = x_n = 0$ *equation* (2) *holds as well.*

Further, in §§5 and 6 it is shown that similar facts hold also for arbitrary curves of $m$-dimensional manifolds, if the latter are differentiable sufficiently many times. Here we may already consider only the case $r_1 = \cdots = r_n = r$. In this case the roles of the partial derivatives (2) are played by the corresponding (mixed) normal derivatives to the manifold.

All of these theorems involve inequalities establishing connections between constants defined by functions entering into the theorem. The significance of these inequalities is not discussed at all in this paper, since it is already long enough without that. We note only that in well-known circumstances they automatically give conditions for the stability of solutions of boundary value problems under appropriate variations of the boundary data.[1)]

Let me discuss for a moment the methods which are applied in this paper.

On the one hand, there is the method of extending functions beyond the limits of the regions where they are given in such a way as to preserve their

1) On this point see [9], [12], [13].

differentiability properties. Here I combine the way in which they are extended (see §3.4) with the already known methods applied in the papers of Hestenes and Whitney (on this point see [10], 1947 ed., pp. 670–674).

On the other hand, we use extensively the methods of the theory of approximation of functions.

Suppose that on some region $G$ of the space we are given a function satisfying certain differential conditions. Then, in order to determine its other differential properties, we extend the function beyond $G_1$, where $\overline{G}_1 \subset G$, to the entire space, preserving all the differential properties. Then it is decomposed into series relative to entire functions $g_{\nu_1 \cdots \nu_n}(x_1, \cdots, x_n)$ of finite degree. Here $\nu_1, \cdots, \nu_n$ are the degrees of $g_{\nu_1 \cdots \nu_n}$ relative to $x_1, \cdots \cdots, x_n$ respectively. Estimates for the terms of the series are obtained using the generalized Jackson inequalities. Further, theorems of the type of the inverse theorems of S. N. Bernšteĭn are applied. Here also an essential role is played not only by the generalized inequality

$$\left\| \frac{\partial g_{\nu_1 \ldots \nu_n}}{\partial x_i} \right\|_p^{(n)} \leqslant \nu_i \| g_{\nu_1 \ldots \nu_n} \|_p^{(n)} \tag{3}$$

of S. N. Bernšteĭn (see [1], Lemma 1), with the aid of which one estimates the norm of the partial derivative, but also by an inequality of another type: [1)]

$$\| g_{\nu_1 \ldots \nu_n} \|_p^{(m)} \leqslant 2^n \left( \prod_{k=m+1}^{n} \nu_k \right)^{\frac{1}{p}} \| g_{\nu_1 \ldots \nu_n} \|_p^{(n)}, \tag{4}$$

which was proved in [1] and using which one can estimate the norm of the function $g_{\nu_1 \cdots \nu_n}$ considered relative to the variables $x_1, \cdots, x_m$ only, with $x_{m+1}, \cdots, x_n$ fixed, in terms of its norm computed in the entire space $R_n$.

## §1. Partial derivatives

1.1. By definition the function $f = f(x_1, \cdots, x_n)$ *has a partial derivative* $\partial^\rho f / \partial x_1^\rho$ *of order* $\rho$ *on the region* $G$, if it may be altered on a set of measure zero in such a way that after that it will have for almost all admissible $(x_2, \cdots, x_n)$ a partial derivative of order $\rho - 1$, absolutely continuous with

1) Inequality (4) is a special case of the following, which holds for $g = g_{\nu_1 \cdots \nu_n}$:

$$\left\{ \int_{-\infty}^{\infty} \cdots \int_{-\infty}^{\infty} \left( \int_{-\infty}^{\infty} \cdots \int_{-\infty}^{\infty} | g(u_1, \ldots, u_m, u_{m+1}, \ldots, u_n) |^p \, du_1 \ldots du_m \right)^{\frac{r}{p}} du_{m+1} \ldots du_n \right\}^{\frac{1}{r}}$$

$$\leqslant 2^n \left( \prod_{k=m+1}^{n} \nu_k \right)^{\frac{1}{p} - \frac{1}{r}} \| g \|_p^{(n)} \quad (1 \leqslant p \leqslant r \leqslant \infty).$$

respect to $x_1$, on any closed segment [1] lying in $G$. In this way the partial derivatives are defined uniquely up to a set of measure zero.

Indeed, suppose that $f$ has already been perturbed on a set of measure zero such that for almost all systems $(x_2, \cdots, x_n)$ forming a set $E$ of measure zero it has an absolutely continuous derivative $\partial^{\rho-1}f/\partial x_1^{\rho-1}$ in $x_1$. Every other alteration of $f$ on a set of measure zero and leading to the same property must encounter certain of the one-dimensional parallels to the $x_1$ axis in cross-sections corresponding to points $(x_2, \cdots, x_n)$ of the set $E$. However the $(n-1)$-dimensional measure of the set of these points must obviously be equal to zero. Such a perturbation leads to a change of $\partial^\rho f/\partial x_1^\rho$ only on a set of $n$-dimensional measure zero.

The mixed derivative

$$\frac{\partial^\rho f}{\partial x_1^{\alpha_1} \ldots \partial x_s^{\alpha_s}} \quad (\alpha_1 + \ldots + \alpha_s = \rho) \tag{1}$$

on $G$ is defined, uniquely up to a set of measure zero, by induction. Indeed, if the partial derivative

$$\psi = \frac{\partial^{\rho - \alpha_s} f}{\partial x_1^{\alpha_1} \ldots \partial x_{s-1}^{\alpha_{s-1}}}$$

is already defined almost everywhere on $G$, then it is possible that, after its alteration on a set of $n$-dimensional measure zero, it will have for almost all $x_1, \cdots, x_{s-1}, x_{s+1}, \cdots, x_n$ an absolutely continuous derivative of order $\alpha_s - 1$ relative to $x_s$ (on any closed segment running parallel to the $x_s$-axis and lying in $G$). We may speak, according to the definition given above, of the derivative

$$\frac{\partial^{\alpha_s} \psi}{\partial x_s^{\alpha_s}}$$

and put

$$\frac{\partial^\rho f}{\partial x_1^{\alpha_1} \ldots \partial x_s^{\alpha_s}} = \frac{\partial^{\alpha_s}}{\partial x_s^{\alpha_s}} \frac{\partial^{\rho-\alpha_s} f}{\partial x_1^{\alpha_1} \ldots \partial x_{s-1}^{\alpha_{s-1}}}.$$

Our task does not include an exposition of the question of the change of

---

1) Here we mean any segment lying in $G$ and lying on a line all of whose points have fixed coordinates $x_2, \cdots, x_n$.

the order of differentiation in general form. Further on (see the corollary in §1.2) it will be clear that additional conditions, which we shall always impose on the functions being considered, lead to the independence of the mixed partial derivatives from the order of differentiation, in every case up to a set of $n$-dimensional measure zero.

We note further that the definition being considered here of the nonmixed derivative $\partial^\rho f/\partial x_1^\rho$ completely coincides with the definition of the corresponding generalized derivative due to Sobolev (see [1], §4), if that derivative and $f$ are summable on the region where the derivative is given. As to the mixed derivative, this is generally not so.

As an example it suffices to consider the function

$$\varphi(x, y) = f_1(x) + f_2(y),$$

analyzed on p. 40 of [6] (transl. pp. 34–35), where $f_1$ and $f_2$ are continuous nowhere differentiable functions. The function $\phi$ does not have Sobolev first partial derivatives $\partial\phi/dx$ and $\partial\phi/\partial y$, and accordingly it does not have them in the sense defined above either. Thus in this sense the question of the derivative $\partial^2\phi/\partial x\partial y$ cannot come up. Nevertheless, the generalized Sobolev derivative $\partial^2\phi/\partial x\partial y$ exists and is identically zero.

However, it is not hard to show that if a function $f(x_1, \cdots, x_n)$ is summable in the region $G$ and if the derivatives

$$\frac{\partial^{\alpha_1} f}{\partial x_1^{\alpha_1}}, \frac{\partial^{\alpha_1+\alpha_2} f}{\partial x_1^{\alpha_1}\partial x_2^{\alpha_2}}, \ldots, \frac{\partial^\rho f}{\partial x_1^{\alpha_1}\ldots\partial x_s^{\alpha_s}} \qquad \left(\rho = \sum_{i=1}^{s} \alpha_i\right) \tag{2}$$

exist and are summable in $G$, then these derivatives are at the same time the corresponding generalized Sobolev derivatives. Conversely, if it is known that a given function $f$ is summable on $G$ and has all the generalized Sobolev derivatives indicated in the series (2), then they are at the same time the corresponding derivatives as defined in this section.

The following situation will always hold in our considerations. Given only that the mixed derivative (1) exists and is summable on $G$, all the partial derivatives appearing in the series (2) are in every case summable on any region $G_1$ whose closure belongs to $G$. Therefore the partial derivatives, mixed and nonmixed, considered in this paper, may be considered as generalized derivatives in the sense of Sobolev.

1.2. Lemma. *Suppose given a sequence of functions*

$$\psi_k = \psi_k(x_1, \ldots, x_n) \quad (k = 1, 2, \ldots),$$

*continuous and integrable to the pth power* ($1 \le p \le \infty$) *along with their partial derivatives to order* $\rho$ *inclusive. Suppose moreover that we are given functions*

$$f, f_{\alpha_1}, f_{\alpha_1\alpha_2}, \ldots, f_{\alpha_1\ldots\alpha_s} \ (\alpha_1 + \alpha_2 + \ldots + \alpha_s \leqslant \rho, \ \alpha_i \text{ positive integers}),$$

*such that*

$$\left.\begin{aligned}
&\lim_{k\to\infty} \| f - \psi_k \|_{L_p(G)} = \lim_{k\to\infty} \Big( \int \ldots \int_G | f - \psi_k |^p dx_1 \ldots dx_n \Big)^{\frac{1}{p}} = 0, \\
&\lim_{k\to\infty} \left\| f_{\alpha_1} - \frac{\partial^{\alpha_1}\psi_k}{\partial x_1^{\alpha_1}} \right\|_{L_p(G)} = 0, \\
&\ldots\ldots\ldots\ldots\ldots\ldots\ldots\ldots\ldots, \\
&\lim_{k\to\infty} \left\| f_{\alpha_1\ldots\alpha_s} - \frac{\partial^{\alpha_1+\ldots+\alpha_s}\psi_k}{\partial x_1^{\alpha_1}\ldots\partial x_s^{\alpha_s}} \right\|_{L_p(G)} = 0.
\end{aligned}\right\} \tag{1}$$

*Then*

$$f_{\alpha_1} = \frac{\partial^{\alpha_1} f}{\partial x_1^{\alpha_1}}, \ f_{\alpha_1\alpha_2} = \frac{\partial^{\alpha_1+\alpha_2} f}{\partial x_a^{\alpha_1}\partial x_2^{\alpha_2}}, \ldots, f_{\alpha_1\ldots\alpha_s} = \frac{\partial^{\alpha_1+\ldots+\alpha_s} f}{\partial x_1^{\alpha_1}\ldots\partial x_s^{\alpha_s}}. \tag{2}$$

For the case $s = 1$ and a right-angled region $G$, this lemma was presented in [1] (Lemma 4). The transition from a right-angled to an arbitrary region $G$ does not present difficulties if we take into account the fact that, as shown above, we require the absolute continuity of $\partial^{\alpha_1-1}f/\partial x_1^{\alpha_1-1}$ on almost all admissible closed segments lying in $G$ and parallel to the $x_1$ axis. The transition from $\alpha_1$ to $\alpha_1\alpha_2, \cdots, \alpha_1 \cdots \alpha_s$ is accomplished without difficulty by induction on applying the lemma for $s = 1$.

**Corollary.** *Since the functions* $\psi_k$ *appearing in the lemma and their partial derivatives to order* $\rho$ *are continuous on* $G$, *the order of differentiation with respect to the various variables in the partial derivatives* (2) *may be changed without destroying their existence, or changing their values outside possibly a set of measure zero.*

1.21. **Lemma.** *Suppose that the conditions of the previous lemma, in particular the relations* (1), *are satisfied for any set of nonnegative numbers* $\alpha_1, \cdots, \alpha_n$ *whose sum does not exceed* $\rho$. *Suppose moreover that the region* $G$ *of the variables* $x_1, \cdots, x_n$ *is mapped in a one-to-one way onto a region* $\tilde{G}$ *of variables* $t_1, \cdots, t_n$, *by means of the functions*

$$x_i = \varphi_i(t_1, \ldots, t_n), \tag{1}$$

*these being continuous and having bounded continuous partial derivatives of orders not exceeding* $p$ *on* $\widetilde{G}$, *and such that the Jacobian*

$$\frac{D(x_1, \ldots, x_n)}{D(t_1, \ldots, t_n)} > K > 0.$$

*Then the function*

$$F(t_1, \ldots, t_n) = f(\varphi_1, \ldots, \varphi_n)$$

*is integrable to the pth power in* $\widetilde{G}$ *along with its partial derivatives of orders up to* $p$ *inclusive, these partial derivatives being computable according to the classical formulas just as if the function* $f$ *had continuous partial derivatives.*

Proof. Under the transformation (1) every measurable set $e \subset G$ goes into a measurable set $\widetilde{e} \subset \widetilde{G}$ and conversely. Hence if $f(x_1, \cdots, x_n)$ is measurable on $G$ then $F(t_1, \cdots, t_n)$ is measurable on $\widetilde{G}$. We have

$$\int \cdots \int_{\widetilde{G}} |f_k(\varphi_1, \ldots, \varphi_n) - f_{k'}(\varphi_1, \ldots, \varphi_n)|^p dt_1 \ldots dt_n \tag{2}$$

$$\leqslant \frac{1}{K} \int \cdots \int_{\widetilde{G}} |f_k(\varphi_1, \ldots, \varphi_n) - f_{k'}(\varphi_1, \ldots, \varphi_n)|^p \frac{D(x_1, \ldots, x_n)}{D(t_1, \ldots, t_n)} dt_1 \ldots dt_n$$

$$= \frac{1}{K} \int \cdots \int_{G} |f_k(x_1, \ldots, x_n) - f_{k'}(x_1, \ldots, x_n)|^p dx_1 \ldots dx_n \to 0 \quad \text{as} \quad k, k' \to \infty.$$

But $f_k(x_1, \cdots, x_n)$ tends to $f(x_1, \cdots, x_n)$ almost everywhere on $G$ for some sequence of values $k = k_1, k_2, \cdots \to \infty$. Accordingly,

$$f_{k_i}(\varphi_1, \ldots, \varphi_n) \to f(\varphi_1, \ldots, \varphi_n) \text{ almost everywhere on } \widetilde{G}, \tag{3}$$

and from (2) and (3) we find that

$$\int \cdots \int_{\widetilde{G}} |f(\varphi_1, \ldots, \varphi_n) - f_k(\varphi_1, \ldots, \varphi_n)|^p dt_1 \ldots dt_n \to 0 \quad \text{as} \quad k \to \infty. \tag{4}$$

Thus we have proved that

$$\int \cdots \int_{\widetilde{G}} |f(\varphi_1, \ldots, \varphi_n)|^p dt_1 \ldots dt_n < \infty.$$

In addition, from the equation

$$\int_G \cdots \int |f_k(x_1, \ldots, x_n)|^p \, dx_1 \ldots dx_n$$
$$= \int_{\widetilde{G}} \cdots \int |f_k(\varphi_1, \ldots, \varphi_n)|^p \frac{D'(x_1, \ldots, x_n)}{D(t_1, \ldots, {}_n)} dt_1 \ldots dt_n,$$

on passing to the limit as $k + \infty$, and from (4) and the assumption that the partial derivatives of the functions $\phi_i$ are bounded on $\widetilde{G}$, we obtain

$$\int_G \cdots \int |f(x_1, \ldots, x_n)|^p \, dx_1 \ldots dx_n$$
$$= \int_{\widetilde{G}} \cdots \int |f(\varphi_1, \ldots, \varphi_n)|^p \frac{D(x_1, \ldots, x_n)}{D(t_1, \ldots, t_n)} dt_1 \ldots dt_n. \tag{5}$$

Put

$$\psi_k(\varphi_1, \ldots, \varphi_n) = F_k(t_1, , \ldots, t_n).$$

The partial derivative $\partial^\alpha F_k / \partial t_1^{\alpha_1} \cdots \partial t_n^{\alpha_n}$ $(\Sigma_{s=1}^n \alpha_s = \alpha \leq \rho)$ is a linear combination of the partial derivatives of the $\psi_k$ with respect to the $x_i$ $(i = 1, 2, \cdots, n)$ of various orders not exceeding $\rho$, calculated according to the well-known classical formulas:

$$\frac{\partial^\alpha F_k}{\partial t_1^{\alpha_1} \ldots \partial t_n^{\alpha_n}} = \lambda_1 \frac{\partial \psi_k}{\partial x_1} + \ldots + \lambda_N \frac{\partial^\alpha \psi_k}{\partial x_n^\alpha}, \tag{6}$$

where the coefficients $\lambda_1, \cdots, \lambda_N$ are certain continuous functions, bounded on the region $\widetilde{G}$.

The following relation holds:

$$\left( \int_{\widetilde{G}} \cdots \int \left| \frac{\partial^\alpha F_k}{\partial t_1^{\alpha_1} \ldots \partial t_n^{\alpha_n}} - \frac{\partial^\alpha F_{k_1}}{\partial t_1^{\alpha_1} \ldots \partial t_n^{\alpha_n}} \right|^p dt_1 \ldots dt_n \right)$$

$$= \int_G \cdots \int \left| \lambda_1 \left( \frac{\partial^\alpha \psi_k}{\partial x_1^\alpha} - \frac{\partial^\alpha \psi_{k_1}}{\partial x_1^\alpha} \right) + \ldots \right|^p \frac{D(t_1, \ldots, t_n)}{D(x_1, \ldots, x_n)} dx_1 \ldots dx_n \to 0$$

$$\text{as } k, k_1 \to \infty,$$

since the Jacobian is bounded on $G$, just as the coefficients $\lambda_1, \cdots$ are bounded (because of the assumption of boundedness of the $\phi_i$).

Accordingly, the partial derivatives

$$\frac{\partial^\alpha F_k}{\partial t_1^{\alpha_1} \ldots \partial t_n^{\alpha_n}}$$

tend as $k \to \infty$ in the $L_p(\widetilde{G})$-norm to the corresponding functions $F_{\alpha_1 \ldots \alpha_n}$ at the same time as $F_k$ tends to $F$ in the $L_p(\widetilde{G})$-norm.

In this case, from Lemma 1.2,

$$F_{\alpha_1 \ldots \alpha_n} = \frac{\partial^\alpha F}{\partial t_1^{\alpha_1} \ldots \partial t_n^{\alpha_n}} \in L_p(\widetilde{G}).$$

Moreover, we may pass to the limit in (6) as $k \to \infty$ (in the sense of $L_p(\widetilde{G})$), so that

$$\frac{\partial^\alpha F}{\partial t_1^{\alpha_1} \ldots \partial t_n^{\alpha_n}} = \lambda_1 \frac{\partial f}{\partial x_1} + \ldots + \lambda_N \frac{\partial^\alpha f}{\partial x_n^\alpha}, \tag{7}$$

and the lemma is proved.

We note further that the equation

$$\int_{\widetilde{G}} \ldots \int \left| \frac{\partial^\alpha F_k}{\partial t_1^{\alpha_1} \ldots \partial t_n^{\alpha_n}} \right|^p dt_1, \ldots dt_n,$$

$$= \int_G \ldots \int \left| \lambda_1 \frac{\partial \psi_k}{\partial x_1} + \ldots + \lambda_N \frac{\partial^\alpha \psi_k}{\partial x_n^\alpha} \right|^p \frac{D(t_1, \ldots, t_n)}{D(x_1, \ldots, x_n)} dx_1 \ldots dx_n$$

$$(k = 1, 2, \ldots)$$

on passing to the limit as $k \to \infty$, directly implies that

$$\int_{\widetilde{G}} \ldots \int \left| \frac{\partial^\alpha F}{\partial t_1^{\alpha_1} \ldots \partial t_n^{\alpha_n}} \right|^p dt_1 \ldots dt_n$$

$$= \int_G \ldots \int \left| \lambda_1 \frac{\partial f}{\partial x_1} + \ldots + \lambda_N \frac{\partial^\alpha f}{\partial x_n^\alpha} \right|^p \frac{D(t_1, \ldots, t_n)}{D(x_1, \ldots, x_n)} dx_1 \ldots dx_n, \tag{8}$$

from which follows, on taking account of the boundedness on $G$ of the coefficients $\lambda$ and of the Jacobian, that

$$\left\| \frac{\partial^\alpha F}{\partial t_1^{\alpha_1} \ldots \partial t_n^{\alpha_n}} \right\|_{L_p(\widetilde{G})}^p \leqslant C \sup_{1 \leqslant \alpha \leqslant \rho} \left\| \frac{\partial^\alpha f}{\partial x_1^{\alpha_1} \ldots \partial x_n^{\alpha_n}} \right\|_{L_p(G)}, \tag{9}$$

where $C$ is a positive constant.

## §2. The classes $H_p^{(r_1,\cdots,r_n)}(G; M_1, \cdots, M_n)$

2.1. Suppose given in the space $R_n$ a region $G$ and an $\eta > 0$. Denote by $G_\eta$ the region lying in $G$ and consisting of the points whose distance from the boundary of $G$ is larger than $\eta$.

We shall say that *the function* $f = f(x_1, \cdots, x_n)$ *lies in the class* $H_{px_1}^{(r)}(G; M)$ if it is bounded on $G$, integrable to the $p$th power on $G$ along with its partial derivatives

$$\frac{\partial^k f}{\partial x_1^k} \quad (k = 1, 2, \ldots, \bar{r}),$$

where $r = \bar{r} + \alpha$ ($\bar{r}$ an integer, $0 < \alpha \leq 1$), and, moreover,

$$\left(\int_{G_\eta}\!\!\cdots\!\int |f_{x_1}^{(\bar{r})}(x_1 + h, x_2, \ldots, x_n) - f_{x_1}^{(\bar{r})}(x_1, \ldots, x_n)|^p\, dx_1 \ldots dx_n\right)^{\frac{1}{p}} < M|h|^\alpha \quad \text{for } \alpha < 1, \tag{1}$$

$$\left(\int_{G_\eta}\!\!\cdots\!\int |f_{x_1}^{(\bar{r})}(x_1 + h, x_2, \ldots, x_n) - 2f_{x_1}^{(\bar{r})}(x_1, \ldots, x_n) + f_{x_1}^{(\bar{r})}(x_1 - h, x_2, \ldots, x_n)|^p\, dx_1 \ldots dx_n\right)^{\frac{1}{p}} < M|h| \quad \text{for } \alpha = 1, \tag{2}$$

for any $h$ satisfying the inequality $|h| < \eta$, where $\eta$ is an arbitrary positive number.

Analogously we define the classes $H_{px_i}^{(r)}(G; M)$ $(i = 2, \cdots, n)$.

Further, if the function $f$ lies simultaneously in the classes $H_{px_i}^{(r_i)}(G; M_i)$ $(i = 1, 2, \cdots, n)$, then we shall say that *it lies in the class* $H_p^{(r_1,\cdots,r_n)}(G; M_1, \cdots, M_n)$.

If $G$ is the entire space $R_n$, we shall take $G_\eta = R_n$. In this case we shall write $H_p^{(r_1,\cdots,r_n)}(M_1, \cdots, M_n)$ instead of $H_p^{(r_1,\cdots,r_n)}(R_n; M_1, \cdots, M_n)$. If we are not interested in the values of the constants which correspond to the given function $f$, then for short we shall write $f \in H_p^{(r_1,\cdots,r_n)}(G)$, or that $f \in H_p^{(r_1,\cdots,r_n)}$ in the case when $G = R_n$. In what follows, if $r_1 = \cdots = r_n = r$ and $M_1 = \cdots = M_n = M$, we shall employ the still shorter notations $H_p^{(r)}(G, M)$, $H_p^{(r)}(G)$, $H_{pn}^{(r)}(M)$, $H_{pn}^{(r)}$ for these classes, instead of $H_p^{(r,\cdots,r)}(G; M, \cdots, M)$, $H_p^{(r,\cdots,r)}(G)$, $H_p^{(r,\cdots,r)}(M, \cdots, M)$, $H_p^{(r,\cdots,r)}$ respectively.

2.11. **Lemma.** *Suppose that the function* $f \in H_{px_1}^{(r)}(G; M)$. *For any* $\eta$ *one has*

$$\left\| \frac{\partial^k f}{\partial x_1^k} \right\|_{L_p(G_\eta)} < c_{1\eta} \| f \|_{L_p(G)} + c_{2\eta} M \qquad (k = 1, \ldots, \bar{r}), \tag{1}$$

*where $c_{1\eta}$ and $c_{2\eta}$ depend possibly on $\eta$ but not on $\|f\|_{L_p(G)}$ or $M$.*

**Proof.** Put $r = \bar{r} + \alpha$, where $\bar{r}$ is an integer and $0 < \alpha \leq 1$.

We first consider the case $0 < \alpha < 1$. For almost all (admissible for $G_\eta$) $x_2, \cdots, x_n$, the function $f$ may be expanded by Taylor's formula, written as follows:

$$f(x_1 + h, x_2, \ldots, x_n) = f(x_1, \ldots, x_n) + h \frac{\partial f}{\partial x_1} + \ldots + \frac{h^{\bar{r}}}{\bar{r}!} \frac{\partial^{\bar{r}} f}{\partial x_1^{\bar{r}}} + R(h), \tag{2}$$

where

$$R(h) = \frac{1}{(\bar{r}-1)!} \int_0^h (h-t)^{\bar{r}-1} \left[ \frac{\partial^{\bar{r}} f(x_1 + t, x_2, \ldots, x_n)}{\partial x_1^{\bar{r}}} - \frac{\partial^{\bar{r}} f(x_1, \ldots, x_n)}{\partial x_1^{\bar{r}}} \right] dt. \tag{3}$$

This equation is therefore valid for almost all $(x_1, \cdots, x_n) \in G$ and all $h$ satisfying the inequality $|h| < \eta$.

If in equation (2) the remainder $R(h)$ is transferred to the left side and if one then substitutes in both sides of the equation the values $h_0$, $h_1, \cdots, h_{\bar{r}}$ in place of $h$, these values being chosen arbitrarily except that they are all distinct and satisfy the inequalities $|h_i| < \eta$, then we obtain a linear system of algebraic equations in the unknowns $f$, $\partial f / \partial x_1, \cdots$ $\cdots, \partial^{\bar{r}} f / \partial x_1^{\bar{r}}$ with a Vandermonde determinant not equal to zero. Hence we conclude that

$$\frac{\partial^k f}{\partial x_1^k} = \sum_{i=0}^{\bar{r}} \gamma_{ik} [f(x_1 + h_i, x_2, \ldots, x_n) - R(h_i)] \quad (k = 0, 1, \ldots, \bar{r}),$$

where $\gamma_{ik}$ $(i, k = 0, 1, \cdots, \bar{r})$ are numbers depending on $h_j$. Accordingly

$$\left\| \frac{\partial^k f}{\partial x_1^k} \right\|_{L_p(G_\eta)}$$

$$\leqslant \sum_{i=0}^{\bar{r}} |\gamma_{ik}| \cdot \left\{ \left( \int \underset{G_\eta}{\cdots} \int |f(x_1 + h_i, x_2, \ldots, x_n)|^p \, dx_1 \ldots dx_n \right)^{\frac{1}{p}} \right.$$

$$\left. + \left( \int \underset{G_\eta}{\cdots} \int |R(h_i)|^p \, dx_1 \ldots dx_n \right)^{\frac{1}{p}} \right\} \leqslant \|f\|_{L_p(G)} \sum_{i=0}^{\bar{r}} |\gamma_{ik}|$$

$$+ \sum_{i=0}^{\bar{r}} |\gamma_{ik}| \frac{1}{(\bar{r}-1)!} \int_0^{h_i} (h_i - t)^{\bar{r}-1} \left( \int \underset{G_\eta}{\cdots} \int \left| \frac{\partial^{\bar{r}} f(x_1 + t, x_2, \ldots, x_n)}{\partial x_1^{\bar{r}}} \right. \right.$$

$$\left. \left. - \frac{\partial^{\bar{r}} f(x_1, \ldots, x_n)}{\partial x_1^{\bar{r}}} \right|^p dx_1 \ldots dx_n \right)^{\frac{1}{p}} \leqslant \|f\|_{L_p(G)} \sum_{i=0}^{\bar{r}} |\gamma_{ik}|$$

$$+ \frac{1}{(\bar{r}-1)!} \sum_{i=0}^{\bar{r}} |\gamma_{ik}| \int_0^{h_i} (h_i - t)^{\bar{r}-1} M t^\alpha dt < c_{1\eta} \|f\|_{L_p(G)} + c_{2\eta} M, \tag{4}$$

where

$$c_{1\eta} = \max_k \sum_{i=0}^{\bar{r}} |\gamma_{ik}|,$$

$$c_{2\eta} = \max_k \frac{1}{(\bar{r}-1)!} \sum_{i=0}^{\bar{r}} |\gamma_{ik}| \int_0^{h_i} (h_i - t)^{\bar{r}-1} t^\alpha dt,$$

and inequality (1) is proved.

The case $\alpha = 1$. Using formulas (2) and (3), and taking into account that $(x_1, \cdots, x_n) \in G_{\eta/2}$ and $|h| < \eta/2$, and then replacing $h$ by $-h$, we get

$$f(x_1 - h, x_2, \ldots, x_n) = f - h \frac{\partial f}{\partial x_1} + \ldots + (-1)^{\bar{r}} \frac{h^{\bar{r}}}{\bar{r}!} \frac{\partial^{\bar{r}} f}{\partial x_1^{\bar{r}}}$$

$$+ \frac{(-1)^{\bar{r}}}{(\bar{r}-1)!} \int_0^h (h - t)^{\bar{r}-1} \left[ \frac{\partial^{\bar{r}} f(x_1 - t, x_2, \ldots, x_n)}{\partial x_1^{\bar{r}}} - \frac{\partial^{\bar{r}} f(x_1, \ldots, x_n)}{\partial x_1^{\bar{r}}} \right] dt. \tag{5}$$

Now suppose that $\bar{r}$ is an even integer. Then, adding (2) and (5), we obtain

$$f(x_1+h, x_2, \ldots, x_n)+f(x_1-h, x_2, \ldots, x_n)$$
$$=2\left(f+\frac{h^2}{2!}\frac{\partial^2 f}{\partial x_1^2}+\ldots+\frac{h^{\bar{r}}}{\bar{r}!}\frac{\partial^{\bar{r}} f}{\partial x_1^{\bar{r}}}\right)+R_*(h), \tag{6}$$

where

$$R_*(h)=\frac{1}{(\bar{r}-1)!}\int_0^h (h-t)^{\bar{r}-1}\left[\frac{\partial^{\bar{r}} f(x_1+t, x_2, \ldots, x_n)}{\partial x_1^{\bar{r}}}-2\frac{\partial^{\bar{r}} f(x_1, \ldots, x_n)}{\partial x_1^{\bar{r}}}\right.$$
$$\left.+\frac{\partial^{\bar{r}} f(x_1-t, x_2, \ldots, x_n)}{\partial x_1^{\bar{r}}}\right]dt. \tag{7}$$

We may transfer $R_*(h)$ to the left side of (6) and replace $h$ in its left and right sides by the numbers $h_k$ $(k=0, 1, \cdots, \bar{r}/2)$, which are arbitrary except that they are distinct and satisfy the inequality $|h_k|<\eta/2$. Then we solve the resulting linear system for $f$, $\partial^2 f/\partial x_1^2, \cdots, \partial^{\bar{r}} f/\partial x_1^{\bar{r}}$. As a result we get

$$\frac{\partial^{2k} f}{\partial x_1^{2k}}=\sum_{i=0}^{\frac{\bar{r}}{2}} \delta_{ik}\,[f(x_1+h_i, x_2, \ldots, x_n)+f(x_1-h_i, x_2, \ldots, x_n)-R_*(h_i)]$$
$$\left(k=0, 1, \ldots, \frac{\bar{r}}{2}\right),$$

so that, integrating both sides over the region $G_{\eta/2}$ and going through the same arguments as used in deriving (4), we get

$$\left\|\frac{\partial^{2k} f}{\partial x_1^{2k}}\right\|_{L_p\left(G_{\frac{\eta}{2}}\right)}<c_{3\eta}\,\|f\|_{L_p(G)}+c_{4\eta}M \tag{8}$$
$$\left(k=0, 1, \ldots, \frac{\bar{r}}{2}\right),$$

where

$$c_{3\eta} = 2 \max_k \sum_{i=0}^{\frac{\bar{r}}{2}} |\delta_{ik}|,$$

$$c_{4\eta} = \max_k \frac{1}{(\bar{r}-1)!} \sum_{i=0}^{\frac{\bar{r}}{2}} |\delta_{ik}| \int_0^{h_i} (h_i - t)^{\bar{r}-1} t\, dt.$$

Since

$$\|\varphi\|_{L_p(G_\eta)} \leqslant \|\varphi\|_{L_p\left(G_{\frac{\eta}{2}}\right)},$$

we see that (8) implies (1), for the time being for even $k$ equal to $2, 4, \cdots, \bar{r}$.

If $k$ is odd and satisfies the inequality $1 < k < \bar{r}$, then on applying Taylor's formula we may write

$$\frac{\partial^{k-1} f(x_1 + h, x_2, \ldots, x_n)}{\partial x_1^{k-1}}$$

$$= \frac{\partial^{k-1} f}{\partial x_1^{k-1}} + h \frac{\partial^k f}{\partial x_1^k} + \int_0^h (h-t) \frac{\partial^{k+1} f(x_1 + t, x_2, \ldots, x_n)}{\partial x_1^{k+1}}\, dt, \tag{9}$$

where we may now suppose that $(x_1, \cdots, x_n) \in G_\eta$ and $|h| < \eta/2$. Transferring the remainder term to the left side and integrating with respect to $x_1, \cdots, x_n$ over the region $G_\eta$, we find on the basis of (8) that

$$|h| \left\| \frac{\partial^k f}{\partial x_1^k} \right\|_{L_p(G_\eta)}$$

$$< \left\| \frac{\partial^{k-1} f}{\partial x_1^{k-1}} \right\|_{L_p\left(G_{\frac{\eta}{2}}\right)} + \int_0^h (h-t) \left( \int_{G_{\frac{\eta}{2}}} \cdots \int \left| \frac{\partial^{k+1} f(x_1, \ldots, x_n)}{\partial x_1^{k+1}} \right|^p dx_1 \ldots dx_n \right)^{\frac{1}{p}} dt$$

$$< (c_{3\eta} \|f\|_{L_p(G)} + c_{4\eta} M) \left(1 + \frac{h^2}{2}\right) = c_{1\eta} \|f\|_{L_p(G)} + c_{2\eta} M.$$

Inequality (1) is thus completely proved for $\alpha = 1$ and even $\bar{r}$. For odd $\bar{r}$ we subtract (5) from (2) and then carry out analogous arguments.

2.2. **Theorem.** *If the function* $f \in H_p^{(r_1, \cdots, r_n)}(G; M_1, \cdots, M_n)$, *then for any* $\eta > 0$ *it is possible to extend it from* $G_\eta$ *to* $R_n$ *with the differential properties preserved. More precisely, there exists a function* $\phi$ *defined on* $R_n$ *and lying in the class* $H_p^{(r_1, \cdots, r_n)}(\bar{M}_1, \cdots, \bar{M}_n)$, *coinciding with* $f$ *on* $G_\eta$, *where*

$$\overline{M}_i < c_{1\eta} \| f \|_{L_p(G)} + c_{2\eta} M_i \quad (i = 1, 2, \ldots, n), \tag{1}$$

*and* $c_{1\eta}$ *and* $c_{2\eta}$ *are positive constants depending only on* $G$, $\eta$, *and* $r_i$ $(i = 1, \cdots, n)$. *In addition we have*

$$\| \varphi \|_p^{(n)} < c_\eta \| f \|_{L_p(G)}, \tag{2}$$

*where* $c_\eta$ *depends only on* $G_\eta$.

Proof. Given an $\eta > 0$, we cover each point of the closure $\overline{G}_\eta$ of $G_\eta$ by disks $\omega$ and $\omega'$ with centers at that point and radii equal to $\eta/3$ and $2\eta/3$ respectively. Now, among the disks $\omega$, we select, using the Borel lemma, a finite number of disks

$$\omega_1, \ldots, \omega_N,$$

covering $G_\eta$. Suppose their centers are

$$P_1, \ldots, P_N$$

respectively.

We define a function $h(t)$, having on the real axis a continuous derivative of order $r > r_k$ $(k = 1, 2, \cdots, n)$ and satisfying the conditions

$$h(t) = 0 \text{ for } t \leqslant 1, \quad h(t) = 1 \text{ for } t \geqslant 2.$$

Further, we put

$$h_i = h_i(P) = h\left(\frac{\overline{PP}_i}{r}\right) \quad \left(r = \frac{\eta}{3}\right) \tag{3}$$

$$(i = 1, 2, \ldots, N),$$

where $\overline{PP}_1$ denotes the distance between the arbitrary point of $P$ of $R_n$ and the point $P_i$. Put

$$H = 1 - h_1 h_2 \ldots h_N.$$

Now we shall suppose that $f = 0$ outside the region $G$, and show that the function

$$\varphi = fH \tag{4}$$

satisfies the assertion of the theorem. [1)]

1) The outline of the argument applied here is analogous to the corresponding outline of Hestenes and Whitney (see [10], 1947 ed., pp. 672–674).

Indeed, if the point $P$ is in $G_\eta$, then it lies also in one of the disks $\omega_1$ and $h_i(0) = 0$, so that $H(P) = 1$. Accordingly

$$\varphi(P) = f(P) \qquad \text{for all} \qquad P \in G_\eta.$$

We note that if $\mu < \eta/3$, then $\Sigma \omega_i' \in G_\mu$, since if $P \in \omega_i'$ and $P_L$ is any point of the boundary $L$ of $G$ we have $\overline{PP}_L \geq \overline{P_iP}_L - \overline{P_iP} \geq \eta - (2/3)\eta = \eta/3$. Therefore every point $P \in G - G_\mu$ lies outside all the disks $\omega_i'$, all the $h_i$ are equal to 1 there, and accordingly $H = 0$. Thus the function $H$, continuous and bounded on $R_n$ along with its partial derivatives, is identically equal to zero on a strip (of positive thickness) of the region $G$ adjacent to its boundary. Hence it follows, since $f \in H_p^{(r_1, \cdots, r_n)}(G)$, that the function $\phi$ is integrable to the pth power on $R_n$ together with its partial derivatives through orders $\overline{r}_i$.

If $0 < h < \mu/2 < 1$, then outside the region $G_{\mu/2}$ not only is $\phi(x_1, \cdots, x_n) = 0$, but also $\phi(x_1 + h, x_2, \cdots, x_n)$ is identically equal to zero. Therefore, since the constant $K$ is larger than the function $H$ and its partial derivatives to order $r > r_k$ $(k = 1, \cdots, n)$ inclusive, for $\alpha_1 < 1$ we will have

$$\Big(\int_{R_n}\!\!\cdots\!\int |\varphi_{x_1}^{(\overline{r}_1)}(x_1+h, x_2, \ldots, x_n) - \varphi_{x_1}^{(\overline{r}_1)}(x_1, \ldots, x_n)|^p\, dx_1 \ldots dx_n\Big)^{\frac{1}{p}}$$

$$= \Big\{\int_{G_{\frac{\mu}{2}}}\!\!\cdots\!\int \Big|\sum_{k=0}^{\overline{r}_1} C_{\overline{r}_1}^k [f_{x_1}^{(k)}(x_1+h, x_2, \ldots) H_{x_1}^{(\overline{r}_1-k)}(x_1+h, x_2, \ldots)$$

$$- f_{x_1}^{(k)}(x_1, \ldots) H_{x_1}^{(\overline{r}_1-k)}(x_1, \ldots)]\Big|^p dx_1 \ldots dx_n\Big\}^{\frac{1}{p}}$$

$$\leqslant \sum_{k=0}^{\overline{r}_1} C_{\overline{r}_1}^k \Big[\Big(\int_{G_{\frac{\mu}{2}}}\!\!\cdots\!\int |H_{x_1}^{(\overline{r}_1-k)}(x_1+h, \ldots)|^p\, |f_{x_1}^{(k)}(x_1+h, x_2, \ldots)$$

$$- f_{x_1}^{(k)}(x_1, x_2, \ldots)|^p\, dx_1 \ldots dx_n\Big)^{\frac{1}{p}}$$

$$+ \Big(\int_{G_{\frac{\mu}{2}}}\!\!\cdots\!\int |f_{x_1}^{(k)}(x_1, \ldots, x_n)|^p\, |H_{x_1}^{(\overline{r}_1-k)}(x_1+h, x_2, \ldots)$$

$$- H_{x_1}^{(\overline{r}_1-k)}(x_1, x_2, \ldots)|^p\, dx_1 \ldots dx_n\Big)^{\frac{1}{p}}\Big] \leqslant$$

$$\leqslant \sum_{k=0}^{\bar r_1} C^k_{\bar r_1} \Big[ K\Big( \int_{G_{\frac{\mu}{2}}} \cdots \int |f^{(k)}_{x_1}(x_1+h, x_2,\ldots) - f^{(k)}_{x_1}(x_1, x_2, \ldots)|^p \, dx_1 \ldots dx_n \Big)^{\frac{1}{p}}$$

$$+ h \Big( \int_{G_{\frac{\mu}{2}}} \cdots \int |f^{(k)}_{x_1}(x_1, \ldots, x_n)|^p \, |H^{(\bar r_1 - k+1)}_{x_1}(x_1 + h\theta, x_2, \ldots, x_n)|^p \, dx_1 \ldots dx_n \Big)^{\frac{1}{p}} \Big]$$

$$\leqslant K \sum_{k=0}^{\bar r_1} C^k_{\bar r_1} \Big[ \Big( \int_{G_{\frac{\mu}{2}}} \cdots \int |f^{(k)}_{x_1}(x_1+h, x_2,\ldots) - f^{(k)}_{x_1}(x_1, x_2, \ldots)|^p \, dx_1 \ldots dx_n \Big)^{\frac{1}{p}}$$

$$+ h \Big( \int_{G_{\frac{\mu}{2}}} \cdots \int |f^{(k)}_{x_1}|^p \, dx_1 \ldots dx_n \Big)^{\frac{1}{p}} \Big]^{1)}$$

$$\leqslant K \Big\{ \sum_{k=0}^{\bar r_1 - 1} C^k_{\bar r_1} \Big( \int_{G_{\frac{\mu}{2}}} \cdots \int \Big| \int_{x_1}^{x_1+h} f^{(k+1)}_{x_1}(u, x_2, \ldots, x_n) \, du \Big|^p dx_1 \ldots dx_n \Big)^{\frac{1}{p}}$$

$$+ M_1 h^{\alpha_1} + 2^{\bar r_1} h \big( c_{1_{\frac{\mu}{2}}} \|f\|_{L_p(G)} + c_{2_{\frac{\mu}{2}}} M_1 \big) \Big\}^{2)}$$

$$\leqslant K \Big\{ \sum_{k=0}^{\bar r_1 - 1} C^k_{\bar r_1} h^{\frac{1}{q}} \Big( \int_{G_{\frac{\mu}{2}}} \cdots \int \int_{x_1}^{x_1+h} |f^{(k+1)}_{x_1}(u, x_2, \ldots, x_n)|^p \, du \, dx_1 \ldots dx_n \Big)^{\frac{1}{p}}$$

$$+ M_1 h^{\alpha_1} + 2^{\bar r_1} h \big( c_{1_{\frac{\mu}{2}}} \|f\|_{L_p(G)} + c_{2_{\frac{\mu}{2}}} M_1 \big)^{3)}$$

$$\leqslant K \Big\{ \sum_{k=0}^{\bar r - 1} C^k_{\bar r_1} h^{\frac{1}{p} + \frac{1}{q}} \Big( \int_{G_{\frac{\mu}{4}}} \cdots \int |f^{(k+1)}_{x_1}(u, x_2, \ldots, x_n)|^p du \, dx_1 \ldots dx_n \Big)^{\frac{1}{p}}$$

$$+ M_1 h^{\alpha_1} + 2^{\bar r_1} h \big( c_{1_{\frac{\mu}{2}}} \|f\|_{L_p(G)} + c_{2_{\frac{\mu}{2}}} M_1 \big) \Big\} \leq {}^{4)}$$

---

1) Because of Lemma 2.11.

2) The Minkowski inequality; $1/p + 1/q = 1$.

3) We change the order of integration with respect to $x_1$ and $u$. We are taking $0 < h < \mu/4$.

4) Lemma 2.11.

$$\leqslant K\Big\{\sum_{k=0}^{\bar{r}-1} C_{\bar{r}_1}^k h\,(c_{1_{\frac{\mu}{4}}}\|f\|_{L_p(G)} + c_{2_{\frac{\mu}{4}}} M_1) + M_1 h^{\alpha_1}$$

$$+\, 2^{\bar{r}_1} h\,(c_{1_{\frac{\mu}{2}}}\|f\|_{L_p(G)} + c_{2_{\frac{\mu}{2}}} M_1)\Big\} \leqslant (c_{1\eta}^{(1)}\|f\|_{L_p(G)} + c_{2\eta}^{(1)} M_1)\, h^{\alpha_1} \qquad (5)$$

($\mu$ may be considered to depend on $\eta$, for example, putting $\mu = \eta/4$).

The resulting inequality is proved so far for $0 < h < \mu/4$. For $h > \mu/4$ we have

$$\Big(\int_{R_n}\!\!\dots\!\int |\,\varphi_{x_1}^{(\bar{r}_1)}(x_1 + h, x_2, \dots, x_n) - \varphi_{x_1}^{(\bar{r}_1)}(x_1, \dots, x_n)\,|^p\, dx_1 \dots dx_n\Big)^{\frac{1}{p}}$$

$$\leqslant 2\,\|\,\varphi_{x_1}^{(\bar{r}_1)}\|_p^{(n)} \leqslant 2\,\|\sum_{k=0}^{\bar{r}_1} C_{\bar{r}_1}^k f_{x_1}^{(k)} H^{\bar{r}_1 - k}\|_{L_p(G_\mu)}{}^{1)}$$

$$\leqslant 2^{\bar{r}_1+1}\,(c_{1\eta}\,\|f\|_{L_p(G)} + c_{2\eta} M_1)\, K\Big(\frac{4h}{\mu}\Big)^{\alpha_1}.$$

Therefore, putting $\mu = \eta/4$, we may say that inequality (5) is satisfied for any $h$, if the constants $c_{1\eta}^{(1)}$ anc $c_{2\eta}^{(2)}$ are increased where necessary so that they become not less than the corresponding coefficients on $\|f\|_{L_p(G)}$ and the $M_1$ on the right side of the last inequality.

We carry out similar arguments for $x_i$. Then we put

$$c_{1\eta} = \max_i c_{1\eta}^{(i)}, \qquad c_{2\eta} = \max_i c_{2\eta}^{(i)}.$$

If $\alpha_1 = 1$ the arguments just presented remain the same up to inequality (5). This last must be replaced by the corresponding inequality for the integral over $R_n$ of the pth power of the second difference

$$\varphi_{x_1}^{(\bar{r}_1)}(x_1 + h,\ x_2, \dots) - 2\varphi_{x_1}^{(\bar{r}_1)}(x_1,\ x_2, \dots) + \varphi_{x_1}^{(\bar{r}_1)}(x_1 - h,\ x_2, \dots).$$

Here the second differences for the functions $f_{x_1}^{(k)} H_{x_1}^{(k)}$ have to be transformed, as is done below for the function

$$f(t+h)\,H(t+h) - 2f(t)\,H(t) + f(t-h)\,H(t-h)$$
$$= [f(t+h) - 2f(t) + f(t-h)]\,H(t+h) + 2f(t)\,[H(t+h) - H(t)]$$
$$- f(t-h)\,[H(t+h) - H(t-h)].$$

The constants $c_{1\eta}$ and $c_{2\eta}$ will of course be different.

We proceed in the same way for the partial derivatives $\phi_{x_2}^{(\bar{r}_2)}, \dots$ $\dots, \phi_{x_n}^{(\bar{r}_n)}$.

---

1) Lemma 2.11.

Inequality (3) follows immediately from equation (4), if we put

$$c_\eta = \max_G |H|$$

(the function $H$ depends on $\eta$). This proves the theorem.

2.3. The arguments preceding inequality (4) imply the following result.

**Theorem.** *If the function $f$ and its partial derivatives of orders $r_i$ respectively ($r_i$ integers, $i = 1, \cdots, n$) are integrable to the pth power on $G$ (and continuous on $G$), then, for any $\eta > 0$, $f$ may be extended beyond the region $G_\eta$ to the entire space $R_n$ in such a way that the extended function will have on $R_n$ derivatives of orders $r_i$ which are integrable to the pth power and continuous.* [1]

Indeed, from equation (4) of §2.2 we get

$$\varphi_{x_i}^{(l)} = \sum_{k=0}^{l} C_l^k f_{x_i}^{(k)} H_{x_i}^{(l-k)},$$

so that, using the notation adopted above, we obtain

$$\| \varphi_{x_i}^{(l)} \|_p^{(n)} \leqslant (c_{1\eta} \| f \|_{L_p(G)} + c_{2\eta} M_i) K \sum_{k=0}^{l} C_l^k.$$

The functions $\phi$ and its partial derivatives are continuous along with $f$.

2.31. If $r_1 = \cdots = r_n = r$ and $M_1 = \cdots = M_n = M$, then we may write

$$H_p^{(r)}(G;\ M), \qquad H_{pn}^{(r)}(M) = H_p^{(r)}(R_n;\ M)$$

for short, instead of

$$H_p^{(r,\ldots,r)}(G;\ M), \qquad H_p^{(r,\ldots,r)}(M) = H_p^{(r,\ldots,r)}(R_n;\ M)$$

respectively.

2.32. **Theorem.** *If the function $f = f(x_1, \cdots, x_n)$ belongs to the class $H_p^{(r)}(G;\ M)$, then for any $\eta > 0$ one has*

$$\left\| \frac{\partial^\lambda f}{\partial x_1^{\lambda_1} \ldots \partial x_n^{\lambda_n}} \right\|_{L_p(G_\eta)} < c_{1\eta} \| f \|_{L_p(G)} + c_{2\eta} M, \qquad \begin{matrix} \lambda_1 + \ldots + \lambda_n = \lambda \\ 0 \leqslant \lambda \leqslant r \end{matrix} \tag{1}$$

*where $c_{1\eta}$ and $c_{2\eta}$ depend on $\eta$ but not on $M$ and $\|f\|_{L_p(G)}$.*

**Proof.** By Theorem 2.2, $f$ may be prolonged from $G_\eta$ to the entire space $R_n$ in such a way that the following properties will hold for the extended function, which we shall denote by $\overline{f}$:

---

1) If the boundary of the region $G$ is sufficiently smooth the theorem may be strengthened (see [13]). The extension can be carried out from $G$ instead of $G_\eta$.

$$1)\qquad \bar f \in H^{(r)}_{pn}(\overline{M}),$$
$$2)\qquad \overline{M} < c'_\eta \|f\|_{L_p(G)} + c''_\eta M,$$
$$3)\qquad \|\bar f\|^{(n)}_p < c_\eta \|f\|_{L_p(G)}.$$

On the basis of Theorem 7 of [1] it follows from these properties that if one denotes by $g_\nu = g_\nu(x_1, \cdots, x_n)$ an entire function of degree $2^\nu$ in each of the variables $x_i$, which is the best approximation to $\bar f$ in the sense of $L_p^{(n)}$, then

$$\|\bar f - g_\nu\|^{(n)}_p < \frac{nd\overline{M}}{2^{r\nu}} \quad (\nu = 1, 2, \ldots), \tag{2}$$

where $d$ depends only on $r$. Thus the function $\bar f$ may be represented as a series of the form

$$\bar f = \sum_{\nu=0}^{\infty} Q_\nu, \quad Q_0 = g_0, \quad Q_\nu = g_\nu - g_{\nu-1} \quad (\nu = 1, 2, \ldots), \tag{3}$$

converging in the sense of $L_p^{(n)}$ to $\bar f$. In addition,

$$\|Q_\nu\|^{(n)}_p \leqslant 2\|\bar f - g_{\nu-1}\|^{n}_p \leqslant \frac{2nd\overline{M}}{2^{r(\nu-1)}} = \frac{c_1\overline{M}}{2^{r\nu}} \quad (\nu = 1, 2, \ldots), \tag{4}$$

$$\|Q_0\|^{(n)}_p \leqslant nd\overline{M} + \|f\|^{(n)}_p. \tag{5}$$

If we apply the generalized Bernšteĭn inequality ([1], Lemma 1)

$$\left|\frac{\partial g_\nu}{\partial x_1}\right|^{(n)}_p \leqslant \nu \|g_\nu\|^{(n)}_p, \tag{6}$$

then we get

$$\left\|\frac{\partial^\lambda Q_\nu}{\partial x_1^{\lambda_1} \ldots \partial x_n^{\lambda_n}}\right\|^{(n)}_p \leqslant \frac{c_1\overline{M} \cdot 2^{\lambda\nu}}{2^{r\nu}} \leqslant \frac{c_1\overline{M}}{2^{(r-\lambda)\nu}} \quad (\nu = 1, 2, \ldots),$$

$$\left\|\frac{\partial^\lambda Q_0}{\partial x_1^{\lambda_1} \ldots \partial x_n^{\lambda_n}}\right\|^{(n)}_p \leqslant nd\overline{M} + \|\bar f\|^{(n)}_p.$$

Thus, for $0 \le \lambda \le \bar r$,

$$\left\| \frac{\partial^\lambda f}{\partial x_1^{\lambda_1} \ldots \partial x_n^{\lambda_n}} \right\|_{L_p(G_\eta)} \leqslant \left\| \frac{\partial^\lambda \bar{f}}{\partial x_1^{\lambda_1} \ldots \partial x_n^{\lambda_n}} \right\|_p^{(n)} \leqslant nd\bar{M} + \| f \|_p^{(r)}$$

$$+ \sum_{\nu=1}^{\infty} \frac{c_1 \bar{M}}{2^{(r-\lambda)\nu}} < \| f \|_p^{(n)} + c_2 \bar{M} < c_\eta \| f \|_{L_p(G)} + c_2 (c'_\eta \| f \|_{L_p(G)} + c''_\eta M)$$
$$< c_{1\eta} \| f \|_{L_p(G)} + c_{2\eta} M,$$

and the theorem is proved.

2.4. **Lemma.** *Suppose that $r > 0$ and $r = \bar{r} + \alpha$, where $\bar{r}$ is an integer and $0 < \alpha \leq 1$. Suppose further that some open region $G$ of the space $R_n$ of points $(x_1, \cdots, x_n)$ maps in a one-to-one way onto a region $G^*$ of the space $R_n^*$ of points $(u_1, \cdots, u_n)$ by means of the functions*

$$u_i = \varphi_i (x_1, \ldots, x_n), \tag{1}$$

*which are continuous and have continuous and bounded partial derivatives on $G$ to orders not exceeding $\bar{r} + 1$, and such that the Jacobian*

$$\frac{D(u_1, \ldots, u_n)}{D(x_1, \ldots, x_n)} > K > 0 \quad (on\ G). \tag{2}$$

*Suppose finally that the function $f(u_1, \cdots, u_n)$ lies in the class $H_p^{(r)}(G^*, M^*)$. Then the following assertions are valid.*

1) *The function*

$$F(x_1, \ldots, x_n) = f(\varphi_1, \ldots, \varphi_n)$$

*of the variables $x_1, \cdots, x_n$ is integrable to the pth power on $G$ and has partial derivatives relative to the $x_i$, computed almost everywhere on $G$ by the same formulas which would hold if the function $f(u_1, \cdots, u_n)$ had continuous partial derivatives relative to the $u_i$.*

2) *For any $\eta > 0$ the function $F$ belongs to the class $H_p^{(r)}(G_\eta; M_\eta)$, where*

$$M_\eta < a_{1\eta} \| f \|_{L_p(G^*)} + a_{2\eta} M^*. \tag{3}$$

*Here the $a_{1\eta}$ and $a_{2\eta}$ are constants depending, aside from $r$ and the functions $\phi_i$, on $G_\eta$.*

*Moreover,*

$$\| F \|_{L_p(G)} \leqslant \frac{1}{K^{\frac{1}{p}}} \| f \|_{L_p(G^*)}. \tag{3'}$$

**Remark.** If $G = R_n$, then by (2) we also must have $G^* = R_n^*$. In this case $G_\eta = G$ and the constants $a_{1\eta}$ and $a_{2\eta}$ do not depend on $\eta$.

Suppose given a $\mu > 0$. We extend the function $f(u_1, \cdots, u_n)$ beyond the limits of $G_\mu^*$ to the entire space $R_n^*$ in such a way that the extended function $\bar{f} \in H_p^{(r)}(\bar{M}^*)$. From Theorem 2.2 we may suppose that,

$$\bar{M}^* < c_{1\mu} \| f \|_{L_p(G^*)} + c_{2\mu} M^*,$$
$$\| \bar{f} \|_p^{(n)} < c_\mu \| f \|_{L_p(G^*)}. \tag{4}$$

Denote by $g_\nu = g_\nu(u_1, \cdots, u_n)$ an entire function of degree $2^\nu$ in each of its variables $u_i$, which is the best approximation to $\bar{f}$ in the sense of $L_p^{(n)}$. Then (see [1], Theorem 7)

$$\| \bar{f} - g_\nu \|_p^{(n)} < \frac{nd\bar{M}^*}{2^{r\nu}} \quad (\nu = 1, 2, \ldots), \tag{5}$$

where $d$ depends only on $r$. One may then represent $\bar{f}$ in the form of a series

$$\bar{f} = \sum_{\nu=0}^{\infty} Q_\nu, \quad Q_0 = g_0, \quad Q_\nu = g_\nu - g_{\nu-1} \quad (\nu = 1, 2, \ldots), \tag{6}$$

converging in the sense of $L_p^{(n)}$ to $f$. In addition,

$$\| Q_\nu \|_p^{(n)} \leqslant 2 \| \bar{f} - g_{\nu-1} \|_p^{(n)} \leqslant \frac{2nd\bar{M}^*}{2^{r(\nu-1)}} = \frac{c_1\bar{M}^*}{2^{r\nu}} \quad (\nu = 1, 2, \ldots),$$
$$\| Q_0 \|_p^{(n)} \leqslant nd\bar{M}^* + \| \bar{f} \|_p^{(n)} = c_2\bar{M}^* + \| \bar{f} \|_p^{(n)}. \tag{7}$$

Suppose that $G_\mu$ is the region of variation of the variables corresponding to $G_\mu^*$.

In view of (7) the series (6) converges in the sense of $L_p(R_n^*)$ to the function $\bar{f}$. And if one recalls also inequality (6) of §2.32, according to which

$$\left\| \frac{\partial^\rho Q_\nu}{\partial x_1^{\alpha_1} \ldots \partial x_n^{\alpha_n}} \right\|_p^{(n)} \leqslant 2^{\rho\nu} \| Q_\nu \|_p^{(n)}, \tag{8}$$

then one may conclude that the series (6), after $\rho$ termwise differentiations relative to any variable, where $\rho \leq \bar{r}$, will still continue to converge in the sense of $L_p(R_n^*)$ and moreover in the sense of $L_p(G_\mu^*)$. Therefore, on the basis of Lemma 1.21, the function $F = f(\phi_1, \cdots, \phi_n)$ has on $G_\mu$ (and, in view of the arbitrary smallness of $\mu$, on $G$ as well) partial derivatives of orders $\rho \leq \bar{r}$, computed according to the ordinary classical rules, as if the function $f$ had ordinary partial derivatives. Thus for almost all $(x_1, \cdots, x_n)$ of $G$ we have

$$\frac{\partial^\rho f(\varphi_1, \ldots, \varphi_n)}{\partial x_1^{\alpha_1} \ldots \partial x_n^{\alpha_n}} = \lambda_1 \frac{\partial f}{\partial u_1} + \ldots + \lambda_N \frac{\partial^\rho f}{\partial u_n^\rho}, \tag{9}$$

where the functions $\lambda_k$, because of the conditions imposed on the $\phi_i$, are continuous and bounded on $G^*$. One also has (see §1.21, formula (5)) the equation [1])

$$\int_G \cdots \int |F|^p \, dx_1 \ldots dx_n = \int_{G^*} \cdots \int |f|^p \frac{D(x_1, \ldots, x_n)}{D(u_1, \ldots, u_n)} du_1 \ldots du_n,$$

from which, taking account of (2), inequality (3′) follows directly.

Thus the first part of the theorem is proved. In order to prove the second part for $\alpha < 1$, we need to prove that for every $\eta > 0$ the partial derivatives $\partial^\rho f / \partial x_i^\rho$ ($\rho \leq \bar{r}$, $i = 1, 2, \cdots, n$) are integrable to the $p$th power on $G_\eta$ and that for every $\eta_1 > \eta$ and $|h| < \eta_1 - \eta$ the inequality

$$\left( \int_{G_{\eta_1}} \cdots \int |F_{x_1}^{(\bar{r})}(x_1 + h, x_2, \ldots, x_n) - F_{x_1}^{(\bar{r})}(x_1, \ldots, x_n)|^p dx_1 \ldots dx_n \right)^{\frac{1}{p}} \leqslant M_\eta |h|^\alpha \tag{10}$$

holds, and also the similar inequality when $x_1$ is replaced by $x_i$ ($i = 2, \cdots, n$). At the same time one can give for $M_\eta$ an estimate of the form (3). Now in the case $\alpha = 1$ we need to show the validity of analogous inequalities, where the second difference appears under the integral sign instead of the first difference $F_{x_i}^{(\bar{r})}$.

The integrability to the $p$th power on $G_\eta$ of the partial derivatives $\partial^\rho F / \partial x_1^\rho$ for $p \leq \bar{r}$ will follow from (9), where we need to assume that $\alpha_s = 0$ for $s \neq i$, as soon as the integrability of the partial derivatives of $f$ with respect to the $u_k$ which appear in the right side of (9) is established over the region $G^*_{(\eta)}$ corresponding to $G_\eta$. The integrability to the $p$th power of those derivatives which are nonmixed was assumed, inasmuch as

---

1) Directly from formula (5) of §1.21 there follows a similar equation for $G_{(\mu)}$ and $G^*_{(\mu)}$. Then we need to pass to the limit as $\mu \to 0$.

$f \in H_p^{(r)}(G^*)$. As to the mixed derivatives, their integrability in the sense of $L_p(G^*_{(\eta)})$ follows from the fact that $f$, as we know, may be extended beyond the limits of $G^*_{(\eta)}$ to $R^*_n$ in such a way that the extended function $f_1 \in H_p^{(r)}$, and then on the basis of Theorem 13 of [1] $f_1$ will have mixed derivatives of all orders to $\bar{r}$ inclusive and lying in $L_p(R_n)$.

The further arguments connected with the establishment of inequality (10) and the corresponding inequality for $\alpha = 1$ will be presented separately for $\alpha < 1$ and $\alpha = 1$.

The case $0 < \alpha < 1$. Suppose given $\eta > 0$. Suppose that $\eta_1 > \eta$. The image of $G_\eta$ under the transformation (1) will be a region $G^*_{(\eta)}$, completely lying in some region $G^*_\mu$, where $\mu = \mu(\eta)$ is a function of $\eta$.

We give to $x_1$ an increment $h$, where $|h| < \eta_1 - \eta$. Thus, if $(x_1, \cdots, x_n) \in G_{\eta_1}$, then $(x_1 + h, x_2, \cdots, x_n) \in G_\eta$, and at such points we may use formula (9), in which we may take $\rho = \bar{r}$. Obviously

$$|\Delta u_i| = |\varphi_i(x_1 + h, x_2, \ldots, x_n) - \varphi_i(x_1, \ldots, x_n)| \leqslant \gamma |h|, \qquad (11)$$

where $\gamma$ is a positive constant. We may write our difference in the form of a sum:

$$F^{(\bar{r})}_{x_1}(x_1 + h, x_2, \ldots) - F^{(\bar{r})}_{x_1}(x_1, x_2, \ldots)$$

$$= \sum_{k=1}^{N} [\lambda_k (u_1 + \Delta u_1, \ldots, u_n + \Delta u_n) f^{(l)}_u (u_1 + \Delta u_1, \ldots, u_n + \Delta u_n)$$

$$- \lambda_k (u_1, \ldots, u_n) f^{(l)}_u (u_1, \ldots u_n)],$$

where $f^{(l)}_u$ denotes one of the partial derivatives (the mixed derivatives included) of order $l \leq \bar{r}$. Each term of this sum in its turn may be represented in the form $\sigma + \sigma'$, where

$$\sigma = \lambda_k (u_1 + \Delta u_1, \ldots) [f^{(l)}_u (u_1 + \Delta u_1, \ldots) - f^{(l)}_u (u_1, \ldots)],$$

$$\sigma' = f^{(l)}_u (u_1, \ldots) [\lambda_k (u_1 + \Delta u_1, \ldots) - \lambda_k (u_1, \ldots)].$$

The functions $\lambda_k$ are simple combinations of the partial derivatives of the functions $\phi_i$ of orders not larger than $\bar{r}$. Since by hypothesis the $\phi_i$ have continuous and bounded derivatives to order $\bar{r} + 1$ inclusive on $G$, we have

$$|\lambda_k| \leqslant c, \quad |\lambda_k (u_1 + \Delta u_1, \ldots) - \lambda_k (u_1, \ldots)| \leqslant c |h|, \qquad (12)$$

and accordingly, for the region $G^*_{(\eta_1)} \subset R^*_n$ corresponding to $G_{\eta_1}$, we will have

$$\left(\int_{G^*_{(\eta)}}\!\!\cdots\int |\sigma'|^p \frac{D(x_1, \ldots, x_n)}{D(u_1, \ldots, u_n)} du_1 \ldots du_n\right)^{\frac{1}{p}}$$

$$\leqslant \frac{1}{K^{\frac{1}{p}}} c\,|h|\left(\int_{G^*_{\mu}}\!\!\cdots\int |f_u^{(l)}|\, du_1 \ldots du_n\right)^{\frac{1}{p}} \text{[1]}$$

$$\leqslant c_1 |h|\,(c_{1\mu} \|f\|_{L_p(G^*)} + c_{2\mu} M^*), \tag{13}$$

$$\left(\int_{G^*_{(\eta_1)}}\!\!\cdots\int |\sigma|^p \frac{D(x_1, \ldots, x_n)}{D(u_1, \ldots, u_n)} du_1 \ldots du_n\right)^{\frac{1}{p}}$$

$$\leqslant c_2\left(\int_{G^*_{(\eta_1)}}\!\!\cdots\int |f_u^{(l)}(u_1 + \Delta u_1, \ldots) - f_u^{(l)}(u_1, \ldots)|^p du_1 \ldots du_n\right)^{\frac{1}{p}} = c_2 I. \tag{14}$$

Below we shall show that each of the quantities $I$ satisfies the inequality

$$I \leqslant (c_{3\mu} \|f\|_{L_p(G^*)} \mid c_{4\mu} M^*)\,|h|^{\alpha} \qquad (0 < \alpha < 1). \tag{15}$$

Therefore it follows from (13)–(15) that

$$\left(\int_{G_{\eta_1}}\!\!\cdots\int |F_{x_1}^{(\overline{r})}(x_1 + h,\ x_2, \ldots, x_n) - F_{x_1}^{(\overline{r})}(x_1, \ldots, x_n)|^p dx_1 \ldots dx_n\right)^{\frac{1}{p}}$$

$$\leqslant \frac{1}{K^{\frac{1}{p}}}\Bigg[\int_{G^*_{(\eta_1)}}\!\!\cdots\int \Big|\sum_{k=1}^{N} [\lambda_k (u_1 + \Delta u_1, \cdots, u_n + \Delta u_n) f_u^{(l)} (u_1 + \Delta u_1, \ldots, u_n + \Delta u_n)$$

$$- \lambda_k (u_1, \ldots, u_n) f_u^{(l)} (u_1, \ldots, u_n]\Big|^p \frac{D(x_1, \ldots, x_n)}{D(u_1, \ldots, u_n)} du_1 \ldots du_n\Bigg]^{\frac{1}{p}}$$

$$< a_{1\mu} \|f\|_{L_p(G^*)} + a_{2\mu} M^*,$$

and thus inequality (3) is proved.

We note that to each point $(u_1, \cdots, u_n) \in G^*_{(\eta_1)}$ there corresponds a unique point $(x_1, \cdots, x_n) \in G_{\eta_1}$, this relation being established by the equations

$$u_i = \varphi_i (x_1, \ldots, x_n) \qquad (i = 1, 2, \ldots, n).$$

---

1) Theorem 2.32.

This last point for a given $h$ ($|h| < \eta_1 - \eta$) corresponds to a unique point $(v_1, \cdots, v_n)$ whose coordinates are expressed by the equations

$$\left.\begin{aligned} v_1 = u_1, \ \ldots, \ v_s &= u_s, \\ v_{s+1} = u_{s+1} + \Delta u_{s+1} &= \varphi_{s+1}(x_1 + h, x_2, \ldots, x_n), \\ \ldots\ldots\ldots\ldots\ldots &\ldots\ldots\ldots\ldots, \\ v_n = u_n + \Delta u_n &= \varphi_n(x_1 + h, x_2, \ldots, x_n). \end{aligned}\right\} \tag{16}$$

Thus we consider $n$ transformations, corresponding to the values $s = 0, \cdots, n-1$, of the point $(u_1, \cdots, u_n)$ into the point $(v_1, \cdots, v_n)$, depending continuously on $h$ and turning into identity when $h = 0$. The Jacobian of each transformation (16) on $G^*_{(\eta_1)}$ depends continuously on $h$ and this dependence is uniform relative to $G^*_{(\eta_1)}$. For $h = 0$ it is equal to 1, and hence for sufficiently small $h_0$ we have

$$\frac{D(v_1, \ldots, v_n)}{D(u_1, \ldots, u_n)} > \frac{1}{2}, \quad (u_1, \ldots, u_n) \in G^*_{(\eta)} \tag{17}$$

for all $|h| < h_0$. One may therefore assert that the transformation (16), for sufficiently small $h$, is not only uniquely defined, but one-to-one, since under the conditions holding here the implicit function theorem implies that there exists a $\rho$ so small that any set of $G^*_{(\eta_1)}$ having diameter $\rho$ is mapped under equations (16) in a one-to-one manner for all $h$ with $|h| < h_0$.

If we suppose the existence of a sequence of numbers $h_k$, for each of which there is a pair of points $(u_1^{(k)}, \cdots, u_n^{(k)})$ and $(u_1^{(k)'}, \cdots, u_n^{(k)'})$ in $G^*_{(\eta_1)}$ going under the transformation (16) into the same point $(v_1^{(k)}, \cdots$ $\cdots, v_n^{(k)})$, then the points of this pair must lie at a distance greater than $\rho$ from one another. But this is impossible, since because of the uniform continuity of (16) relative to $h$ one would have to conclude that three of our points for some sequence of values of $k$ would tend respectively to the points $(u_1, \cdots, u_n)$, $(u_1', \cdots, u_n')$ and $(v_1, \cdots, v_n)$, the first two being distinct, and at the same time all three must coincide, since if $h = 0$ our transformation is the identity.

From what has been said it follows that it is legitimate to apply the ordinary formulas of change of variables $(u_1, \cdots, u_n)$ to $(v_1, \cdots, v_n)$ in the integral, as will be done in what follows.

On the basis of (16), for points of the region $G^*_\mu$ lying in $G^*_{(\eta)}$ one may write

$$f = \overline{f} = \sum_{\nu=0}^{\infty} Q_\nu,$$

so that after differentiating this equation termwise $l$ times we obtain

$$f_u^{(l)} = \overline{f}_u^{(l)} = \sum_{\nu=0}^{\infty} Q_{\nu u}^{(l)}, \quad l \leqslant \overline{r}, \tag{18}$$

the symbol $f_u^{(l)}$ again expressing any partial derivative of the $l$th order, mixed derivatives included. Differentiation $l$ times is legitimate here, since, from (7), on the basis of the generalized Bernšteĭn inequality, we have

$$\| Q_{\nu u}^{(l)} \|_p^n \leqslant \frac{c_1 \overline{M}^*}{2^{(r-l)\nu}}, \quad r - l > 0 \quad (\nu = 1, 2, \ldots). \tag{19}$$

Suppose $h$ is given. We choose an integer $k$ such that

$$2^{-k} \leqslant |h| < 2^{-(k-1)}.$$

We write $f_u^{(l)}$ in the form of a sum:

$$f_u^{(l)} = S' + S'',$$

where

$$S' = \sum_{\nu=0}^{k-1} Q_{\nu u}^{(l)}, \quad S'' = \sum_{\nu=k}^{\infty} Q_{\nu u}^{(l)}.$$

We have

$$\Big(\int_{G^*_{(\eta_1)}}\!\!\ldots\int | S'(u_1 + \Delta u_1, \ldots, u_n + \Delta u_n) - S'(u_1, \ldots, u_n)|^p \, du_1, \ldots, du_n\Big)^{\frac{1}{p}}$$

$$\leqslant \sum_{\nu=0}^{k-1} \sum_{s=1}^{n} \Big\{\int_{G^*_{(\eta_1)}}\!\!\ldots\int | Q_{\nu u}^{(l)}(u_1, \ldots, u_{s-1}, u_s + \Delta u_s, \ldots, u_n + \Delta u_n)$$

$$- Q_{\nu u}^{(l)}(u_1, \ldots, u_s, u_{s+1} + \Delta u_{s+1}, \ldots, u_{n+1} + \Delta u_{n+1}) |^p \, du_1 \ldots du_n\Big\}^{\frac{1}{p}}$$

$$= \sum_{\nu=0}^{k-1} \sum_{s=1}^{n} I_{\nu s} < (c' \|\overline{f}\|_p^{(n)} + c''\overline{M}^*) |h|^\alpha,$$

since

$$I_{\nu s}=\Big(\int\limits_{G^*_{(\eta_1)}}\!\!\dots\!\int | Q^{(l)}_{\nu u}(u_1,\dots,u_{s-1},u_s+\Delta u_s,\dots,u_n+\Delta u_n)$$

$$-Q^{(l)}_{\nu u}(u_1,\dots,u_{s+1}+\Delta u_{s+1},\dots,u_n+\Delta u_n)|^p\,du_1\dots du_n\Big)^{\frac{1}{p}}.$$

$$=\Big\{\int\limits_{G^*_{(\eta_1)}}\!\!\dots\!\int \Big|\int\limits_{u_s}^{u_s+\Delta u_s} Q^{(l+1)}_{\nu u u_s}(u_1,\dots,u_{s-1},t,u_{s+1}+\Delta u_{s+1},\dots,u_n$$

$$+\Delta u_n)\,dt\Big|^p du_1\dots du_n\Big\}^{\frac{1}{p}}\,{}^{1)}$$

$$\leqslant\Big\{\int\limits_{G^*_{(\eta_1)}}\!\!\dots\!\int\Big(\int\limits_{-\gamma|h|}^{\gamma|h|} | Q^{(l+1)}_{\nu u u_s}(u_1,\dots,u_{s-1},u_s+\tau,u_{s+1}+\Delta u_{s+1},\dots,u_n$$

$$+\Delta u_n)|\,d\tau\Big)^p du_1\dots du_n\Big\}^{\frac{1}{p}}.$$

$$\leqslant\int\limits_{-\gamma|h|}^{\gamma|h|}\Big(\int\limits_{G^*_{(\eta_1)}}\!\!\dots\!\int | Q^{(l+1)}_{\nu u u_s}(u_1,\dots,u_{s-1},u_s+\tau,u_{s+1}+\Delta u_{s+1},\dots,u_n$$

$$+\Delta u_n)|^p du_1\dots du_n\Big)^{\frac{1}{p}}d\tau\,{}^{2)}$$

$$\leqslant\int\limits_{-\gamma|h|}^{\gamma|h|}\Big(2\int\limits_{G^*_{(\eta_1)}}\!\!\dots\!\int | Q^{(l+1)}_{\nu u u_s}(u_1,\dots,u_{s-1},u_s+\tau,u_{s+1}+\Delta u_{s+1}\dots,u_n$$

$$+\Delta u_n)|^p\frac{D(v_1,\dots,v_n)}{D(u_1,\dots,u_n)}du_1\dots du_n\Big)^{\frac{1}{p}}d\tau$$

$$=\int\limits_{-\gamma|h|}^{\gamma|h|}\Big(2\int\limits_{R_n}\!\!\dots\!\int | Q^{(l+1)}_{\nu u u_s}(v_1,\dots,v_{s-1},v_s+\tau,v_{s+1},\dots,v_n)|^p dv_1\dots dv_n\Big)^{\frac{1}{p}}d\tau$$

$$=2^{1+\frac{1}{p}}\gamma\,\|Q^{(l+1)}_{\nu u u_s}\|^{(n)}_p|h|\,{}^{3)}$$

$$\leqslant\begin{cases} c|h|2^{(l+1-r)\nu}\overline{M}^* & (\nu=1,2,\dots),\\ |h|(\|\overline{f}\|^{(n)}_p+c_2\overline{M}^*) & (\nu=0),\end{cases}$$

and thus

1) In view of inequality (11).

2) See inequality (17).

3) In view of relations (7) and the generalized Bernšteĭn inequality.

$$\sum_{\nu=0}^{k-1}\sum_{s=1}^{n} I_{\nu s} \leqslant n\,|h|\,(\|f\|_p^{(n)} + c_2\overline{M}^* + \sum_{\nu=1}^{k-1} 2^{(l+1-r)\nu}\overline{M}^*)$$

$$< \begin{cases} (c_3\|\overline{f}\|_p^{(n)} + c_4\overline{M}^*)\,|h|, & \text{if } l < \overline{r}, \\ |h|\,(c_5\|\overline{f}\|_p^{(n)} + c_6\overline{M}^* + c_7 2^{(1-\alpha)k}\overline{M}^*) \\ < (c_5\|\overline{f}\|_p^{(n)} + c_6\overline{M}^*)\,|h| + c_8|h|^\alpha, & \text{if } l = \overline{r}. \end{cases}$$

In both cases, for $|h| \leq 1$

$$\sum_{\nu=0}^{k-1}\sum_{s=1}^{n} I_{\nu s} < (c_9\|\overline{f}\|_p^{(n)} + c_{10}\overline{M}^*)\,|h|^\alpha.$$

Further,

$$\Big(\int\limits_{G^*_{(\eta)}}\!\!\dots\!\int |S''(u_1+\Delta u_1,\dots,u_n+\Delta u_n) - S''(u_1,\dots,u_n)|^p\,du_1\dots du_n\Big)^{\frac{1}{p}}$$

$$\leqslant \sum_{\nu=k}^{\infty}\Big(\int\limits_{G^*_{(\eta_1)}}\!\!\dots\!\int |Q_{\nu u}^{(l)}(u_1+\Delta u_1,\dots,u_n+\Delta u_n)|^p\,du_1\dots du_n\Big)^{\frac{1}{p}}$$

$$+ \sum_{\nu=k}^{\infty}\Big(\int\limits_{G^*_{(\eta_1)}}\!\!\dots\!\int |Q_{\nu u}^{(l)}(u_1,\dots,u_n)|^p\,du_1\dots du_n\Big)^{\frac{1}{p}}$$

$$\leqslant \sum_{\nu=k}^{\infty}\Big[2\int\limits_{G^*_{(\eta_1)}}\!\!\dots\!\int |Q_{\nu u}^{(l)}(u_1+\Delta u_1,\dots,u_n+\Delta u_n)|^p\,\frac{D(v_1,\dots,v_n)}{D(u_1,\dots,u_n)}\,du_1\dots du_n\Big]^{\frac{1}{p}}$$

$$+ \sum_{\nu=k}^{\infty}\|Q_{\nu u}^{(l)}\|_p^{(n)} \leqslant \sum_{\nu=k}^{\infty}\Big(2\int\limits_{R_n}\!\!\dots\!\int |Q_{\nu u}^{(l)}(v_1,\dots,v_n)|^p\,dv_1\dots dv_n\Big)^{\frac{1}{p}}$$

$$+ \sum_{\nu=k}^{\infty}\|Q_{\nu u}^{(l)}\|_p^{(n)} \leqslant \big(2^{\frac{1}{p}}+1\big)\sum_{\nu=k}^{\infty}\|Q_{\nu u}^{(l)}\|_p^{(n)} \leqslant c_4\overline{M}^*\sum_{\nu=k}^{\infty}\frac{1}{2^{\nu(r-l)}}$$

$$< c_5\cdot\frac{\overline{M}^*}{2^{(r-l)k}} < c_6\overline{M}^*|h|^{r-l} \leqslant c_7\overline{M}^*|h|^\alpha.$$

Thus we have proved that

$$I = \Big(\int\limits_{G^*_{(\eta)}}\!\!\dots\!\int |f_u^{(l)}(u_1+\Delta u_1,\dots) - f_u^{(l)}(u_1,\dots)|^p\,du_1\dots du_n\Big)^{\frac{1}{p}}$$

$$< (c_8\|\overline{f}\|_p^{(n)} + c_9\overline{M}^*)\,|h|^\alpha$$

and, in view of inequality (4),

$$I \leqslant (c_{1\mu} \| f \|_{L_p(G^*)} + c_{2\mu} M^*) \,|\, h \,|^\alpha.$$

Thus inequality (15) is proved for $0 < \alpha < 1$.

We turn to the proof of inequality (15) in the case $\alpha = 1$.

Put

$$\begin{aligned} u_i + \Delta u_i &= \varphi_i (x_1 + h, x_2, \ldots, x_n), \\ u_i - \Delta u_i' &= \varphi_i (x_1 - h, x_2, \ldots, x_n), \end{aligned} \qquad (i = 1, 2, \ldots, n).$$

We have the following inequality, analogous to (8):

$$|\Delta u_i'| < |\varphi_i (x_1 - h, x_2, \ldots) - \varphi_i (x_1, x_2, \ldots)| \leqslant \gamma |h|. \qquad (11')$$

We represent our second difference as follows:

$$F_{x_1}^{(\bar{r})} (x_1 + h, x_2, \ldots) - 2F_{x_1}^{(\bar{r})} (x_1, x_2, \ldots) + F_{x_1}^{(\bar{r}_1)} (x_1 - h, x_2, \ldots)$$

$$= \sum_{k=1}^{N} [\lambda_k (u_1 + \Delta u_1, \ldots, u_n + \Delta u_n) f_u^{(l)}(u_1 + \Delta u_1, \ldots, u_n + \Delta u_n) \cdot$$

$$- 2\lambda_k (u_1, \ldots, u_n) f_u^{(l)} (u_1, \ldots, u_n)$$

$$+ \lambda_k (u_1 - \Delta u_1', \ldots, u_n - \Delta u_n') f_u^{(l)} (u_1 - \Delta u_1', \ldots, u_n - \Delta u_n')]$$

$$= \sum_{k=1}^{N} \{[f_u^{(l)} (u_1 + \Delta u_1, \ldots) - 2f_u^{(l)} (u_1, \ldots)$$

$$+ f_u^{(l)} (u_1 - \Delta u_1', \ldots)] \lambda_k (u_1 + \Delta u_1, \ldots)$$

$$+ 2f_u^{(l)} (u_1, \ldots) [\lambda_k (u_1 + \Delta u_1, \ldots) - \lambda_k (u_1, \ldots)]$$

$$- f_u^{(l)} (u_1 - \Delta u_1', \ldots) [\lambda_k (u_1 + \Delta u_1, \ldots) - \lambda_k (u_1 - \Delta u_1', \ldots)]\}$$

$$= \sum_{k=1}^{N} \left\{ \overset{(k)}{S}_* + \overset{(k)}{S}{}_*' + \overset{(k)}{S}{}_*'' \right\}. \qquad (20)$$

Just as in the derivation of (13), we find from (12) that

$$\left( \int \cdots \int_{G^*_{(\eta_1)}} |\overset{(k)}{S}{}_*'|^p \frac{D(x_1, \ldots, x_n)}{D(u_1, \ldots, u_n)} \, du_1 \ldots du_n \right)^{\frac{1}{p}}$$

$$< (c_{1\mu} \| f \|_{L_p(G^*)} + c_{2\mu} M^*) \,|\, h \,|. \qquad (21)$$

Further, since $|\lambda_k(u_1 + \Delta u_1, \cdots) - \lambda_k(u_1 - \Delta u_1', \cdots)| \leq c\,|h|$, the following relation holds:

$$\Big(\int\limits_{G^*_{(\eta_1)}}\!\!\dots\!\int |\overset{(k)}{S}{}''_{*}|^p \frac{D(x_1,\dots,x_n)}{D(u_1,\dots,u_n)}\,du_1\dots du_n\Big)^{\frac{1}{p}}$$

$$\leqslant \frac{c\,|h|}{K^{\frac{1}{p}}}\Big(\int\limits_{G^*_{(\eta_1)}}\!\!\dots\!\int |f_u^{(l)}(u_1-\Delta u_1', \dots|^p\,du_1\dots du_n\Big)^{\frac{1}{p}}$$

$$\leqslant c_1|h|\Big(2\int\limits_{G^*_{(\eta_1)}}\!\!\dots\!\int |f_u^{(l)}(u_1-\Delta u_1', \dots)|^p \frac{D(v_1,\dots,v_n)}{D(u_1,\dots,u_n)}\,du_1\dots du_n\Big)^{\frac{1}{p}} \quad (22)$$

$$= c_2|h|\Big(\int\limits_{R_n}\!\!\dots\!\int |\overline{f}_u^{(l)}(v_1,\dots,v_n)|^p\,dv_1\dots dv_n\Big)^{\frac{1}{p}} \leqslant (c_3\|\overline{f}\|_p^{(n)} + c_4\overline{M}^*)\,|h|.$$

This last inequality follows from the fact that the function $\overline{f}_u^{(l)}$ is representable in the form of the series (18), and also from (19) and the inequality

$$\|Q_{0u}^{(l)}\|_p^{(n)} \leqslant \|\overline{f}\|_p^{(n)} + c_5\overline{M}^*,$$

obtained via the generalized Bernšteĭn inequality from the second inequality (7).

Finally, later on we shall prove the inequality

$$\beta_\nu = \Big(\int\limits_{G^*_{(\eta_1)}}\!\!\dots\!\int | Q_{\nu u}^{(l)}(u_1+\Delta u_1, \dots) - 2Q_{\nu u}^{(l)}(u_1, \dots)$$

$$+ Q_{\nu u}^{(l)}(u_1-\Delta u_1', \dots)|^p\,du_1\dots du_n\Big)^{\frac{1}{p}}$$

$$< \begin{cases} \dfrac{c_1 h^2 \overline{M}^*}{2^{\nu(r-l-2)}} & (\nu = 1, 2, \dots), \\ (c_2\overline{M}^* + c_3\|\overline{f}\|_p^{(n)})\,h^2 & (\nu = 0), \end{cases} \quad (23)$$

in view of which, on the basis of the inequality

$$|\lambda_k| \leqslant c_4 \quad (k = 1, 2, \dots, N),$$

choosing the natural number $m = m(h)$ so as to satisfy the relations

$$2^{-m} \leqslant |h| < 2^{-(m-1)},$$

we will have

$$\left(\int\limits_{G^*_{(\eta_1)}}\cdots\int |\overset{(k)}{S}_*|^p \frac{D(x_1, \ldots, x_n)}{D(u_1, \ldots, u_n)} du_1 \ldots du_n\right)^{\cdot}$$

$$\leqslant c \int\limits_{G^*_{(\eta_1)}}\cdots\int | f_u^{(l)}(u_1+\Delta u_1, \ldots) - 2f_u^{(l)}(u_1, \ldots)$$

$$+ f_u^{(l)}(u_1 - \Delta u_1, \ldots)|^p du_1 \ldots du_n\Big)^{\frac{1}{p}}$$

$$\leqslant c \sum_{\nu=0}^{\infty}\left(\int\limits_{G^*_{(\eta_1)}}\cdots\int | Q_{\nu u}^{(l)}(u_1 + \Delta u_1, \ldots) - 2Q_\nu^{(l)}(u_1, \ldots)\right.$$

$$\left. + Q_{\nu u}^l(u_1 - \Delta u_1', \ldots)|^p du_1 \ldots du_n\right)^{\frac{1}{p}}$$

$$\leqslant c \sum_{\nu=0}^{m} \beta_\nu + c_2 \sum_{\nu=m+1}^{\infty} 4 \| Q_{\nu u}^{(l)} \|_p^{(n)\,1)}$$

$$\leqslant c(c_2 \overline{M}^* + c_3 \|\overline{f}\|_p^{(n)}) h^2 + c_4 \left(\sum_{\nu=0}^{m} \frac{h^2}{2^{\nu(r-l-2)}} + \sum_{\nu=m+1}^{\infty} \frac{1}{2^{\nu(r-l)}}\right) \overline{M}^*$$

$$< (c_5 \overline{M}^* + c_6 \|\overline{f}\|_p^{(n)}) h^2 + c_4 \left(\sum_{\nu=0}^{m} \frac{h^2}{2^{\nu(r-l-2)}} + |h|^{r-l}\right) \overline{M}^*$$

$$< (c_7 \|\overline{f}\|_p^{(n)} + c_8 \overline{M}) |h|, \tag{24}$$

since

$$h^2 \sum_{\nu=0}^{m} \frac{1}{2^{\nu(r-l-2)}} < \begin{cases} c_4 |h| & \text{for } l = \overline{r}\,(= r-1), \\ c_4 h^2 \lg \frac{1}{|h|} & \text{for } l = \overline{r} - 1, \\ c_4 |h|^2 & \text{for } l < \overline{r} - 1. \end{cases}$$

On taking account of the fact that $\mu = \mu(\eta)$ and using (20)–(22), (24) and (4), we obtain the required inequality (15):

---

1) This estimate of $\beta_\nu (\nu > m)$ is obtained by replacing the variables $u_1 + \Delta u_1, u_2 + \Delta u_2, \cdots$ by $v_1, v_2, \cdots$, and correspondingly replacing $u_1 - \Delta u_1'$ by $v_1, \cdots$, which was justified above, and passing from integration over $G^*_{(\eta_1)}$ to integration over $R^*_n$.

$$\Big(\int_{G_\eta}\!\!\cdots\!\int |F_{x_1}^{(\bar r)}(x_1+h, x_2, \ldots, x_n) - 2F_{x_1}^{(\bar r)}(x_1, \ldots, x_n)$$

$$+ F_{x_1}^{(\bar r)}(x_1-h, x_2, \ldots, x_n)|^p\, dx_1 \ldots dx_n\Big)^{\frac{1}{p}}$$

$$= \Big(\int_{G^*_{(\eta)}}\!\!\cdots\!\int \Big|\sum_{k=1}^{N} (\overset{(k)}{S}_* + \overset{(k)}{S}{}'_* + \overset{(k)}{S}{}''_*)\Big|^p \frac{D(x_1, \ldots, x_n)}{D(u_1, \ldots, u_n)}\, du_1 \ldots du_n\Big)^{\frac{1}{p}}$$

$$< a_{1\eta}\|f\|_{L_p(G^*)} + a_{2\eta}M^*.$$

Proof of inequality (23). For conciseness we put $Q^{(l)}_{\nu u} = q$. We observe that

$$\Delta u_i = \varphi_i(x_1+h, x_2, \ldots) - \varphi_i(x_1, x_2, \ldots) = \frac{\partial \varphi_i}{\partial x_1} h + O(h^2),$$

$$-\Delta u_i' = \varphi_i(x_1-h, x_2, \ldots) - \varphi_i(x_1, x_2, \ldots) = -\frac{\partial \varphi_i}{\partial x_1} h + O(h^2),$$

so that

$$|\Delta u_i - \Delta u_i'| < \omega h^2,$$

where $\omega$ is a constant not depending on $h$ and which is the same for all the points $(x_1, \cdots, x_n) \in G_{\eta_1}$.

We begin with the inequality

$$\Big(\int_{G^*_{(\eta_1)}}\!\!\cdots\!\int |q(u_1+\Delta u_1, \ldots, u_n+\Delta u_n)$$

$$- q(u_1+\Delta u_1', \ldots, u_n+\Delta u_n')|^p\, du_1 \ldots du_n\Big)^{\frac{1}{p}}$$

$$\leqslant \sum_{s=1}^{n} \Big(\int_{G^*_{(\eta_1)}}\!\!\cdots\!\int |q(u_1+\Delta u_1', \ldots, u_{s-1}+\Delta u_{s-1}', u_s+\Delta u_s, \ldots, u_n+\Delta u_n)$$

$$- q(u_1+\Delta u_1', \ldots, u_s+\Delta u_s', u_{s+1}+\Delta u_{s+1}, \ldots, u_n+\Delta u_n)|^p\, du_1 \ldots du_n\Big)^{\frac{1}{p}}$$

$$= \sum_{s=1}^{n} \Big(\int_{G^*_{(\eta_1)}}\!\!\cdots\!\int \Big|\int_0^{\Delta u_s - \Delta u_s'} q'_{u_s}(u_1+\Delta u_1', \ldots, u_{s-1}+\Delta u_{s-1}', u_s+\Delta u_s'+\tau, u_{s+1}$$

$$+ \Delta u_{s+1}, \ldots, u_n+\Delta u_n)\, d\tau\Big|^p du_1 \ldots du_n\Big)^{\frac{1}{p}} \leq$$

$$\leqslant \sum_{s=1}^{n} \Big( \int \dots \int_{G^*_{(\eta_1)}} \Big| \int_{-\omega h^2}^{\omega h^2} q'_{u_s} (u_1 + \Delta u'_1, \dots, u_{s-1} + \Delta u'_{s-1}, u_s + \Delta u'_s + \tau,$$

$$u_{s+1} + \Delta u_{s+1}, \dots, u_n + \Delta u_n)\, d\tau \Big|^p du_1 \dots du_n \Big)^{\frac{1}{p}}$$

$$\leqslant \sum_{s=1}^{n} \int_{-\omega h^2}^{\omega h^2} \Big[ \int \dots \int_{G^*_{(\eta_1)}} \Big| q'_{u_s} (u_1 + \Delta u'_1, \dots, u_{s-1} + \Delta u'_{s-1},$$

$$u_s + \Delta u'_s + \tau, u_{s+1} + \Delta u_{s+1}, \dots, u_n + \Delta u_n) \Big|^p du_1 \dots, du_n \Big]^{\frac{1}{p}} d\tau$$

$$\leqslant \sum_{s=1}^{n} \int_{-\omega h^2}^{\omega h^2} \Big( \int \dots \int_{G^*_{(\mu)}} \Big| q_{u_s} (v_1, \dots, v_{s-1}, v_s + \tau, v_{s+1}, \dots, v_n) \Big|^p$$

$$\times \frac{D(u_1, \dots, u_n)}{D(v_1, \dots, v_n)} dv_1 \dots dv_n \Big)^{\frac{1}{p}} d\tau$$

$$\leqslant c \sum_{s=1}^{n} \int_{-\omega h^2}^{\omega h^2} \Big( \int \dots \int_{R_n} | q'_{u_s} (v_1, \dots, v_{s-1}, v_s + \tau, v_{s+1}, \dots, v_n) |^p dv_1 \dots dv_n \Big)^{\frac{1}{p}} d\tau$$

$$= 2c\omega h^2 \sum_{s=1}^{n} \| q'_{u_s} \|_p^{(n)} = c_1 h^2 \sum_{s=1}^{n} \| Q^{(l+1)}_{\nu u u_s} \|_p^{(n)} \quad {}^{1)}$$

$$\leqslant \begin{cases} c_2 h^2 \displaystyle\sum_{s=1}^{n} \frac{\overline{M}^*}{2^{(r-l-1)\nu}} < \frac{c_3 h^2 \overline{M}^*}{2^{(r-l-2)\nu}} & (\nu = 1, 2, \dots), \\ (c_4 \overline{M}^* + c_5 \| f \|_p^{(n)})\, h^2 & (\nu = 0). \end{cases}$$

We obtain a similar inequality if in it we replace $\Delta u_i$ and $\Delta u'_i$ throughout by $-\Delta u$ and $-\Delta u_i$ respectively. Therefore inequality (23) will be proved if we prove the inequality obtained from (23) by replacing each $\Delta u'_i$ by $\Delta u_i$.

We have

$$q(u_1 + \Delta u_1, \dots, u_n + \Delta u_n) - 2q(u_1, \dots, u_n)$$

$$+ q(u_1 - \Delta u_1, \dots, u_n - \Delta u_n) =$$

1) In view of relations (7).

$$= \sum_{s=1}^{n} [q(u_1, \ldots, u_{s-1}, u_s + \Delta u_s, \ldots, u_n + \Delta u_n) - q(u_1, \ldots, u_s, u_{s+1}$$

$$+ \Delta u_{s+1}, \ldots, u_n + \Delta u_n) + q(u_1, \ldots, u_{s-1}, u_s - \Delta u_s, \ldots, u_n - \Delta u_n)$$
$$- q(u_1, \ldots, u_s, u_{s+1} - \Delta u_{s+1}, \ldots, u_n - \Delta u_n)]$$

$$= \sum_{s=1}^{n} \int_0^{\Delta u_s} [q'_{u_s}(u_1, \ldots, u_{s-1}, u_s + \tau, u_{s+1} + \Delta u_{s+1}, \ldots, u_n + \Delta u_n)$$
$$- q'_{u_s}(u_1, \ldots, u_{s-1}, u_s - \tau, u_{s+1} - \Delta u_{s+1}, \ldots, u_n - \Delta u_n)]\, d\tau$$

$$= \sum_{s=1}^{n} \int_0^{\Delta u_s} \Big\{ [q'_{u_s}(u_1, \ldots, u_{s-1}, u_s + \tau, u_{s+1} + \Delta u_{s+1}, \ldots, u_n + \Delta u_n)$$
$$- q'_{u_s}(u_1, \ldots, u_{s-1}, u_s - \tau, u_{s+1} + \Delta u_{s+1}, \ldots, u_n + \Delta u_n)]$$
$$+ \sum_{k=1}^{n-s} [q'_{u_s}(u_1, \ldots, u_{s-1}, u_s - \tau, u_{s+1} - \Delta u_{s+1}, \ldots$$
$$\ldots, u_{s+k-1} - \Delta u_{s+k-1}, u_{s+k} + \Delta u_{s+k}, \ldots, u_n + \Delta u_n)$$
$$- q'_{u_s}(u_1, \ldots, u_{s-1}, u_s - \tau, u_{s+1} - \Delta u_{s+1}, \ldots, u_{s+k} - \Delta u_{s+k}, \ldots, u_n + \Delta u_n)] \Big\}\, d\tau$$

$$= \sum_{s=1}^{n} \int_0^{\Delta u_s} \Big\{ \int_{-\tau}^{\tau} q''_{u_s u_s}(u_1, \ldots, u_{s-1}, u_s + t, u_{s+1} + \Delta u_{s+1}, \ldots, u_n + \Delta u_n)$$
$$+ \sum_{k=1}^{n-s} \int_{-\Delta u_{s+k}}^{\Delta u_{s+k}} q''_{u_s u_{s+k}}(u_1, \ldots, u_{s-1}, u_s - \tau, u_{s+1} - \Delta u_{s+1}, \ldots$$
$$\ldots, u_{s+k-1} - \Delta u_{s+k-1}, u_{s+k} + t, u_{s+k+1}$$
$$+ \Delta u_{s+k+1}, \ldots, u_n + \Delta u_n)\, dt \Big\}\, d\tau.$$

Hence, in view of (11),

$$| q(u_1 + \Delta u_1, \ldots, u_n + \Delta u_n) - 2q(u_1, \ldots, u_n)$$
$$+ q(u_1 - \Delta u_1, \ldots, u_n - \Delta u_n) |$$

$$\leqslant \sum_{s=1}^{n} \int_{-\gamma|h|}^{\gamma|h|} \int_{-\gamma|h|}^{\gamma|h|} \Big\{ | q''_{u_s u_s}(u_1, \ldots, u_{s-1}, u_s + t, u_{s+1}$$
$$+ \Delta u_{s+1}, \ldots, u_n + \Delta u_n)$$

$$+ \sum_{k=1}^{n-s} | q''_{u_s u_{s+k}}(u_1, \ldots, u_{s-1}, u_s - \tau, u_{s+1} - \Delta u_{s+1}, \ldots$$

$$\ldots, u_{s+k-1}-\Delta u_{s+k-1}, u_{s+k}+t, u_{s+k+1}$$

$$+\Delta u_{s+k+1}, \ldots, u_n+\Delta u_n)|\Big\} \, dt \, d\tau,$$

and accordingly, applying the generalized Minkowski inequality, we get

$$\Big(\int_{G^*_{(\eta_1)}}\!\!\ldots\int |q(u_1+\Delta u_1, \ldots)-2q(u_1, \ldots)+q(u_1-\Delta u_1, \ldots)|^p du_1 \ldots du_n\Big)^{\frac{1}{p}}$$

$$\leqslant \sum_{s=1}^{n}\int_{-\gamma|h|}^{\gamma|h|}\int_{-\gamma|h|}^{\gamma|h|}\Big\{\Big(\int_{G^*_{(\eta_1)}}\!\!\ldots\int |q''_{u_s u_s}(u_1, \ldots, u_{s-1}, u_s+t, u_{s+1}+\Delta u_{s+1}, \ldots, u_n+\Delta u_n)$$

$$+\sum_{k=1}^{n-s} q''_{u_s u_{s+k}}(u, \ldots, u_{s-1}, u_s-\tau, u_{s+1}-\Delta u_{s+1}, \ldots, u_{s+k-1}-\Delta u_{s+k-1}, u_{s+k}$$

$$+t, u_{s+k+1}+\Delta u_{s+k+1}, \ldots, u_n+\Delta u_n)|^p du_1 \ldots du_n\Big\}^{\frac{1}{p}} dt \, d\tau$$

$$< c\sum_{s=1}^{n}\int_{-\gamma|h|}^{\gamma|h|}\int_{-\gamma|h|}^{\gamma|h|}\Big\{\Big(\int_{R_n}\!\!\ldots\int |q''_{u_s u_s}(v_1, \ldots, v_{s-1}, v_s+t, v_{s+1}, \ldots, v_n)|^p dv_1 \ldots dv_n\Big)^{\frac{1}{p}}$$

$$+\sum_{k=1}^{n-s}\Big(\int_{R_n}\!\!\ldots\int |q''_{u_s u_{s+k}}(v_1, \ldots, v_{s-1}, v_s-\tau, v_{s+1}, \ldots$$

$$\ldots, v_{s+k-1}, v_{s+k}+t, v_{s+k+1}, \ldots, v_n)|^p dv_1, \ldots dv_n\Big)^{\frac{1}{p}}\Big\} dt \, d\tau$$

$$< c_1 4\gamma^2 h^2 \sum_{s=1}^{n}\Big\{\| Q^{(l+2)}_{\nu u u_s u_s}\|_p^{(n)} + \sum_{k=1}^{n-s}\| Q^{(l+2)}_{\nu u u_s u_{s+k}}\|_p^{(n)}\Big\}^{1)}$$

$$\leqslant \begin{cases} c_2 h^2 \overline{M}^* \sum_{s=1}^{n}\Big\{\frac{1}{2^{\nu(r-l-2)}}+\frac{(n-s)}{2^{\nu(r-l-2)}}\Big\} < \frac{c_3 h^2 \overline{M}^*}{2^{\nu(r-l-2)}} & (\nu=1, 2, \ldots), \\ (c_4 \overline{M}^* + c_5 \|\overline{f}\|_p^{(n)}) h^2 & (\nu = 0), \end{cases}$$

and inequality (23) is proved.

Thus the proof of our lemma is complete.

## §3. Behavior of functions of the classes $H_p^{(r_1, \cdots, r_n)}$ on planes parallel to the coordinate axes. Direct and inverse theorems

3.11. **Lemma.** *The entire function* $g = g_{\nu_1 \cdots \nu_n}(x_1, \cdots, x_n)$ *of degrees* $\nu_i$ *relative to the variables* $x_i$ $(i = 1, \cdots, n)$ *respectively, satisfying the inequality*

1) See relation (7).

$$\| g_{\nu_1 \ldots \nu_n} \|_p^{(n)} \leqslant M,$$

*lies in the class* $H_p^{(r_1, \cdots, r_n)} (4M \nu_1^{r_1}, \cdots, 4M \nu_n^{r_n})$.

Proof. Put $r_i = \bar{r}_i + \alpha_i$, where $0 < \alpha_i \leq 1$ and the $\bar{r}_i$ are integers. Denote the first and second differences of the function $f$ relative to the variable $x_i$ respectively by $\Delta_i f$ and $\Delta_i^2 f$. On the basis of Lemmas 1 and 3 of [1] we find that

$$\| \Delta_i g_{x_i}^{(\bar{r}_i)} \|_p^{(n)} = \| g_{x_i}^{(\bar{r}_i + 1)} \|_p^{(n)} |h| \leqslant M \nu_i^{\bar{r}_i + 1} |h| \leqslant M \nu_i^{r_i} |h|^{\alpha_i},$$

if $|\nu_i h| \leq 1$. On the other hand, if $|\nu_i h| > 1$ we obtain

$$\| \Delta_i g_{x_i}^{(\bar{r}_i)} \|_p^{(n)} \leqslant 2 \| g_{x_i}^{(\bar{r}_i)} \|_p^{(n)} \leqslant 2 M \nu_i^{\bar{r}_i} (\nu_i h)^{\alpha_i} = 2 M \nu_i^{r} |h|^{\alpha_i}.$$

In a similar way for $\alpha_i = 1$ we get

$$\| \Delta_i^2 g_{x_i}^{(\bar{r}_i)} \|_p^{(n)} \leqslant M \nu_i^{(\bar{r}_i + 2)} h^2 \leqslant M \nu_i^{r_i} |h|,$$

if $\nu_i |h| \leq 1$, and if $\nu_i |h| > 1$, then

$$\| \Delta_i^2 g_{x_i}^{(\bar{r}_i)} \|_p^{(n)} \leqslant 4 \| g_{x_i}^{(\bar{r}_i)} \|_p^{(n)} \leqslant 4 \nu_i^{\bar{r}_i} \leqslant 4 \nu_i^{r_i} |h|.$$

3.12. **Lemma.** *Suppose that the function* $F = F(x_1, \cdots, x_n)$ *is the sum of the series*

$$F(x_1, \ldots, x_n) = \sum_{s=0}^{\infty} Q_s (x_1, \ldots, x_n), \tag{1}$$

*converging to it in* $L_p^{(n)}$, *where the* $Q_s$ $(s = 0, 1, \cdots)$ *are entire functions of degrees* $\nu_k = 2^{s/r_k}$ $(k = 1, 2, \cdots, n)$ *respectively in the variables* $x_k$, *and suppose that*

$$\| Q_0 \|_p^{(n)} \leqslant A, \quad \| Q_s \|_p^{(n)} \leqslant \frac{B}{2^{\alpha s}} \quad (s = 1, 2, \ldots), \tag{2}$$

*where* $A$, $B$ *and* $\alpha$ *are positive constants not depending on* $s$. *Then the following properties hold*:

1) $F \in H_p^{(\alpha r_1, \cdots, \alpha r_n)} (K, \cdots, K)$, *where* $K < 4A + cB$,

2) $\|F\|_p^{(n)} \leq A + cB$,

*where the constant* $c$ *does not depend on* $A$ *and* $B$.

Proof. Put

$$S_{\nu_1 \ldots \nu_n} = \sum_{s=0}^{\mu-1} Q_s.$$

This sum, in view of the conditions imposed on the $Q_s$, is an entire function of degrees

$$\nu_k = 2^{\frac{\mu}{r_k}} \quad (\mu = 1, 2, \ldots) \tag{3}$$

in the $x_k$ respectively ($k = 1, 2, \cdots, n$). Therefore, from (1),

$$\| F - S_{\nu_1 \ldots \nu_n} \|_p^{(n)} \leqslant \sum_{s=\mu}^{\infty} \| Q_s \|_p^{(n)} < \sum_{s=\mu}^{\infty} \frac{B}{2^{\alpha s}} = \frac{B}{1-2^{\alpha}} \frac{1}{2^{\alpha\mu}}$$

$$< \frac{c_3 B}{2^{\alpha\mu}} < \frac{c_3 B}{\nu_k^{\alpha r_k}} < \sum_{k=1}^{n} \frac{c_3 B}{\nu_k^{\alpha r_k}}. \tag{4}$$

Put

$$F = Q_0 + \Phi.$$

Because of inequality (4), where the numbers $\nu_k$ run through the geometrically increasing progression (3), on the basis of Theorem 10 of [1], the function $\Phi$ lies in the class $H^{(\alpha r_1, \cdots, \alpha r_n)}(K_1, \cdots, K_1)$, where

$$K_1 < c_4 B.$$

As to the function $Q_0$, it is an entire function of degree 1 in all the variables $x_k$ ($k = 1, \cdots, n$), satisfying the first of inequalities (2). Therefore on the basis of Lemma 3.11 it may also be considered as belonging to the class $H_p^{(\alpha r_1, \cdots, \alpha r_n)}(4A, \cdots, 4A)$. In such a case the function $F$ lies in the class $H^{(\alpha r_1, \cdots, \alpha r_n)}(K, \cdots, K)$, where

$$K \leqslant 4A + c_4 B.$$

Thus the first property of the function $F$ is proved.

The second property follows directly from the inequality

$$\| F \|_p^{(n)} \leqslant A + \sum_{s=1}^{\infty} \frac{B}{2^{\alpha s}} = A + cB.$$

3.13. **Lemma.** *If, aside from the conditions imposed on the function $F$ in the preceding lemma, the inequality*

$$\gamma = \alpha - \frac{1}{p} \sum_{k=m+1}^{n} \frac{1}{r_k} > 0, \tag{1}$$

*is satisfied, where m is an integer with* $1 \leq m < n$, *then the function F may be altered on a set of measure zero in such a way that the series* (1) *of* §3.12 *will converge in the variables* $x_1, \cdots, x_m$ *for any fixed* $x_{m+1}, \cdots, x_n$ *in the sense of* $L_p^{(m)}$ *to F, and in addition the function F of the variables* $x_1, \cdots, x_m$ *will lie in the class* $H_p^{(\gamma r_1, \cdots, \gamma r_n)}(\overline{K}, \cdots, \overline{K})$, *where*

$$\overline{K} \leqslant c_1 A + c_2 B. \tag{2}$$

*Moreover,*

$$\| F \|_p^{(m)} \leqslant c_1 A + c_2 B, \tag{3}$$

$$\lim_{h_k \to 0} \int_{-\infty}^{\infty} \cdots \int_{-\infty}^{\infty} | F(x_1, \ldots, x_m, x_{m+1} + h_{m+1}, \ldots, x_n + h_n)$$

$$- F(x_1, \ldots, x_m, x_{m+1}, \ldots, x_n) |^p \, dx_1 \ldots dx_n = 0 \tag{4}$$

$$(k = m+1, \; m+2, \; \ldots, \; n).$$

Proof. On the basis of inequality (4) of the Introduction, we have

$$\| Q_0 \|_p^{(m)} \leqslant 2^n \| Q_0 \|_p^{(n)} \leqslant 2^n A,$$

$$\| Q_s \|_p^{(m)} \leqslant 2^n \left( \prod_{k=m+1}^{n} \nu_k \right)^p \| Q_s \|_p^{(n)} \leqslant 2^n 2^{\frac{s}{p} \sum_{k=m+1}^{n} \frac{1}{r_k}} \frac{B}{2^{\alpha s}} \tag{5}$$

$$= \frac{2^n B}{2^{\left(\alpha - \frac{1}{p} \sum_{k=m+1}^{n} \frac{1}{r_k}\right) s}} = \frac{2^n B}{2^{\gamma s}}.$$

In view of (1) it is obvious that the series

$$\sum_{s=0}^{\infty} Q_s \tag{6}$$

converges for any fixed $x_{m+1}, \cdots, x_n$ in the variables $x_1, \cdots, x_m$ in the sense of $L_p^{(m)}$ to some function $F_1$. On the other hand, as we know from the preceding lemma, it converges in the sense of $L_p^{(n)}$ to some function $F$. It is easy to see that $F$ differs from $F_1$ almost nowhere. Obviously,

$$\| F \|_p^{(m)} \leqslant 2^n \left( A + \sum_{s=1}^{\infty} \frac{B}{2^{\gamma s}} \right) \leqslant c_1 A + c_2 B,$$

and we have obtained relation (3).

The fact that the function $F$, relative to the variables $x_1, \cdots, x_m$, lies in the class $H_p^{(\gamma r_1, \cdots, \gamma r_n)}(\overline{K}, \cdots, \overline{K})$, and the truth of inequality (2), follow directly on the basis of (5) from the preceding lemma, if this is applied to the series (6).

Now we shall prove relation (4). We have

$$\| F(x_1, \ldots, x_m, x_{m+1} + h_{m+1}, \ldots, x_n + h_n) - F(x_1, \ldots, x_m, x_{m+1}, \ldots, x_n) \|_p^{(m)}$$

$$\leqslant \sum_{s=0}^{\mu} \| Q_s(x_1, \ldots, x_m, x_{m+1} + h_{m+1}, \ldots, x_n + h_n) - Q_s(x_1, \ldots, x_m, x_{m+1}, \ldots, x_n) \|_p^{(m)} + 2 \sum_{s=\mu+1}^{\infty} \| Q_s \|_p^{(m)} = \sigma' + \sigma''.$$

In view of inequalities (5) it is obvious that $\mu$ may be chosen so large that

$$\sigma'' < \varepsilon$$

uniformly relative to $h_k$ $(k = m+1, m+2, \cdots, n)$.

Now we take into consideration the fact that for each entire function $g = g_{\nu_1, \cdots, \nu_n}$ of degree $\nu_1, \cdots, \nu_n$ in the variables $x_1, \cdots, x_n$ respectively, the following inequality, which was proved in [1] (Theorem 5), holds:

$$\| g \|_p^{(m)} = \left( \int_{-\infty}^{\infty} \cdots \int_{-\infty}^{\infty} | g(u_1, \ldots, u_m, x_{m+1}, \ldots, x_n) |^p du_1, \ldots, du_m \right)^{\frac{1}{p}} \leqslant 2^n \left( \prod_{k=m+1}^{n} \nu_k \right)^{\frac{1}{p}} \| g \|_p^{(n)}.$$

Applying this inequality to each of the differences appearing in the sum $\sigma'$ we find that

$$\| Q_s(x_1, \ldots, x_m, x_{m+1}+h_{m+1}, \ldots, x_n+h_n) - Q_s(x_1, \ldots, x_n) \|_p^{(m)}$$
$$\leqslant 2^n \left(\prod_{k=m+1}^{n} \nu_k\right)^{\frac{1}{p}} \| Q_s(x_1, \ldots, x_m, x_{m+1}+h_{m+1}, \ldots, x_n+h_n)$$
$$- Q_s(x_1, \ldots, x_n)\|_p^{(n)} \leqslant 2^n \left(\prod_{k=m+1}^{n} \nu_k\right)^{\frac{1}{p}} \sum_{k=0}^{n-m-1} \| Q_s(x_1 \ldots, x_{m+k}, x_{m+k+1}$$
$$+ h_{m+k+1}, \ldots, x_n+h_n) - Q_s(x_1, \ldots, x_{m+k+1}, x_{m+k+2}+h_{m+k+2}, \ldots, x_n+h_n)\|_p^{(n)} \text{ 1)}$$
$$\leqslant 2^n \left(\prod_{k=m+1}^{n} \nu_k\right)^{\frac{1}{p}} \sum_{k=m+1}^{n} h_k \| Q'_{sx_k} \|_p^{(n)} \to 0 \text{ as } \sum_{k=m+1}^{n} h_k \to 0.$$

And since the sum $\sigma$ has a finite number of terms, for sufficiently small $|h_k|$ it, too, will be less than $\epsilon$, which proves property (4).

3.14. Remark. The function $F = F(x_1, \cdots, x_n)$, given on $R_n$ and belonging to one or another class, is generally defined only almost everywhere on $R_n$. Therefore, formally, the function $F(x_1, \cdots, x_m, x_{m+1}, \cdots, x_n)$ of the variables $x_1, \cdots, x_m$ for fixed $x_{m+1}, \cdots, x_n$ is undefined.

We shall always understand by this function a function for which equation (4) of §3.13 holds. The basis of the validity of this equation for the functions which will be encountered in what follows will, as a rule, reduce to the preceding lemma.

3.2. Theorem. *Suppose that the function $f$ lies in the class $H_p^{(r_1,\cdots,r_n)}(M, \cdots, M)$, and suppose given nonnegative numbers $\rho_1, \cdots, \rho_n$ satisfying the inequality*

$$\sum_{k=1}^{n} \frac{\rho_k}{r_k} < 1. \tag{1}$$

*Then there exists on $R_n$ a partial derivative, $\partial^{\rho_1+\cdots+\rho_n} f/\partial x_1^{\rho_1} \cdots \partial x_n^{\rho_n}$, belonging to the class $H_p^{(r_1',\cdots,r_n')}(\overline{M}, \cdots, \overline{M})$, where*

$$r_i' = r_i\left(1 - \sum_{k=1}^{n} \frac{\rho_k}{r_k}\right) \quad (i = 1, 2, \ldots, n), \tag{2}$$

$$\overline{M} \leqslant 4\|f\|_p^{(n)} + cM. \tag{3}$$

*In addition,*

$$\left\| \frac{\partial^{\rho_1+\cdots+\rho_n} f}{\partial x_1^{\rho_1} \ldots \partial x_n^{\rho_n}} \right\|_p^{(n)} \leqslant \|f\|_p^{(n)} + cM. \tag{4}$$

1) See [1], Lemma 3.

*In inequalities* (3) *and* (4) *the constant* $c$ *depends on* $p$, $r_k$ *and* $\rho_k$, *but not on* $\|f\|_p^{(n)}$ *or* $M$.

**Proof.** We denote by $g_s = g_s(x_1, \cdots, x_n)$ $(s = 1, 2, \cdots)$ an entire function of degrees $\nu_k = \nu_k(s)$ relative to the variables $x_1, \cdots, x_n$ respectively, which is a best approximation in the sense of $L_p^{(n)}$ to the function $f$. Here we will suppose that

$$\nu_k = 2^{\frac{s}{r_k}} \quad (s = 1, 2, \ldots).$$

From Theorem 7 of [1] we have the inequality

$$\|f - g_s\| \leqslant \frac{dnM}{2^s},$$

where $d$ is a constant depending only on $p$ and $r_k$. The function $f$ decomposes into a series converging to it in the sense of $L_p^{(n)}$:

$$f = Q_0 + \sum_{s=1}^{\infty} Q_s, \quad Q_0 = g_0, \quad Q_s = g_s - g_{s-1} \ (s = 1, 2, \ldots),$$

where

$$\|Q_s\|_p^{(n)} \leqslant \|g_s - f\|_p^{(n)} + \|f - g_{s-1}\|_p^{(n)} \leqslant \frac{3dnM}{2^s} \quad (s = 1, 2, \ldots),$$
$$\|Q_0\|_p^{(n)} \leqslant \|f\|_p^{(n)} + dnM.$$

After termwise differentiation of this series we get

$$\frac{\partial^{\rho_1+\ldots+\rho_n} f}{\partial x_1^{\rho_1} \ldots \partial x_n^{\rho_n}} = \frac{\partial^{\rho_1+\ldots+\rho_n} Q_0}{\partial x_1^{\rho_1} \ldots \partial x_n^{\rho_n}} + \sum_{s=1}^{\infty} \frac{\partial^{\rho_1+\ldots+\rho_n} Q_s}{\partial x_1^{\rho_1} \ldots \partial x_n^{\rho_n}}, \tag{5}$$

where

$$\left\| \frac{\partial^{\rho_1+\ldots+\rho_n} Q_s}{\partial x_1^{\rho_1} \ldots \partial x_n^{\rho_n}} \right\|_p^{(n)} \leqslant \frac{3dnM}{2^s} 2^{s \sum_{k=1}^{n} \frac{\rho_k}{r_k}} = \frac{3dnM}{2^{s\left(1 - \sum_{k=1}^{n} \frac{\rho_k}{r_k}\right)}} \quad (s = 1, 2, \ldots),$$

$$\left\| \frac{\partial^{\rho_1+\ldots+\rho_n} Q_0}{\partial x_1^{\rho_1} \ldots \partial x_n^{\rho_n}} \right\|_p^{(n)} \leqslant \|f\|_p^{(n)} + dnM.$$

In view of inequality (1), the differentiation is legitimate.

Now our theorem follows from Lemma 3.12, on applying it to the series (5).

3.3. **Theorem.** *Given the function* $f = f(x_1, \cdots, x_n)$, *lying in the class*

$H_p^{(r_1,\cdots,r_n)}(M, \cdots, M)$, *an integer* $1 \le m < n$, *and a collection of nonnegative integers* $\lambda_{m+1}, \cdots, \lambda_n$, *forming a system* $(\lambda)$, *suppose that the inequality*

$$\rho_i^{(\lambda)} = r_i\left(1 - \sum_{j=m+1}^{n} \frac{\lambda_i}{r_j} - \frac{1}{p}\sum_{j=m+1}^{n} \frac{1}{r_j}\right) > 0 \tag{1}$$

*is satisfied, which in particular implies the inequality*

$$\sum_{j=m+1}^{n} \frac{\lambda_j}{r_j} < 1. \tag{2}$$

*Then the partial derivative* [1]

$$\psi(x_1, \ldots, x_m) = \frac{\partial^{\lambda_{m+1}+\cdots+\lambda_n} f(x_1, \ldots, x_m, x_{m+1}, \ldots, x_n)}{\partial x_{m+1}^{\lambda_{m+1}} \cdots \partial x_n^{\lambda_n}},$$

*as a function of* $x_1, \cdots, x_m$ *for any fixed* $x_{m+1}, \cdots, x_n$ *lies in the class* $H_p^{\left(\rho_1^{(\lambda)},\ldots,\rho_m^{(\lambda)}\right)}(K, \ldots, K)$, *where*

$$K < c_1 \|f\|_p^{(n)} + c_2 M. \tag{3}$$

*Moreover,*

$$\lambda_{m+1}, \ldots, \lambda_n. \tag{4}$$

*The constants* $c$ *entering in* (3) *and* (4) *depend on* $p$, $r_k$, $\lambda_j$ *and* $m$, *but not on* $\|f\|_p^{(n)}$ *or* $M$.

**Proof.** On the basis of Theorem 3.2 the partial derivative

$$\frac{\partial^{\lambda_{m+1}+\cdots+\lambda_n} f(x_1, \ldots, x_n)}{\partial x_{m+1}^{\lambda_{m+1}} \cdots \partial x_n^{\lambda_n}},$$

in view of inequality (2), belongs to the class $H_p^{(r_1',\cdots,r_n')}(M_1, \cdots, M_1)$, where

$$r_i' = r_i\left(1 - \sum_{k=m+1}^{n} \frac{\lambda_k}{r_k}\right) \quad (i = 1, 2, \ldots, n), \tag{5}$$

$$M_1 < 4\|f\|_p^{(n)} + cM. \tag{6}$$

---

1) If $\sum_{j=m+1}^{n} \lambda_j = 0$ we are dealing with the function $f$ itself.

Therefore, from Theorem 7 of [1], we have the equation

$$\frac{\partial^{\lambda_{m+1}+\cdots+\lambda_n} f(x_1, \ldots, x_n)}{\partial x_{m+1}^{\lambda_{m+1}} \ldots \partial x_n^{\lambda_n}} = g_0 + \sum_{s=1}^{\infty} g_s, \tag{7}$$

where $g_s = g_s(x_1, \cdots, x_n)$ are entire functions of degrees

$$\nu_k = 2^{\frac{s}{r_k'}} \quad (s = 0, 1, \ldots)$$

relative to the variables $x_k$ $(k = 1, \cdots, n)$ respectively, satisfying the inequalities

$$\| g_s \|_p^{(n)} < \frac{3dnM_1}{2^s} < \frac{c_1 \| f \|_p^{(n)} + c_2 M}{2^s} \quad (s = 1, 2, \ldots),$$

$$\| g_0 \|_p^{(n)} \leqslant \left\| \frac{\partial^{\lambda_{m+1}+\cdots+\lambda_n} f}{\partial x_1^{\lambda_{m+1}} \ldots \partial x_{m+1}^{\lambda_n}} \right\|_p^{(n)} + dnM_1 < c_3 \| f \|_p^{(n)} + c_4 M.$$

We may apply Lemma 3.13 to the series (7), with $\alpha$ and $r_k$ equal to 1 and $r_k'$ respectively, since it follows from (1) that the quantity

$$\gamma = 1 - \frac{1}{p} \sum_{k=m+1}^{n} \frac{1}{r_k'} > 0.$$

The assertion of the theorem now follows from it, if one takes into consideration that

$$r_i' \gamma = r_i \left(1 - \sum_{j=m+1}^{n} \frac{\lambda_j}{r_j} - \frac{1}{p} \sum_{j=m+1}^{n} \frac{1}{r_j}\right).$$

3.4. **Theorem** 1) **(inverse to Theorem 3.3).** *Suppose given positive numbers $r_i$ $(i = 1, \cdots, n)$, and all possible systems $(\lambda)$ of nonnegative integers*

$$\lambda_{m+1}, \ldots, \lambda_n,$$

*for which*

$$\rho_i^{(\lambda)} = r_i \left(1 - \sum_{j=m+1}^{n} \frac{\lambda_j}{r_j} - \frac{1}{p} \sum_{j=m+1}^{n} \frac{1}{r_j}\right) > 0. \tag{1}$$

*Suppose moreover that to each system $(\lambda)$ there is assigned a function $\varphi_{(\lambda)}(x_1, \cdots, x_m)$ of $m$ variables, lying in the class*

$$H_p^{\left(\rho_1^{(\lambda)}, \ldots, \rho_n^{(\lambda)}\right)} (M^{(\lambda)}, \ldots, M^{(\lambda)}). \tag{2}$$

1) A special case of this theorem was proved in [2]. A summary of its contents was published in [3].

*Then it is possible to construct a function* $f(x_1, \cdots, x_n)$ *of* $n$ *variables, having the following properties:* [1]

1) $$f \in H_p^{(r_1, \ldots, r_n)}(K, \ldots, K),$$

*where*

$$K < c \sum_{(\lambda)} (\| \varphi_{(\lambda)} \|_p^{(m)} + M^{(\lambda)}), \tag{3}$$

$$\| f \|_p^{(n)} < c \sum_{(\lambda)} (\| \varphi_{(\lambda)} \|_p^{(n)} + M^{(\lambda)}), \tag{3'}$$

2) $$\frac{\partial^{\lambda_{m+1} + \cdots + \lambda_n} f(x_1, \ldots, x_m, 0, \ldots 0)}{\partial x_{m+1}^{\lambda_{m+1}} \ldots \partial x_n^{\lambda_n}} = \varphi_{(\lambda)}(x_1, \ldots, x_m). \tag{4}$$

**Proof.** Put

$$r_i^{(\lambda)} = r_i \left(1 - \sum_{k-m+1}^{n} \frac{\lambda_k}{r_k}\right) \quad (i = 1, 2, \ldots, n), \tag{5}$$

$$\varkappa^{(\lambda)} = 1 - \frac{1}{p} \sum_{i=m+1}^{n} \frac{1}{r_i^{(\lambda)}}. \tag{6}$$

Then obviously

$$\rho_i^{(\lambda)} = r_i^{(\lambda)} \varkappa^{(\lambda)}. \tag{7}$$

Let

$$\varphi_{(\lambda)} \in H_p^{\left(\rho_1^{(\lambda)}, \ldots, \rho_n^{(\lambda)}\right)} (M^{(\lambda)}, \ldots, M^{(\lambda)}).$$

Denote by

$$g_{s(\lambda)} = g_{s(\lambda)}(x_1, \ldots, x_m) \quad (s = 1, 2, \ldots),$$

an entire function of degrees $2^{s/r_i^{(\lambda)}}$ relative to the $x_i$ respectively, which is a best approximation for $\varphi_{(\lambda)}$ in the sense of $L_p^{(m)}$. Then

$$\| \varphi_{(\lambda)} - g_{s(\lambda)} \|_p^{(m)} \leqslant \frac{d m M^{(\lambda)}}{2^{s \varkappa^{(\lambda)}}}.$$

---

1) If $\sum_{j=m+1}^{n} \lambda_j = 0$ the left side of (4) is the function $f$ itself.

Accordingly the function $\varphi_{(\lambda)}$ decomposes into a series

$$\varphi_{(\lambda)}=\sum_{s=0}^{\infty} Q_{s\,(\lambda)}, \quad Q_{0\,(\lambda)}=g_{0\,(\lambda)}, \quad Q_{s\,(\lambda)}=g_{s\,(\lambda)}-g_{s-1\,(\lambda)},$$

converging to it in the sense of $L_p^{(m)}$, with

$$\| Q_{s\,(\lambda)} \|_p^{(m)} \leqslant \frac{3d\,mM^{(\lambda)}}{2^{s\varkappa(\lambda)}} \quad (s=1,2,\ldots), \quad \| Q_{0\,(\lambda)} \|_p^{(m)} \leqslant \| \varphi_{(\lambda)} \|_p^{(m)} + dmM^{(\lambda)}. \tag{8}$$

Now we introduce trigonometrical polynomials $T_\nu(x)$, $\nu = 0, 1, \cdots, l$, where $l$ denotes the largest of the numbers $\lambda_j$ among all of those occurring in the various systems $\lambda_{m+1}, \cdots, \lambda_n$ which we are considering. Suppose that these trigonometric polynomials have the following properties: the function

$$\Phi_\nu(x)=\frac{T_\nu(x)}{x^2}$$

is finite for $x = 0$, and moreover it satisfies the equation

$$\Phi_\nu^{(\nu)}(0)=\frac{d^\nu}{dx^\nu}\Phi_\nu(x)\Big|_{x=0}=1,$$

$$\Phi_\nu^{(k)}(0)=0 \qquad (k=0,1,\ldots,\nu-1,\nu+1,\ldots,l).$$

We are not concerned with the magnitude of the degree of the trigonometric polynomial $T_\nu(x)$, and thus the conditions indicated above do not define it uniquely. We shall suppose that we have chosen completely defined polynomials, having degrees $\mu(\nu)$ respectively. Then $\Phi_\nu(x)$ is an entire function of the same degree, and $\Phi_\nu(nx/\mu(\nu))$ is an entire function of degree $n$. Obviously, also,

$$\left\| \Phi_\nu\left(\frac{n}{\mu(\nu)}x\right)\right\|_p^{(1)}=\left(\int_{-\infty}^{\infty}\left|\Phi_\nu\left(\frac{n}{\mu(\nu)}x\right)\right|^p dx\right)^{\frac{1}{p}}$$

$$=\left(\frac{\mu(\nu)}{n}\right)^{\frac{1}{p}}\left(\int_{-\infty}^{\infty}|\Phi_\nu(u)|^p\,du\right)^{\frac{1}{p}}=\frac{A_\nu}{n^{\frac{1}{p}}}, \tag{9}$$

where $A_\nu$ is a constant depending only on $\nu$.

Now we define functions $f_{(\lambda)}(x_1, \ldots, x_n)$, corresponding to the various systems $(\lambda)$, in terms of the series

$$f_{(\lambda)}(x_1, \ldots, x_n)$$

$$= \sum_{s=0}^{\infty} Q_{s(\lambda)} \prod_{j=m+1}^{n} \left( \frac{\mu(\lambda_j)}{2^{\frac{s}{r_j^{(\lambda)}}}} \right)^{\lambda_j} \Phi_{\lambda_j} \left( 2^{\frac{s}{r_j^{(\lambda)}}} \frac{x_j}{\mu(\lambda_j)} \right) = \sum_{s=0}^{\infty} R_{s(\lambda)}.$$

Using (5)–(9) one obtains the following estimate for $R_{s(\lambda)}$:

$$\| R_{s(\lambda)} \|_p^{(\lambda)}$$

$$\leqslant \frac{3dmM^{(\lambda)}}{2 \cdot 2^{s\varkappa^{(\lambda)}}} \frac{\prod_{j=m+1}^{n} [\mu(\lambda_j)]^{\lambda_j} A_{\lambda_j}}{2^{s \cdot \sum_{j=m+1}^{n} \frac{\lambda_j}{r_j^{(\lambda)}}} \cdot 2^{\frac{s}{p} \sum_{j=m+1}^{n} \frac{1}{r_j^{(\lambda)}}}} = \frac{c_1 M^{(\lambda)}}{2^{\frac{s r_i}{r_i^{(\lambda)}}}} \quad (s = 1, 2, \ldots),$$

$$\| R_{0(\lambda)} \|_p^{(n)}$$

$$\leqslant (\| \varphi_{(\lambda)} \|_p^{(m)} + dmM^{(\lambda)}) \prod_{j=m+1}^{n} [\mu(\lambda_j)]^{\lambda_j} A_{\lambda_j} \leqslant c_2 \| \varphi_{(\lambda)} \|_p^{(m)} + c_3 M^{(\lambda)}, \tag{10}$$

where the $c$ are constants not depending on $s$ or on the $M^{(\lambda)}$, and since

$$\sum_{s=1}^{k-1} R_{s(\lambda)}$$

is an entire function of degree $2^{k/r_i^{(\lambda)}}$ in the $x_i$ respectively, then on the basis of Lemma 3.12

$$f_{(\lambda)} \in H_p^{(r_1, \ldots, r_n)} (K^{(\lambda)}, \ldots, K^{(\lambda)}), \tag{11}$$

where

$$K^{(\lambda)} < c_2 \| \varphi_{(\lambda)} \|_p^{(m)} + c_4 M^{(\lambda)}$$

and

$$\| f_{(\lambda)} \|_p^{(n)} < c_5 \| \varphi_{(\lambda)} \|_p^{(m)} + c_6 M^{(\lambda)}. \tag{12}$$

We note further that it follows from the properties of the functions $\Phi_\nu$ that for any system $(\lambda)$ which is admissible, i.e. which satisfies condition (4), the following equation holds for the corresponding function $f_{(\lambda)}$:

$$\frac{\partial^{\lambda_1+\cdots+\lambda_m} f_{(\lambda)}(x_1, \ldots, x_m, 0, \ldots, 0)}{\partial x_{m+1}^{\lambda_{m+1}} \cdots \partial x_n^{\lambda_n}} = \varphi_{(\lambda)}(x_1, \ldots, x_n). \tag{13}$$

If now $(\lambda')$ is another admissible system of numbers $(\lambda'_{m+1}, \cdots, \lambda'_n)$, then *a fortiori*

$$\frac{\partial^{\lambda'_{m+1}+\cdots+\lambda'_n} f_{(\lambda)}(x_1, \ldots, x_m, 0, \ldots, 0)}{\partial x_{m+1}^{\lambda'_{m+1}} \cdots \partial x_n^{\lambda'_n}} \equiv 0. \tag{14}$$

In this case the function

$$f(x_1, \ldots, x_n) = \sum_{(\lambda)} f_{(\lambda)}(x_1, \ldots, x_n),$$

where the summation is extended over all possible admissible systems $(\lambda)$, satisfies all the requirements of the theorem.

Indeed, in view of (10), it lies in the class $H_p^{(r_1,\cdots,r_n)}(K, \cdots, K)$ if one takes

$$K = \sum_{(\lambda)} K_{(\lambda)} < c \sum_{(\lambda)} (\| \varphi_{(\lambda)} \|_p^{(m)} + M^{(\lambda)}).$$

Moreover

$$\| f \|_g^{(n)} \leqslant \sum_{(\lambda)} \| f_{(\lambda)} \|_p^{(n)} < c \sum_{(\lambda)} (\| \varphi_{(\lambda)} \|_p^{(m)} + M^{(\lambda)})$$

and, in view of (13) and (14), equation (4) holds.

3.5. Theorems 3.3 and 3.4 proved above will in the sequel be applied in the case $r_1 = \cdots = r_n = r$. Therefore it is appropriate to formulate them in this special case.

We agree for short to denote the class $H_p^{(r,\cdots,r)}(M, \cdots, M)$ by $H_{pn}^{(r)}(M)$.

3.51. **Theorem.** *Given a function $f(x_1, \cdots, x_n) \in H_{pn}^{(r)}(M)$ and a system $(\lambda)$ of nonnegative numbers $\lambda_{m+1}, \cdots, \lambda_n$ for which*

$$\lambda = \sum_{k=m+1}^{n} \lambda_k < r, \tag{1}$$

*suppose that*

$$\rho^{(\lambda)} = r - \lambda - \frac{n-m}{p} > 0. \tag{2}$$

*Then for any fixed* $x_{m+1}, \cdots, x_n$

$$\psi(x_1, \ldots, x_m) = \frac{\partial^\lambda f(x_1, \ldots, x_m, x_{m+1}, \ldots, x_n)}{\partial x_{m+1}^{\lambda_{m+1}} \ldots \partial x_n^{\lambda_n}} \in H_{pm}^{(\rho^{(\lambda)})}(K), \tag{3}$$

*where*

$$K < c_1 \|f\|_p^{(n)} \mid c_2 M, \tag{4}$$

$$\|\psi\|_p^{(m} < c_1 \|f\|_p^{(n)} + c_2 M. \tag{5}$$

*The constants* $c_1$ *and* $c_2$ *depend on* $p$, $r$ *and* $(\lambda)$, *but not on* $\|f\|_p^{(n)}$ *and* $M$.

*For* $\lambda = 0$ *the partial derivative coincides with the function* $f$ *itself.*

3.52. **Theorem.** *Suppose given a positive number* $r$ *and integers* $m$ *and* $n$, *with* $1 \leq m < n$. *Moreover suppose given all possible systems* $(\lambda)$ *of nonnegative numbers*

$$\lambda_{m+1}, \ldots, \lambda_n,$$

*for which*

$$\lambda = \sum_{j=m+1}^{n} \lambda_j < r, \tag{1}$$

*and numbers*

$$\rho^{(\lambda)} = r - \lambda - \frac{n-m}{p} > 0. \tag{2}$$

*Suppose further that to each system* $(\lambda)$ *there has been assigned a function* $\phi_{(\lambda)}(x_1, \cdots, x_m)$ *of* $m$ *variables, lying in the class* $H_{pm}^{(\rho^{(\lambda)})}(M^{(\lambda)})$, *where the* $M^{(\lambda)}$ *are certain positive numbers.*

*Then it is possible to construct a function* $f$ *of* $n$ *variables, having the following properties:*

1) $$f \in H_{pn}^{(r)}(K),$$

where

$$K < c \sum_{(\lambda)} \left(\|\varphi_{(\lambda)}\|_p^{(m)} + M^{(\lambda)}\right),$$

2) $$\|f\|_p^{(n)} < c \sum_{(\lambda)} (\|\varphi_{(\lambda)}\|_p^{(m)} + M^{(\lambda)}),$$

3) $$\frac{\partial^\lambda f(x_1, \ldots, x_m, 0, \ldots, 0)}{\partial x_{m+1}^{\lambda_{m+1}} \ldots \partial x_n^{\lambda_n}} = \varphi_{(\lambda)}(x_1, \ldots, x_m).$$

*For* $(\lambda) = 0$ *the partial derivative coincides with the function itself.*

## §4. Values of functions of the classes $H_{pn}^{(r)}$ on differentiable manifolds

4.1. In this section we shall consider functions $f(x_1, \cdots, x_m)$ belonging to the classes $H_{pn}^{(r)}$, on various $m$-dimensional manifolds.

An $m$-dimensional manifold $S$ is defined as follows. This is a closed bounded set lying in $R$, such that each of its points $P_0$ may be surrounded by a sphere with center at $P_0$ and of radius sufficiently small that it excises from $S$ a piece $\sigma$ defined under an appropriate enumeration of the coordinate axes by the equations

$$\begin{array}{l} x_{m+1} = \varphi_{m+1}(x_1, \ldots, x_m), \\ \ldots\ldots\ldots\ldots\ldots\ldots, \\ x_n = \varphi_n(x_1, \ldots, x_m), \end{array} \tag{1}$$

when the point $(x_1, \cdots, x_m)$ runs through some $m$-dimensional region $G$. Here the functions $\phi_k$ $(k = m + 1, \cdots, n)$ have on $G$ continuous partial derivatives of orders at least $\bar{r} + 1$. Moreover, it is supposed that it is possible at each point $P \in \sigma$ to choose $n - m$ unit vectors

$$\bar{N}_{m+1}, \ldots, \bar{N}_n,$$

orthogonal to one another and such that each of them is orthogonal to the linear manifold tangent to $S$ at the point $P$. Here the projections

$$\bar{N}_j = (\alpha_1^{(j)}, \ldots, \alpha_n^{(j)}) \quad (j = m + 1, m + 2, \ldots, n) \tag{2}$$

of these vectors are functions of $x_1, \cdots, x_m$, having continuous partial derivatives of orders up to $\bar{r} + 1$ inclusive.

If some function $f \in H_{pn}^{(r)}$ is continuous, then it induces some function $f|_S$ on the manifold $S$. On each of the pieces $\sigma$ this function may be written in terms of the function

$$F(x_1, \ldots, x_m) = f(x_1, \ldots, x_m, \varphi_{m+1}, \ldots, \varphi_n) \tag{3}$$

of the variables $x_1, \cdots, x_m$, defined on an appropriate $m$-dimensional region $G$. Of course, here we may clearly make the convention of expressing this function explicitly in terms of the first $m$ coordinates. In fact, on each different piece $\sigma$ these $m$ coordinates are generally speaking different.

In the general case, when the function $f$ is not necessarily continuous, it is defined on $R_n$ only up to a set of $n$-dimensional measure zero, and therefore the function $f|_S$ remains undefined. Now we shall define what we shall mean by $f|_S$ in this case.

Suppose we are given arbitrary numbers

$$h_{m+1}, \ldots, h_n.$$

Define on $G$ the function

$$f\left(x_1 + \sum_{j=m+1}^{n} \alpha_1^{(j)} h_j, \ldots, x_m + \sum_{j=m+1}^{n} \alpha_m^{(j)} h_j, \varphi_{m+1}\right.$$

$$\left. + \sum_{j=m+1}^{n} \alpha_{m+1}^{(j)} h_j, \ldots, \varphi_n + \sum_{j=m+1}^{n} \alpha_n^{(j)} h_j\right)$$

$$= F(x_1, \ldots, x_m, h_{m+1}, \ldots, h_n) = F_{(h)} \tag{4}$$

of the variables $x_1, \cdots, x_m$. By the expression (3) on $G$ we will understand a function of $x_1, \cdots, x_m$ for which the relation

$$\lim_{\sum_{k=m+1}^{n} |h_k| \to 0} \int \cdots \int_{G_1} |F_{(h)} - F|^p \, dx_1 \ldots dx_m = 0, \tag{5}$$

holds for any region $G_1$ whose closure lies in $G$.

In the following subsection we shall show that equation (5) always holds if

$$\rho = r - \frac{n-m}{p} > 0.$$

In §§4.12 and 4.13 it will be proved, moreover, that the function $F$ does not depend on the choice of the normal vectors (2), and in the case when the piece $\sigma$ of the manifold $S$ is given explicitly in terms of other coordinates distinct from $x_1, \cdots, x_m$, the function $F$ transforms to the new coordinates in the same way as in the case of a function $f(x_1, \cdots, x_m, \phi_{m+1}, \cdots, \phi_n)$, understood in the usual sense of this term.

4.11. Suppose we are given an arbitrary region $G_1$ such that $\overline{G}_1 \subset G$, and suppose that $S'$ is the portion of $S$ corresponding to $G_1$. The equations

$$\begin{aligned} u_i &= x_i + \sum_{j=m+1}^{n} \alpha_i^{(j)} h_j \quad (i = 1, 2, \ldots, m), \\ u_k &= \varphi_k + \sum_{j=m+1}^{n} \alpha_k^{(j)} h_j \quad (k = m+1, m+2, \ldots, n) \end{aligned} \tag{1}$$

define a transformation of the points $(x_1, \cdots, x_m, h_{m+1}, \cdots, h_n)$ of some space $R_n^*$ into points $(u_1, \cdots, u_n)$ of the space $R_n$. The Jacobian of this transformation, at the point $(x_1, \cdots, x_m, 0, \cdots, 0)$, has the form

$$D = \begin{vmatrix} 1 & 0 & \ldots & 0 & \alpha_1^{(m+1)} & \ldots & \alpha_1^{(n)} \\ 0 & 1 & \ldots & 0 & \alpha_2^{(m+1)} & \ldots & \alpha_2^{(n)} \\ \cdot & \cdot & \cdot & \cdot & \cdot & \cdot & \cdot \\ 0 & 0 & \ldots & 1 & \alpha_m^{(m+1)} & \ldots & \alpha_m^{(n)} \\ \frac{\partial \varphi_{m+1}}{\partial x_1} & & \ldots & \frac{\partial \varphi_{m+1}}{\partial x_m} & \alpha_{m+1}^{(m+1)} & \ldots & \alpha_{m+1}^{(n)} \\ \cdot & \cdot & \cdot & \cdot & \cdot & \cdot & \cdot \\ \frac{\partial \varphi_n}{\partial x_1} & & \ldots & \frac{\partial \varphi_n}{\partial x_m} & \alpha_n^{(m+1)} & \ldots & \alpha_n^{(n)} \end{vmatrix} \tag{2}$$

The first $m$ columns of this determinant may be considered as vectors tangent to $S'$ at the point $P$, not linearly dependent on one another, since the matrix consisting of these $m$ columns has rank $m$. The remaining columns define unit normal vectors $\overline{N}_{m+1}, \cdots, \overline{N}_n$.

Thus all $n$ columns of $D$ are linearly independent and $D \neq 0$ at any point of the $n$-dimensional region $\overline{G}_1^*$ consisting of the points $(x_1, \cdots, x_m, 0, \cdots, 0)$ for which $(x_1, \cdots, x_m) \in \overline{G}_1$.

Moreover, at these points $|D| \geq 1$. In fact, because of the orthogonality of the first $m$ columns of the determinant to the remaining $n - m$ columns, the magnitude of $|D|$ is equal to the product of the ($m$-dimensional) volume of the parallelepiped constructed on the first $m$ columns by the ($(n - m)$-dimensional) volume of the cube constructed on the remaining columns. The second volume is obviously equal to unity, and the first is not less than unity, since the corresponding matrix consisting of $m$ columns contains a square $m$-dimensional unit matrix. Therefore each such point $(x_1, \cdots, x_m, 0, \cdots, 0)$ may be encircled by a ball $V \subset R_n^*$ with center at that point of a radius so small that it transforms under the transformations (1) in a 1-1 manner into some region $\Delta \subset R_n^*$ containing the point

$$u_1 = x_1, \ldots, u_m = x_m, u_{m+1} = \varphi_{m+1}, \ldots, u_n = \varphi_n.$$

But, in view of the 1-1 property of the relationship between the points $(x_1, \cdots, x_m, 0, \cdots, 0)$ and the points $(x_1, \cdots, x_m, \phi_{m+1}, \cdots, \phi_n)$, one may assert even more. In fact, let us denote by $W^{(\epsilon)}$ the region consisting of points of $R_n^*$ lying at a distance less than $\epsilon$ from $\overline{G}_1^*$. Then one may assert that for $\epsilon > 0$ sufficiently small the entire region $W^{(\epsilon)}$ maps under the transformations (1) into some region $R_n$ in a one-to-one manner. In fact, if this were not so, then there would exist in $R_n^*$ two sequences of points $Q_k$ and $Q_k'$, $k = 1, 2, \cdots$, tending as $k \to \infty$ to points $Q_0$ and $Q_0'$ of the set $\overline{G}_1^*$ and such that for each $k = 1, 2, \cdots$, the points $Q_k$ and $Q_k'$ are distinct and go under the transformation (1) into the same point $P_k \in R_n$, while, obviously, $P_k$ tends to some point $P_0 \in \sigma$. But this is not possible, since if $Q_0 = Q_0'$ then no small neighborhood of the point $Q_0$ would transform in a one-to-one manner under (1), and if $Q_0 \neq Q_0'$ one obtains a contradiction with the fact that the set of $G_1^*$ itself maps into some piece of $\sigma$ in a one-to-one manner. Here we may suppose that on $W^{(\epsilon)}$ the Jacobian of the transformation (1) is larger in absolute value than some positive constant.

Since the function $f(u_1, \cdots, u_n)$ lies in the class $H_{pn}^{(r)}$ and the functions $u_j$, $j = 1, \cdots, n$, defining the transformation (1) have continuous partial derivatives of order $\overline{r} + 1$, one may state on the basis of Lemma 2.4 that the function $F_{(h)}$ defined by equation (4) of §4.1 lies in the class $H_p^{(r)}(W^{(\epsilon')})$, where $0 < \epsilon' < \epsilon$. Finally, we may continue the function $F_{(h)}$ beyond the boundary of $W^{(\epsilon'')}$, where $0 < \epsilon'' < \epsilon' < \epsilon$, in such a way that the extended function

$$\Phi_{(h)} = \Phi(x_1, \ldots, x_m, h_{m+1}, \ldots, h_n)$$

will lie in the class $H_{pn}^{(r)}$ (Theorem 2.2).

In view of inequality (6) of §4.1, on the basis of Theorem 3.51, which we apply for $\lambda = 0$, the function

$$\Phi_{(0)} = \Phi(x_1, \ldots, x_m, 0, \ldots, 0) \in H_{pm}^{(\rho)},$$

while

$$\lim_{\sum_{k=m+1}^{n} |h_k| \to 0} \| \Phi_{(h)} - \Phi_{(0)} \|_p^{(m)} = 0.$$

But, in view of the fact that $\Phi_{(h)} = F_{(h)}$ on $W^{(\epsilon'')}$, we find that

$$\lim_{\sum_{k=m+1}^{n} |h_k| \to 0} \int \dots \int_{G_1} | F_{(h)} - F|^p \, dx_1 \dots dx_m = 0. \tag{3}$$

4.12. The function $F(x_1, \cdots, x_m) = f[x_1, \cdots, x_m, \phi_{m+1}, \cdots, \phi_n]$ was defined by us on $G$ using the normal vectors $\overline{N}_{m+1}, \cdots, \overline{N}_n$ to the manifold $S$ introduced for this purpose. We wish to show that the function $F$ in fact does not depend on the way in which these vectors were chosen.

Suppose that we have associated with each point of $\sigma$ another system of unit vectors $\overline{N}'_{m+1}, \cdots, \overline{N}'_n$ with projections

$$\overline{N}'_k = \left(\alpha_1'^{(k)}, \dots, \alpha_n'^{(k)}\right) \quad (k = m+1, m+2, \dots, n),$$

which as functions of $x_1, \cdots, x_m$ have continuous partial derivatives to order $\overline{r}+1$ inclusive.

Suppose given $n-m$ numbers $h_{m+1}, \cdots, h_n$. We construct the corresponding function

$$f\left(x_1 + \sum_{j=m+1}^{n} \alpha_1'^{(j)} h_j, \dots, x_m + \sum_{j=m+1}^{n} \alpha_m'^{(j)} h_j, \varphi_{m+1} \right.$$
$$\left. + \sum_{j=m+1}^{n} \alpha_{m+1}'^{(j)} h_j, \dots, \varphi_n + \sum_{m+1}^{n} \alpha_n'^{(j)} h_j \right)$$
$$= F'(x_1, \dots, x_m, h_{m+1}, \dots, h_n) = F'_{(h)}.$$

The scalar products

$$c_{ks} = (\overline{N}'_k, \overline{N}_s) = \sum_{i=1}^{n} \alpha_i'^{(k)} \alpha_i^{(s)} \quad (k, s = m+1, \dots, n)$$

are functions of $x_1, \cdots, x_m$, having continuous partial derivatives to order $\overline{r}+1$ inclusive. Obviously

$$\alpha_i'^{(k)} = \sum_{s=m+1}^{n} c_{ks} \alpha_i^{(s)}.$$

Therefore it follows from equations (1) of §4.11 that

$$F'_{(h)} = f\Big(x_1 + \sum_j \sum_s c_{js}\alpha_1^{(s)} h_j, \ldots, x_m + \sum_j \sum_s c_{js}\alpha_m^{(s)} h_j, \varphi_{m+1} + \sum_j \sum_s c_{js}\alpha_{m+1}^{(s)} h_j, \ldots, \varphi_n + \sum_j \sum_s \alpha_n^{(s)} h_j\Big)$$

$$= f\Big(x_1 + \sum_s \alpha_1^{(s)} H_s, \ldots, x_m + \sum_s \alpha_m^{(s)} H_s, \varphi_{m+1} + \sum_s \alpha_{m+1}^{(s)} H_s, \ldots, \varphi_n + \sum_s \alpha_n^{(s)} H_s\Big),$$

where

$$H_s = \sum_{j=m+1}^{n} c_{js} h_j \quad (s = m+1, m+2, \ldots, n). \tag{1}$$

But in this case, if we put

$$F'_{(h)} = F(x_1, \ldots, x_m, h_{m+1}, \ldots, h_n), \tag{2}$$

then

$$F'_{(h)} = F(x_1, \ldots, x_m, H_{m+1}, \ldots, H_n). \tag{3}$$

Our problem consists in the following: we must prove that both the functions $F_{(h)}$ and $F'_{(h)}$ of the variables $x_1, \cdots, x_m$ tend in the sense of $L_p(\overline{G}_1)$ to the same (up to a set of measure zero) function $F(x_1, \cdots, x_m)$ on every closed set $\overline{G}_1 \subset G$, when $\sum_{k=m+1}^{n} h_k^2 \to 0$.

Suppose we are given any point $(x_1^*, \cdots, x_m^*) \in \overline{G}_1$. In the space $R_n^*$ of points $(x_1, \cdots, x_m, H_{m+1}, \cdots, H_n)$ we introduce the set $\overline{G}_1^*$ of points of the form $(x_1, \cdots, x_m, 0, \cdots, 0)$, where $(x_1, \cdots, x_m) \in \overline{G}_1$, and the region $W^{(\epsilon)}$ enclosing it consisting of points lying at a distance less than $\epsilon$ from $\overline{G}_1^*$. As was explained in §4.11, for sufficiently small $\epsilon$ the function $F(x_1, \cdots, x_m, H_{m+1}, \cdots, H_n)$ lies in the class $H_p^{(r)}(W^{(\epsilon)})$. Now we extend the function $F(x_1, \cdots, x_m, H_{m+1}, \cdots, H_n)$ beyond the boundary of $W^{(\epsilon')}$, where $0 < \epsilon' < \epsilon$, to the entire space $R_n^*$ in such a way that the extended function $\Phi(x_1, \cdots, x_m, H_{m+1}, \cdots, H_n)$ lies in $H_{pn}^{(r)}$, which is possible in view of Theorem 2.2. (See Theorem 7 of [1].) We now decompose the resulting function $\Phi$ into a series

$$\Phi(x_1, \ldots, x_m, H_{m+1}, \ldots, H_n) = \sum_{s=0}^{\infty} Q_s(x_1, \ldots, x_m, H_{m+1}, \ldots, H_n), \tag{4}$$

where the $Q_s$ $(s = 0, 1, \cdots)$ are entire functions of degrees $2^{s/r}$ relative to each of the variables $x_1, \cdots, x_m, H_{m+1}, \cdots, H_n$. In addition we may suppose that the norm of $Q_s$ satisfies the inequality

$$\| Q_s \|_p^{(n)} \leqslant \frac{B}{2^s} \quad (s = 1, 2, \ldots), \tag{5}$$

where $B$ is a constant not depending on $s$. Put

$$\Phi = \Phi_\mu + R_\mu, \tag{6}$$

where

$$\Phi_\mu = \sum_{s=0}^{\mu-} Q_s, \quad R_\mu = \sum_{s=\mu}^{\infty} Q_s. \tag{7}$$

We have supposed that

$$r - \frac{n-m}{p} > 0. \tag{8}$$

We choose a positive number $r'$ such that

$$r' - \frac{n-m}{p} > 0, \quad \bar{r} < r' < r. \tag{9}$$

Since the function $Q_s$ is of degree $2^{s/r}$ relative to each of the variables and satisfies inequality (5), by Lemma 3.11 it lies in the class

$$H_p^{(r')}\left(\frac{4B}{2^{\frac{r-r'}{r}s}}\right).$$

Hence

$$R_\mu \in H_p^{(r')}(M_\mu),$$

where

$$M_\mu = 4B \sum_{s=\mu}^{\infty} \frac{1}{2^{\frac{r-r'}{r}s}} = \frac{B_1}{2^{\frac{r-r'}{r}\mu}}. \tag{10}$$

Moreover, it follows from (5) and (7) that

$$\| R_\mu \|_p^{(n)} \leqslant \frac{B}{2^{\mu-1}}. \tag{11}$$

In view of the orthogonality of the matrix $\|c_{js}\|$, the equation

$$\sum_{j=m+1}^{n} h_j^2 = \sum_{j=m+1}^{n} H_j^2$$

holds, and therefore with the aid of the transformations

$$x_i = x_i \quad (i = 1, 2, \ldots, m),$$

$$H_s = \sum_{j=m+1}^{n} c_{js} h_j = \sum_{j=m+1}^{n} c_{js}(x_1, \ldots, x_m) h_j \quad (s = m+1, m+2, \ldots, n)$$

the points $(x_1, \cdots, x_m, h_{m+1}, \cdots, h_n)$ of the region $W^{(\epsilon')}$ are brought into one-to-one correspondence with the points $(x_1, \cdots, x_m, H_{m+1}, \cdots, H_n)$ of the same region. The Jacobian of this transformation is equal to unity. Hence, because the function

$$R_\mu = R_\mu(x_1, \ldots, x_m, H_{m+1}, \ldots, H_n),$$

considered as a function of $x_1, \cdots, x_m, H_{m+1}, \cdots, H_n$ on $W^{(\epsilon')}$, lies in the class $H_p^{(r')}(W^{(\epsilon')}, M_\mu)$, and because of (11) and Lemma 2.4, this same function, considered as a function of the variables $x_1, \cdots, x_m$, $h_{m+1}, \cdots, h_n$, lies in the class $H_p^{(r')}(W^{(\epsilon'')}, \overline{M}_\mu)$, where $0 < \epsilon'' < \epsilon$, and

$$\overline{M}_\mu < c_1 \| R_\mu \|_{L_p(W^{(\varepsilon')})} + c_2 M_\mu \leqslant c_1 \| R_\mu \|_p^{(n)} + c_2 M_\mu.$$

Moreover,

$$\| R_\mu \|_{L_p\left(W_h^{(\varepsilon')}\right)} < c_3 \| R_\mu \|_{L_p(W^{(\varepsilon')})} < c_3 \| R_\mu \|_p^{(n)},$$

where the symbol $W_h^{(\epsilon')}$ indicates that the norm is taken relative to the variables $x_1, \cdots, x_m, h_{m+1}, \cdots, h_n$.

Finally, on the basis of Theorem 2.2, we may extend the function $R_\mu$ of the variables $x_1, \cdots, x_m, h_{m+1}, \cdots, h_n$ beyond the limits of the region $W^{(\epsilon''')}$ $(0 < \epsilon''' < \epsilon'')$ in such a way that the extended function, which we denote by

$$\widetilde{R}_\mu = \widetilde{R}_\mu(x_1, \ldots, x_m, h_{m+1}, \ldots, h_n),$$

will lie in the class $H_{pn}^{(r')}(\widetilde{M}_\mu)$, where

$$\widetilde{M}_\mu < c_4 \| R_\mu \|_{L_p\left(W_h^{(\varepsilon'')}\right)} + c_5 \overline{M}_\mu < c_6 \| R_\mu \|_p^{(n)} + c_7 M_\mu,$$

$$\| \widetilde{R}_\mu \|_p^{(n)} < c_8 \| R_\mu \|_{L_p\left(W_h^{(\varepsilon'')}\right)} < c_9 \| R_\mu \|_p^{(n)}.$$

Since the number $r'$ satisfies inequality (9), then from Theorem 3.51, applied for $\lambda = 0$, we find for any fixed $h_{m+1}, \cdots, h_n$ that

$$\left(\int_{-\infty}^{\infty} \dots \int_{-\infty}^{\infty} | \tilde{R}_\mu (x_1, \dots, x_\mu, h_{m+1}, \dots, h_n) |^p dx_1 \dots dx_m\right)^{\frac{1}{p}}$$

$$< c_1 \| \tilde{R}_\mu \|_p^{(n)} + \tilde{M}_\mu < c_{10} \| R_\mu \|_p^{(n)} + c_{11} M_\mu < \varepsilon, \tag{12}$$

given that $\mu$ is sufficiently large, uniformly relative to $h_{m+1}, \cdots, h_n$, as follows from (10) and (11).

Now we shall start from the fact that the function $F_{(h)}$ on the region $W^{(\epsilon''')}$ decomposes into the series (4):

$$F_{(h)} = F(x_1, \dots, x_m, h_{m+1}, \dots, h_n)$$

$$= \sum_{s=0}^{\infty} Q_s (x_1, \dots x_m, h_{m+1}, \dots, h_n),$$

converging to it, on the basis of the estimate (5), not only in the sense of $L_p(W^{(\epsilon'')})$ but also in the variables $x_1, \cdots, x_m$, in the sense of $L_p(G_1)$, for each fixed set of admissible $h_{m+1}, \cdots, h_n$.

From (8) and Lemma 3.13 (see also Theorem 3.51), one has the relation

$$\lim_{\sum_{k=m+1}^{n} h_k^2 \to 0} \int_{G_1}\dots\int | F_{(h)} - F_{(0)} | \, dx_1 \dots dx_m = 0,$$

where

$$F_{(0)} = \sum_{s=0}^{\infty} Q_s (x_1, \dots, x_m, 0, \dots, 0). \tag{13}$$

On the other hand, in view of (6),

$$\| F_{(h)}' - F_{(0)} \|_{L_p(G_1)} \leqslant \sum_{s=0}^{\mu-1} \| Q_s (x_1, \dots, x_m, H_{m+1}, \dots, H_n)$$

$$- Q_s (x_1, \dots, x_m, 0, \dots, 0) \|_{L_p(G_1)}$$

$$+ \| \tilde{R}_\mu \|_{L_p(G_*)} + \sum_{s=\mu}^{\infty} \| Q_s (x_1, \dots, x_m, 0, \dots, 0) \|_{L_p(G_*)} = \gamma_1 + \gamma_2 + \gamma_3.$$

But if $\mu$ is sufficiently large, then from (12) we have $\gamma_2 < \epsilon$; similarly $\gamma_3 < \epsilon$, in view of the convergence of the series (13) in the sense of $L_p^{(m)}$. As to the quantity $\gamma_1$, it satisfies the equation

$$\lim_{\sum_{k=m+1}^{n} h_k^2 \to 0} \gamma_1 = 0,$$

since it is the sum of a finite number of terms. Here we must take into account the fact that the functions $Q_s(x_1, \cdots, x_m, H_{m+1}, \cdots, H_n)$ are uniformly continuous on $G_1$ and, moreover, that $\sum_{k=m+1}^{n} H_k^2 = \sum_{k=m+1}^{n} h_k^2 \to 0$.

Thus both the functions $F_{(h)}$ and $F'_{(h)}$ tend to the same function $F_{(0)}$ in the sense of $L_p(G_1)$.

This proves our assertion.

4.13. Now we shall suppose that one and the same piece $\sigma$ of the manifold $S$ is defined, on the one hand, by means of the equations

$$\begin{aligned} x_{m+1} &= \varphi_{m+1}(x_1, \ldots, x_m), \\ &\cdots\cdots\cdots\cdots, \qquad (x_1, \ldots, x_m) \in G, \\ x_n &= \varphi_n(x_1, \ldots, x_m), \end{aligned} \tag{1}$$

and on the other hand by the equations

$$\begin{aligned} x_{i_{m+1}} &= \overset{*}{\varphi}_{m+1}(x_{i_1}, \ldots, x_{i_m}), \\ &\cdots\cdots\cdots\cdots, \qquad (x_{i_1}, \ldots, x_{i_m}) \in G^*, \\ x_{i_n} &= \overset{*}{\varphi}_n(x_{i_1}, \ldots, x_{i_m}), \end{aligned} \tag{2}$$

which give an explicit expression in terms of other coordinates $x_{i_1}, \cdots, x_{i_m}$, in some way or another different from $x_1, \cdots, x_m$. Then obviously

$$\begin{aligned} x_{i_1} &= \psi_1(x_1, \ldots, x_m), \\ &\cdots\cdots\cdots\cdots, \qquad (x_1, \ldots, x_m) \in G, \\ x_{i_m} &= \psi_m(x_1, \ldots, x_m), \end{aligned} \tag{3}$$

where the functions $\psi_k$ have continuous partial derivatives to order $\bar{r}+1$ inclusive, since the functions $\phi_k$ and $\overset{*}{\phi}_k$ have this property.

For us it will be important to establish that the functions on $\sigma$ defined in §4.1,

$$f(x_1, \ldots, x_m, \varphi_{m+1}, \ldots, \varphi_n) = F(x_1, \ldots, x_m), \tag{4}$$

$$f(\overset{*}{\varphi}_{m+1}, \ldots, \overset{*}{\varphi}_n, x_{i_1}, \ldots, x_{i_m}) = F_*(x_{i_1}, \ldots, x_{i_m}) \tag{5}$$

(on the left side we have arranged the $\overset{*}{\phi}_k$ and $x_{i_k}$ in a purely conventional way) are connected by the usual relation

$$F(x_1, \ldots, x_m) = F_*[\psi_1, \ldots, \psi_m], \quad (x_1, \ldots, x_m) \in F, \tag{6}$$

in every case almost everywhere.

Indeed, it is easy to see that if the function $F_{(h)}$ defined in §4.1 by equation (4), corresponding to the system (1) of §4.12, has the form

$$F_{(h)} = F(x_1, \ldots, x_m, h_{m+1}, \ldots, h_n), \tag{7}$$

and another such function, corresponding to system (2) of §4.12, has the form

$$F^*_{(h)} = F_*(x_{i_1}, \ldots, x_{i_m}, h_{m+1}, \ldots, h_n), \tag{8}$$

then evidently

$$F(x_1, \ldots, x_m, h_{m+1}, \ldots, h_n) = F_*(\psi_1, \ldots, \psi_m, h_{m+1}, \ldots, h_n) \tag{9}$$

identically relative to $x_1, \cdots, x_m, h_{m+1}, \cdots, h_n$.

Since both functions $F_{(h)}$ and $F'_{(h)}$ converge in the sense of $L_p(G_1)$ and $L_p(G_1^*)$ respectively, where $\overline{G}_1 \subset G_1$ and $G_1^*$ is the region corresponding to $G_1$, then it is possible to indicate a subsequence of systems

$$(h^{(k)}) = (h_{m+1}^{(k)}, \ldots, h_n^{(k)}) \quad (k = 1, 2, \ldots),$$

satisfying the condition $\sum_{j=m+1}^{n} |h_j^{(k)}| \longrightarrow 0 \ (k \longrightarrow \infty)$ and such that

$$\begin{aligned} &\lim_{k\to\infty} F_{(h^{(k)})} = F(x_1, \ldots, x_m) \text{ almost everywhere on } G_1, \\ &\lim_{k\to\infty} F^*_{(h^{(k)})} = F^*(x_{i_1}, \ldots, x_{i_m}) \text{ almost everywhere on } G_1^*. \end{aligned} \tag{10}$$

If we take equation (9) into account and use the fact that equations (3) realize a one-to-one transformation of $G$ into $G^*$, such that every measurable set goes into a measurable set, then (6) follows from (10) and our assertion is proved.

## §5. Behavior of the partial derivatives of functions of the classes $H_{pn}^{(r)}$ on differentiable manifolds. Direct theorems

5.1. Again, as in §4.1, we are given an $m$-dimensional differentiable manifold $S$. Suppose as before that a piece $\sigma$ of this manifold is defined by the equations

$$\begin{aligned} x_{m+1} &= \varphi_{m+1}(x_1, \ldots, x_m), \\ \cdot \cdot \cdot \cdot & \cdot \cdot \cdot \cdot \cdot \cdot \cdot \cdot \cdot, \qquad (x_1, \ldots, x_m) \in G, \\ x_n &= \varphi_n(x_1, \ldots, x_m), \end{aligned} \tag{1}$$

where the $\phi_j$ $(j = m+1, \cdots, n)$ are functions having continuous partial derivatives to orders at least $\bar{r}+1$. Moreover we suppose that to each point $P \in \sigma$ we may associate $n-m$ unit vectors

$$\overline{N}_j = (\alpha_1^{(j)}, \ldots, \alpha_n^{(j)}) \quad (j = m+1, m+2, \ldots, n), \tag{2}$$

normal to $S$, whose projections $\alpha_k^{(j)}$ have continuous partial derivatives to orders $\bar{r}+1$ inclusive.

If the function $f \in H_{pn}^{(r)}$ and $r > 1$, then it has a partial derivative $\partial f/\partial x_i \in H_p^{(\rho)}$, where $\rho = r-1$. In turn, if

$$\rho - \frac{n-m}{p} > 0,$$

then we may define our partial derivative on the manifold $S$, i.e. define a function $\partial f/\partial x_i|_S$, which is obtained from $\partial f/\partial x_i$ to the same exactness as that to which $f|_S$ was determined from $f$.

5.11. Put

$$\begin{aligned} F_{(h)} &= F(x_1, \ldots, x_m, h_{m+1}, \ldots, h_n) \\ &= f\Big(x_1 + \sum_{j=m+1}^{n} \alpha_1^{(j)} h_j, \ldots, x_m + \sum_{j=m+1}^{n} \alpha_m^{(j)} h_j, \\ &\qquad \varphi_{m+1} + \sum_{j=m+1}^{n} \alpha_{m+1}^{(j)} h_j, \ldots, \\ &\qquad \varphi_n + \sum_{j=m+1}^{n} \alpha_n^{(j)} h_j\Big). \end{aligned} \tag{1}$$

If the function $f$ has continuous partial derivatives, then its derivative along the direction $\overline{N}_{m+1}$ on the piece $\sigma$ of the manifold $S$ is evidently equal to

$$\frac{\partial f}{\partial N_{m+1}}\Big|_\sigma = \frac{\partial F_{(h)}}{\partial h_{m+1}}\Big|_0,$$

where the subscript $|_0$ means that after the indicated operation has been carried out we put $h_{m+1} = \cdots = h_n = 0$.

Similarly, if $f$ has a sufficient number of continuous partial derivatives, then one may express in terms of $F_{(h)}$ any mixed derivative of $f$ in the directions $\overline{N}_j$ $(j = m+1, m+2, \cdots, n)$. In fact,

$$\left.\frac{\partial^\lambda f}{\partial \overline{N}_{m+1}^{\lambda_{m+1}} \ldots \partial \overline{N}_n^{\lambda_n}}\right|_\sigma = \left.\frac{\partial^\lambda F_{(h)}}{\partial h_{m+1}^{\lambda_{m+1}} \ldots \partial h_n^{\lambda_n}}\right|_0 \quad \left(\lambda = \sum_{j=m+1}^{n} \lambda_j\right). \tag{2}$$

We shall employ equation (2) as the basis of the definition of the mixed derivative of the function $f$ on $S$. Thus the mixed derivative of $f$ indicated on the left side of (2) on the piece $\sigma$ of the manifold $S$ is defined as a function of $x_1, \cdots, x_m$ to which the corresponding mixed derivative of $F_{(h)}$ indicated on the right side of (2) converges in the mean on $\overline{G}_1 \subset G$ as $\sum_{k=m+1}^n h_k^2 \to 0$.

If to the points of the piece $\sigma$ there has been associated another system of normal vectors

$$\overline{N}'_k = (\alpha_1'^{(k)}, \ldots, \alpha_n'^{(k)}) \quad (k = m+1, m+2, \ldots, n)$$

with transformation formulas

$$c_{ks} = (\overline{N}'_k, \overline{N}_s) \quad (k, s = m+1, m+2, \ldots, n),$$

then, as we know from §4.12, the function $F'_{(h)}$ corresponding to this new system may be expressed in terms of $F_{(h)}$ as follows:

$$F'_{(h)} = F(x_1, \ldots, x_m, H_{m+1}, \ldots, H_n), \tag{3}$$

$$H_s = \sum_{j=m+1}^{n} c_{js} h_j \quad (s = m+1, m+2, \ldots, n). \tag{4}$$

We shall prove the following theorem.

5.12. **Theorem.** *Suppose that the function* $f = f(x_1, \cdots, x_n)$ *is in the class* $H_{pn}^{(r)}(M)$ *and that there is given a system* $(\lambda)$ *of nonnegative integers* $\lambda_{m+1}, \cdots, \lambda_n$, *for which*

$$\lambda = \sum_{k=m+1}^{n} \lambda_k < r \tag{1}$$

*and*

$$\rho^{(\lambda)} = r - \lambda - \frac{n-m}{p} > 0. \tag{2}$$

*Then the following assertions are valid.*

1) *There exists a function* [1)]

$$\left.\frac{\partial^\lambda f}{\partial \overline{N}_{m+1}^{\lambda_{m+1}} \ldots \partial \overline{N}_n^{\lambda_n}}\right|_\sigma = \left.\frac{\partial F_{(h)}}{\partial h_{m+1}^{\lambda_{m+1}} \ldots \partial h_n^{\lambda_n}}\right|_0, \tag{3}$$

*lying in the class* $H_p^{(\rho(\lambda))}(G_1; M_{G_1})$, *where* $G_1$ *is any region whose closure lies in* $G$. *In addition*

$$M_{G_1} < c_{1G_1} M + c_{2G_1} \|f\|_p^{(n)}, \tag{4}$$

$$\left\| \left.\frac{\partial^\lambda F_{(h)}}{\partial h_{m+1}^{\lambda_{m+1}} \ldots \partial h_n^{\lambda_n}}\right|_0 \right\|_{L_p(G_1)} \leqslant c_{1G_1} M + c_{2G_1} \|f\|_p^{(n)}, \tag{5}$$

*where the constants* $c_{1G_1}$ *and* $c_{2G_1}$ *depend on* $\overline{G}_1$ *and on the choice of the system of normal vectors* $\overline{N}_{m+1}, \cdots, \overline{N}_n$, *but do not depend on* $M$ *and* $\|f\|_p^{(n)}$.

2) *The derivatives of* $f$ *of order* $\lambda$ *in the directions of the other normal vectors* $\overline{N}'_{m+1}, \cdots, \overline{N}'_n$ *are completely defined in terms of the derivatives of order* $\lambda$ *in the directions* $\overline{N}_{m+1}, \cdots, \overline{N}_n$, *and also in terms of the functions* $\phi_1$ *and* $c_{js}$. *Here the transformation formulas (see formula (6) of this subsection, below) exactly coincide with the classical formulas deducible in the case when the partial derivatives of* $f$ *of order* $\lambda$ *are continuous.*

**Proof.** Suppose given a function $f \in H_{pn}^{(r)}(M)$ in a region $G_1$ for which $\overline{G}_1 \subset G$. In the space $R_n^*$ of points $(x_1, \cdots, x_m, h_{m+1}, \cdots, h_n)$ we define a set $\overline{G}_1^*$ of points of the form $(x_1, \cdots, x_m, 0, \cdots, 0)$, where $(x_1, \cdots, x_m) \in \overline{G}_1$, and a region $W^{(\epsilon)}$ consisting of the points of $R_n^*$ distant from $\overline{G}_1^*$ by less than $\epsilon$. In §4.11 we proved that the function $F_{(h)}$ defined by equation (1) of §5.11 lies in the class $H_p^{(r)}(M_1; W^{(\epsilon')})$, where $\epsilon'$ is a sufficiently small number. As to the constant $M_1$, it is estimated on the basis of Lemma 2.4 using the inequality

$$M_1 < c_1 M + c_2 \|f\|_p^{(n)}.$$

In view of the same lemma,

$$\|F_{(h)}\|_{L_p(W^{(\epsilon')})} < c_1 M + c_2 \|f\|_p^{(n)}.$$

Here the numbers $c_1$ and $c_2$ do not depend on $M$ and $\|f\|_p^{(n)}$, but depend, of course, on the region $G_1$ and the choice of the system of vectors $\overline{N}_{m+1}, \cdots, \overline{N}_n$.

1) If $\lambda = 0$ we mean $f|_\sigma$.

Now we extend the function $F_{(h)}$ by means of Theorem 2.2 beyond the limits of the region $W^{(\epsilon'')}$ $(0 < \epsilon'' < \epsilon')$ in such a way that the extended function, which we shall again denote by $F_{(h)}$, lies in the class $H_{pn}^{(r)}(M_2)$, where

$$M_2 < c_3 M + c_4 \| f \|_p^{(n)},$$

$$\| F_{(h)} \|_p^{(n)} < c_3 M + c_4 \| f \|_p^{(n)}.$$

Here $c_3$ and $c_5$ are determined by the region $G_1$ and the choice of the system $\overline{N}_{m+1}, \cdots, \overline{N}_n$.

Now we may apply Theorem 3.51 to the resulting function $F_{(h)}$, which leads to the first assertion of the theorem under proof.

Now we turn to the proof of the second assertion. We already know that the function

$$F_{(h)} = F(x_1, \ldots, x_m, H_{m+1}, \ldots, H_n)$$

of the variables $(x_1, \cdots, x_m, H_{m+1}, \cdots, H_n)$ lies in the class $H_p^{(r)}(W^{(\epsilon')})$. The substitution

$$x_i = x_i \quad (i = 1, 2, \ldots, m),$$

$$H_j = \sum_{k=m+1}^{n} c_{kj} h_k \quad (j = m+1, m+2; \ldots, n), \tag{6}$$

in view of the fact that $\Sigma_{s=m+1}^n H_s^2 = \Sigma_{s=m+1}^n h_s^2$, transforms $W^{(\epsilon')}$ into $W^{(\epsilon')}$ in a one-to-one manner. In addition the Jacobian of this transformation is equal to $\pm 1$. In such a case Lemma 2.4 is applicable to the function $F_{(h)}$ on the region $W^{(\epsilon')}$ and to the transformations (6), so that the function

$$F'_{(h)} = F(x_1, \ldots, x_m, H_{m+1}, \ldots, H_n)$$

of the variables $x_1, \cdots, x_m, h_{m+1}, \cdots, h_n$ may be differentiated $\overline{r}$ times relative to these variables and moreover according to the usual classical rules. By this lemma,

$$F'_{(h)} \in H_p^{(r)}(W^{(\epsilon')}),$$

while we may suppose also that the function $F'_{(h)}$ has been extended with preservation of the class to the entire space $R_n^*$ of the variables $(x_1, \cdots, x_m, h_{m+1}, \cdots, h_n)$, and

$$\frac{\partial^\lambda F'_{(h)}}{\partial h_{m+1}^{\lambda_{m+1}} \ldots \partial h_n^{\lambda_n}} = \sum_{\substack{m+1 \leqslant s_i \leqslant n \\ 1 \leqslant i \leqslant \lambda}} \cdots \sum \frac{\partial^\lambda F(x_1, \ldots, x_m, H_{m+1}, \ldots, H_n)}{\partial H_{s_1} \ldots \partial H_{s_\lambda}} c_{m+1, s_1} \ldots$$

$$\ldots c_{m+1, s_{\lambda_{m+1}}} c_{m+2, s_{\lambda_{m+1}+1}} \ldots c_{m+2, s_{\lambda_{m+1}+\lambda_{m+2}}} \ldots c_{n, s_{\lambda-\lambda_n+1}} \ldots c_{n, s_\lambda}.$$

Hence, passing to the limit in the sense of $L_p(G_1)$ as $\Sigma_{k=m+1}^n h_k^2 \to 0$, we obtain

$$\left.\frac{\partial^\lambda f}{\partial \overline{N}'^{\lambda_{m+1}}_{m+1} \ldots \partial \overline{N}'^{\lambda_n}_n}\right|_\sigma = \sum_{\substack{m+1 \leqslant s_i \leqslant n \\ 1 \leqslant i \leqslant \lambda}} \cdots \sum \left.\frac{\partial^\lambda F}{\partial h_{s_1} \ldots \partial h_{s_\lambda}}\right|_0 c_{m+1, s_1} \ldots c_{m+1, s_{\lambda_{m+1}}} \ldots c_{n, s_\lambda}$$

$$= \sum_{\substack{m+1 \leqslant s_i \leqslant n \\ 1 \leqslant i \leqslant \lambda}} \cdots \sum \left.\frac{\partial^\lambda f}{\partial \overline{N}^{\lambda_{m+1}}_{m+1} \ldots \partial \overline{N}^{\lambda_n}_n}\right|_\sigma c_{m+1, s_1} \ldots c_{m+1, s_{\lambda_{m+1}}} \ldots c_{n, s_\lambda} \tag{7}$$

(the basis for the passage to the limit is in §4.12).

5.13. Let us cover our manifold $S$ with a finite number of open pieces $\sigma_1', \cdots, \sigma_\mu'$, each of which is explicitly represented in terms of the coordinates $(x_1, \cdots, x_m)$ (or other coordinates) by means of the equations

$$x_j = \varphi_j(x_1, \ldots, x_m), \quad (x_1, \ldots, x_m) \in G_k'$$

$$(k = 1, 2, \ldots, \mu; \; j = m+1, m+2, \ldots, n).$$

For each $k$ we replace the piece $\sigma_k'$ by an open piece $\sigma_k$ such that $\overline{\sigma}_k \subset \sigma_k'$, but the system $\sigma_1, \cdots, \sigma_\mu$ nevertheless covers $S$ completely. Further, we associate to the points of each piece $\sigma_k$ unit vectors $\overline{N}_{m+1}, \cdots, \overline{N}_n$ normal to $S$, subjected to the condition stated in §5.1. The manifold $S$ along with the covering $\sigma_1, \cdots, \sigma_k$ thus formed will be denoted by $S^*$.

We agree to write

$$\left.\frac{\partial^\lambda f}{\partial \overline{N}^{\lambda_{m+1}}_{m+1} \ldots \partial \overline{N}^{\lambda_n}_n}\right|_{S^*} \in H_p^{(\rho)}(M; S^*),$$

if on each piece $\sigma_k$, for the normals $\overline{N}_{m+1}, \cdots, \overline{N}_n$ associated with it, the corresponding function of the variables $(x_1, \cdots, x_m)$ (or other variables) has the property that

$$\left.\frac{\partial^\lambda f}{\partial N^{\lambda_{m+1}}_{m+1} \ldots \partial N^{\lambda_n}_n}\right|_{\sigma_k} \in H_p^{(\rho)}(M; G_k),$$

where $G_k$ is the projection of $\sigma_k$ on the $x_1, \cdots, x_m$ subspace.

Further, we put

$$\left\| \frac{\partial^\lambda f}{\partial \overline{N}_{m+1}^{\lambda_{m+1}} \cdots \partial \overline{N}_n^{\lambda_n}} \Bigg|_{S^*} \right\|_{L_p(S^*)} = \max_k \left\| \frac{\partial^\lambda f}{\partial \overline{N}_{m+1}^{\lambda_{m+1}} \cdots \partial \overline{N}_n^{\lambda_n}} \Bigg|_{\sigma_k} \right\|_{L_p(G_k)}.$$

Theorem 5.12 leads to the following theorem.

5.14. **Fundamental theorem.** *Suppose that the function $f = f(x_1, \cdots, x_n)$ lies in the class $H_{pn}^{(r)}(M)$ and that we are given a system $(\lambda)$ of nonnegative integers $\lambda_{m+1}, \cdots, \lambda_n$, for which*

$$\lambda = \sum_{k=m+1}^{n} \lambda_k < r \tag{1}$$

*and*

$$\rho^{(\lambda)} = r - \lambda - \frac{n-m}{p} > 0. \tag{2}$$

*Then*

$$\frac{\partial^\lambda f}{\partial \overline{N}_{m+1}^{\lambda_{m+1}} \cdots \partial \overline{N}_n^{\lambda_n}} \Bigg|_{S^*} \in H_p^{(\rho^{(\lambda)})}(M^*; S^*),$$

$$M^* < c_1 M + c_2 \| f \|_p^{(n)},$$

$$\left\| \frac{\partial^\lambda f}{\partial \overline{N}_{m+1}^{\lambda_{m+1}} \cdots \partial \overline{N}_n^{\lambda_n}} \Bigg|_{S^*} \right\|_{L_p(S^*)} < c_1 M + c_2 \| f \|_p^{(n)},$$

*where the numbers $c_1$ and $c_2$ depend on r, p and $S^*$, but not on M and $\|f\|_p^{(n)}$.*

**Proof.** The proof of this theorem follows directly from Theorem 5.12, if one takes into account that $c_1$ and $c_2$ are positive numbers satisfying the conditions

$$c_1 = \max_k c_{1G_k}, \quad c_2 = \max_k c_{2G_k},$$

where $c_{1G_k}$ and $c_{2G_k}$ are the constants figuring in inequalities (4) and (5) of §5.12, and the $G_k$ are regions of variation of $x_1, \cdots, x_m$ (or the other variables) on which the pieces $\sigma_k$ are projected.

## §6. Behavior of functions of the classes $H_{pn}^{(r)}$ and their partial derivatives on differentiable manifolds. Inverse theorems

6.1. **Theorem** (inverse of Theorem 5.12). *Suppose given a positive number $r$ and all possible (admissible) systems $(\lambda)$ of nonnegative integers $\lambda_{m+1}, \cdots, \lambda_n$, for which*

$$\lambda = \sum_{j=m+1}^{n} \lambda_j < r, \tag{1}$$

$$\rho^{(\lambda)} = r - \lambda - \frac{n-m}{p} > 0. \tag{2}$$

*Moreover, on the region $G$ of points $(x_1, \cdots, x_m)$, suppose that we are given functions $F_{(\lambda)}(x_1, \ldots, x_m)$, corresponding to each system $(\lambda)$ and lying respectively in the classes $H_p^{(\rho^{(\lambda)})}(G; M)$. To this region there is connected a piece $\sigma$ of a differentiable manifold of dimension $m$, defined by the equations*

$$\begin{aligned} x_{m+1} &= \varphi_{m+1}(x_1, \ldots, x_m), \\ &\cdots\cdots\cdots\cdots\cdots, \qquad (x_1, \ldots, x_m) \in G. \\ x_n &= \varphi_n(x_1, \ldots, x_m), \end{aligned}$$

*In addition, to each point of $\sigma$ there corresponds a definite set of unit vectors $\overline{N}_{m+1}, \cdots, \overline{N}_n$ such that the smoothness conditions formulated in §4.1 are satisfied.*

*Then for any piece $\sigma$ of the manifold whose closure lies in $\sigma$, it is possible to construct on the entire space $R_n$ a function $f(x_1, \cdots, x_n)$ satisfying the conditions*

1) $$f \in H_{pn}^{(r)}(M^*),$$

$$M^* < c_1 \sum_{(\lambda)} \| F_{(\lambda)} \|_{L_p(G)} + c_2 M, \tag{3}$$

$$\| f \|_p^{(n)} < c_1 \sum_{(\lambda)} \| F_{(\lambda)} \|_{L_p(G)} + c_2 M, \tag{4}$$

*where the sums are extended over all admissible systems $(\lambda)$;*

2) $$\left. \frac{\partial^\lambda f}{\partial \overline{N}_{m+1}^{\lambda_{m+1}} \cdots \partial \overline{N}_n^{\lambda_n}} \right|_{\sigma'} = F_{(\lambda)}(x_1, \ldots, x_m) \tag{5}$$

*for all admissible systems $(\lambda)$.*

Proof. Denote by $G_1$ the projection of the piece $\sigma'$ on the linear space $R_m$ of variables $x_1, \cdots, x_m$, and extend the functions $F_{(2)}$ beyond the limits of $G_1$ in such a way that the extended functions $\widetilde{F}_{(\lambda)}$ lie respectively in $H_{pm}^{(\rho(\lambda))}(M_1)$. By Theorem 2.2 we may suppose that

$$M_1 < c_3 M + c_4 \| F_{(\lambda)} \|_{L_p(G)}, \tag{6}$$

$$\| \widetilde{F}_{(\lambda)} \|_p^{(m)} < c_5 \| F_{(\lambda)} \|_{L_p(G)}, \tag{7}$$

where the constants $c_3, c_4, c_5$ depend on $\sigma'$.

Now, on the basis of Theorem 3.52, we construct on $R_n$ a function $\Phi(x_1, \cdots, x_m, h_{m+1}, \cdots, h_n)$ of $n$ variables, having the property $\Phi \in H_{pn}^{(r)}(M_2)$, where, in view of (6) and (7),

$$M_2 < c \sum_{(\lambda)} (\| \widetilde{F}_{(\lambda)} \|_p^{(m)} + M_1) < c \sum_{(\lambda)} (c_5 \| F_{(\lambda)} \|_{L_p(G)} + c_3 M + c_4 \| F_{(\lambda)} \|_{L_p(G)})$$

$$< c_6 \sum_{(\lambda)} \| F_{(\lambda)} \|_{L_p(G)} + c_7 M; \tag{8}$$

$$\| \Phi \|_p^{(n)} < c_6 \sum_{(\lambda)} \| F_{(\lambda)} \|_{L_p(G)} + c_7 M; \tag{9}$$

$$\frac{\partial^\lambda \Phi(x_1, \ldots, x_m, 0, \ldots, 0)}{\partial h_{m+1}^{\lambda_{m+1}} \cdots \partial h_n^{\lambda_n}} = \widetilde{F}_{(\lambda)}(x_1, \ldots, x_m). \tag{10}$$

Consider the transformation of the points $(x_1, \cdots, x_m, h_{m+1}, \cdots, h_n)$ of the space $R_n^*$ into the points $(u_1, \cdots, u_n)$ of the space $R_n$ defined by the equations

$$\begin{aligned} u_i &= x_i + \sum_{j=m+1}^{n} \alpha_i^{(j)} h_j \quad (i = 1, 2, \ldots, m), \\ u_i &= \varphi_i + \sum_{j=m+1}^{n} \alpha_i^{(j)} h_j \quad (i = m+1, m+2, \ldots, n). \end{aligned} \tag{11}$$

Now we introduce a region $G_\eta$, where $\eta$ is so small that $G_1 \subset G_\eta$. Further, we denote by $G_\eta^*$ the region of points $(x_1, \cdots, x_m, 0, \cdots, 0)$ of the space $R_n$ where $(x_1, \cdots, x_m) \subset G_\eta$, and by $\overline{G}_\eta^*$ the closure of this region in $R_n^*$.

In §4.11 we establish the existence of an $\epsilon$ such that the whole region $W^{(\epsilon)}$, consisting of the points of $R_n^*$ lying at a distance less than $\epsilon$ from $\overline{G}_\eta^*$, maps under the transformation (11) into the corresponding region $\Omega$ of the space $R_n$ in a one-to-one way. Thus the transformation (11) may be inverted:

$$\begin{aligned} x_i &= \psi_i(u_1, \ldots, u_n) \quad (i = 1, 2, \ldots, m), \\ h_i &= \psi_i(u_1, \ldots, u_n) \quad (i = m+1, m+2, \ldots, n), \end{aligned} \tag{12}$$

the functions $\psi_i$ having on $\Omega$ continuous partial derivatives to order $\bar{r}+1$ inclusive.

Now put

$$f(u_1, \ldots, u_n) = \Phi(\psi_1, \ldots, \psi_n), \quad (u_1, \ldots, u_n) \in \Omega. \tag{13}$$

Thus Lemma 2.4 is applicable to equation (13), and, denoting by $\Omega'$ a region of $R_n$ whose closure lies in $\Omega$, we may state that the function $f \in H_p^{(r)}(\Omega'; M_3)$, where, taking relations (8) and (9) into account, we have

$$M_3 < c_7 \|\Phi\|_p^{(n)} + c_8 M_2 < c_9 \sum_{(\lambda)} \|F_{(\lambda)}\|_{L_p(G)} + c_{10} M, \tag{14}$$

$$\|f\|_{L_p(\Omega)} < c_{11} \|\Phi\|_p^{(n)} + c_{12} M_2 < c_{13} \sum_{(\lambda)} \|F_{(\lambda)}\|_{L_p(G)} + c_{14} M. \tag{15}$$

Finally, using Lemma 2.2, we may extend $f$ beyond the limits of a region $\Omega''$ whose closure lies in $\Omega'$ in such a way that the extended function, which we will again denote by $f$, lies in $H_{pn}^{(r)}(M_4)$. Here we choose the regions $\Omega'$ and $\Omega''$ in such a way that they contain the piece $\sigma'$. This is possible since $\Omega$ contains $\sigma'$. In this connection, as a consequence of (14) and (15), we will have

$$M_4 < c_{15} \|f\|_{L_p(\Omega)} + c_{16} M_3 < c_{17} \sum_{(\lambda)} \|F_{(\lambda)}\|_{L_p(G)} + c_{18} M, \tag{16}$$

$$\|f\|_p^{(n)} < c_{19} \|f\|_{L_p(\Omega)} < c_{20} \sum_{(\lambda)} \|F_{(\lambda)}\|_{L_p(G)} + c_{21} M. \tag{17}$$

The function $f$ constructed on the space $R_n$ satisfies all the requirements of Theorem 6.1. In fact, it lies in the class $H_{pn}^{(r)}(M^*)$, if one puts $M^* = M_4$. Further, it follows from (16) and (17) that inequalities (3) and (4) hold with the appropriate choice of the constants $c_1$ and $c_2$.

If we again replace $u_i$ in (13) by their expressions (11), we obtain identically

$$\Phi(x_1, \ldots, x_m, h_{m+1}, \ldots, h_n) = f\Big(x_1 + \sum_{j=m+1}^{n} \alpha_1^{(j)} h_j, \ldots, x_m$$
$$+ \sum_{j=m+1}^{n} \alpha_m^{(j)} h_j, \varphi_{m+1} + \sum_{j=m+1}^{n} \alpha_{m+1}^{(j)} h_j, \ldots, \varphi_n + \sum_{j=m+1}^{n} \alpha_n^{(j)} h_j\Big), \tag{18}$$

which is valid in every case for all points $(x_1, \cdots, x_m, h_{m+1}, \cdots, h_n)$ whose images $(u_1, \cdots, u_n)$ lie in $\Omega''$.

Equations (5) follow directly from (10) and (18) and the fact that by the definition itself (see §5.11)

$$\left.\frac{\partial^\lambda f}{\partial \overline{N}_{m+1}^{\lambda_{m+1}} \ldots \partial \overline{N}_n^{\lambda_n}}\right|_{\sigma'} = \left.\frac{\partial^\lambda \Phi}{\partial h_{m+1}^{\lambda_{m+1}} \ldots \partial h_n^{\lambda_n}}\right|_0 .$$

**6.2. Fundamental theorem (inverse of Theorem 5.14).** *Suppose that in the space $R_n$ we are given a finite system of pairwise nonintersecting manifolds $S_1, \cdots, S_\mu$, of dimensions*

$$m_1, m_2, \ldots, m_\mu \quad \left(r - \frac{n - m_k}{p} > 0\right)$$

*respectively. Let each manifold be covered by a finite number of overlapping pieces $\sigma$ lying on it. We express these pieces by some or other $m_k$ coordinates using equations (1) of §4.1 and we attach to their points definite systems of vectors $\overline{N}_{m_k+1}, \cdots, \overline{N}_n$ normal to $\sigma$.*

*Further, suppose given a positive number $r$ and all possible (admissible) systems $(\lambda)_k$ of nonnegative integers $\lambda_{m_k+1}, \cdots, \lambda_n$ associated respectively with the manifolds $S_k$, for which*

$$\lambda = \sum_{k=m_k+1}^{n} \lambda_j < r \tag{1}$$

*and*

$$\rho^{(\lambda)_k} = r - \lambda - \frac{n - m_k}{p} > 0. \tag{2}$$

*Suppose that there is connected to each piece $\sigma$ lying on the manifold $S_k$ and to each admissible system $(\lambda)_k$ a function $F_{(\lambda)_k\sigma}$ defined on that piece, which may also be denoted as $F_{(\lambda)_\sigma\sigma}$. If $\sigma$ is expressed explicitly in terms of the coordinates $x_1, \cdots, x_{m_k}$ (or $m_k$ other coordinates), then the function $F_{(\lambda)_k\sigma}$, expressed in terms of these coordinates, is supposed to* *lie in the class* $H_p^{(\rho^{(\lambda)_k})}(G_\sigma;\ M)$, *where* $G_\sigma$ *is the projection of the piece* $\sigma$ *onto the linear subspace of variables* $x_1, \cdots, x_{m_k}$ *(or other variables).*

*On the overlapping portions of pieces* $\sigma$ *of one and the same manifold* $S_k$ *the functions* $F_{(\lambda)_k\sigma}$ *are assumed to be consistent with one another in the sense that they are subjected to the appropriate formulas (see formulas (6) of §5.12) for transforming from one system of normal vectors* $\overline{N}_{m_k}, \cdots, \overline{N}_n$ *to another and from one set of* $m_k$ *variables to another, necessary in order that the equations*

$$\left.\frac{\partial^\lambda f}{\partial \overline{N}_{m_k+1}^{\lambda_{m_k+1}} \cdots \partial \overline{N}_n^{\lambda_n}}\right|_\sigma = F_{(\lambda)_k\sigma} \tag{3}$$

*should be possible for all admissible systems* $(\lambda)_k$ *and all the pieces covering one and the same manifold, for one and the same function* $f(x_1, \cdots, x_n)$.

*In such a case one may construct a function* $f(x_1, \cdots, x_n)$ *in the space* $R_n$ *satisfying the following conditions.*

1).
$$f \in H_{pn}^{(r)}(M^*),$$

$$M^* < c' \sum_\sigma \sum_{(\lambda)_\sigma} \| F_{(\lambda)_\sigma\sigma} \|_{L_p(G_\sigma)} + c''M, \tag{4}$$

$$\| f \|_p^{(n)} < c' \sum_\sigma \sum_{(\lambda)_\sigma} \| F_{(\lambda)_\sigma\sigma} \|_{L_p(G_\sigma)} + c''M; \tag{5}$$

*here the second sum is extended over all possible admissible systems* $\lambda_{m_k+1}, \cdots, \lambda_n$ *corresponding to the piece* $\sigma$ *in accordance with its dimension, and the first sum is extended over all possible pieces* $\sigma$ *covering our manifold according to the conditions of the theorem. The constants* $c'$ *and* $c''$ *do not depend on the numbers*

$$\sum_\sigma \sum_{(\lambda)_\sigma} \|F_{(\lambda)_\sigma\sigma}\|_{L_p(G)} \text{ and } M.$$

2) *On the corresponding pieces* $\sigma$ *equation* (3) *is valid for the normals* $\overline{N}_{m_k+1}, \cdots, \overline{N}_n$ *defined on them.*

**Proof.** Since our manifolds are bounded, closed and do not intersect one another, there exists a number $\delta > 0$ so small that a ball of radius $\delta$ with center at any point of any manifold does not contain points of other manifolds.

Suppose given any point $P_0$ of any manifold $S_k$. It lies in one of the pieces $\sigma$ covering $S_k$. If one describes a ball of radius $r_{P_0} > \delta$ with $P_0$

as center, then it will contain some piece of the manifold $S_k$ on which the point $P_0$ lies. Decreasing $r_{P_0}$ if necessary, we may arrange that this piece of $S_k$ along with its closure lies completely in $\sigma$. Thus, to each point $P_0$ of any manifold $S_k$ we have attached a ball $V_{P_0}$ with center at that point which excises a piece $\sigma_{P_0}$ of that manifold containing $P_0$ and lying completely in one of the pieces $\sigma$ covering, according to the theorem, our manifold. Within $V_{P_0}$ there are no points of other manifolds. We denote the radius of the sphere by $r_{P_0}$.

Along with $V_{P_0}$ we shall consider also spheres $V'_{P_0}$ and $V''_{P_0}$ with the same center $P_0$ and with radii, respectively, $r_{P_0}/3$ and $r_{P_0}/2$. We shall suppose that these spheres excise on the corresponding manifold $S_k$ pieces $\sigma'_{P_0}$ and $\sigma''_{P_0}$ respectively.

The piece $\sigma_{P_0}$ (since by the hypothesis of the theorem this is true for $\sigma$) is defined explicitly in terms of the coordinates $x_1, \cdots, x_m$ (or other coordinates) in the form of equations

$$\begin{array}{ll} x_{m_k+1} = \varphi_{m_k+1}(x_1, \ldots, x_m), & \\ \ldots\ldots\ldots\ldots\ldots\ldots, & (x_1, \ldots, x_{m_k}) \in G_{P_0}. \\ x_n = \varphi_n(x_1, \ldots, x_m), & \end{array}$$

Moreover, on the piece $\sigma_{P_0}$ (since this is true for $\sigma$) there are given functions $F_{(\lambda)_k \sigma_{P_0}}$, connected with the systems $(\lambda)_k$ and lying respectively in the classes $H_p^{(\rho^{(\lambda)_k})}(M;\, G_{P_0})$, where $G_{P_0}$ is the projection of $G_{P_0}$ on the corresponding linear manifold of points $(x_1, \cdots, x_{m_k})$.

On the basis of Theorem 6.1 we may construct on the entire space $R_n$ a function $f_{P_0} \in H_{pn}^{(r)}(M_{P_0})$ having the following properties:

$$\left. \frac{\partial^\lambda f_{P_0}}{\partial \overline{N}_{m_{k+1}}^{\lambda_{m_k+1}} \cdots \partial \overline{N}_n^{\lambda_n}} \right|_{\sigma''_{P_0}} = F_{(\lambda)_k \sigma_{P_0}}, \tag{6}$$

for any admissible system $(\lambda)_k$,

$$M_{P_0} < c'_{P_0} \sum_{(\lambda)_k} \| F_{(\lambda)_k \sigma_{P_0}} \|_{L_p(G_{P_0})} + c''_{P_0} M, \tag{7}$$

$$\| f_{P_0} \|_p^{(n)} < c'_{P_0} \sum_{(\lambda)_k} \| F_{(\lambda)_k \sigma_{P_0}} \|_{L_p(G_{P_0})} + c''_{P_0} M. \tag{8}$$

With the aid of the Heine-Borel Theorem we now choose a finite number of balls $V'_{P_0}$ completely covering our manifolds $S_k$. We enumerate them in the order $V_1', \cdots, V_l'$. Their radii are respectively equal to $r_i/3$ and their centers are taken to be the points $P_i$, $i = 1, \cdots, l$. Along with them we shall consider the balls $V_i''$ and $V_i$ with the same centers $P_i$ and radii $r_i/2$, $r_i$. The balls $V_i'$, $V_i''$, $V_i$ excise pieces from the corresponding manifold which we denote by $\sigma_i'$, $\sigma_i''$, $\sigma_i$.

To each $P_i$ we assign a function $f_i = f_{P_i}(x_1, \cdots, x_n)$ lying in the class $H_{pn}^{(r)}(M_i)$ $(M_i = M_{p_i})$ and extending the given boundary conditions from the piece $\sigma_i''$. Thus for the functions $f_i$ the relations

$$\left.\frac{\partial^\lambda f_i}{\partial \overline{N}_{m_{k_i}+1}^{\lambda_{m_{k_i}+1}} \cdots \partial \overline{N}_n^{\lambda_n}}\right|_{\sigma_i''} = F_{(\lambda)_{k_i}\sigma_i''} \tag{9}$$

are satisfied for any admissible systems $(\lambda)_{k_i}$, and we have

$$M_i < c_i' \sum_{(\lambda)_{k_i}} \| F_{(\lambda)_{k_i}\sigma_i} \|_{L_p(G_i)} + c_i'' M, \tag{10}$$

$$\| f_i \|_p^{(n)} < c_i' \sum_{(\lambda)_{k_i}} \| F_{(\lambda)_{k_i}\sigma_i} \|_{L_p(G_i)} + c_i'' M, \tag{11}$$

where $G_i$ is the projection of $\sigma_i$ onto the linear manifold of variables $(x_1, \cdots, x_{m_{k_i}})$ (or $m_{k_i}$ other variables).

We further define on $R_n$ functions $u_i(P)$ $(i = 1, \cdots, l)$, as follows.

1) The function $u_i(P)$ is continuous along with its partial derivatives to order $\bar{r} + 1$ inclusive.

2) $u_i(P) = 0$ for $\overline{PP_i} \leq r_i/3$, where $\overline{PQ}$ denotes the distance between the points $P$ and $Q$ of the space $R_n$.

3) $u_i(P) = 1$ for $\overline{PP_i} \geq r_i/2$.

4) Along the manifold $S_k$ on which the point $P_i$ lies, the function $u_i(P)$ has derivatives normal to that manifold of orders up to $\bar{r} + 1$ inclusive, mixed and unmixed, equal to zero.

The possibility of constructing the functions $u_i(P)$ subject to the four requirements enumerated above will be established in §6.21.

Now we introduce the functions

$$U_1 = U_1(P) = 1 - u_1,$$

$$\cdots\cdots\cdots\cdots\cdots$$

$$U_i = U_i(P) = u_1 \ldots u_{i-1}(1 - u_i) \quad (i = 1, 2, \ldots, l),$$

and show that the function $f$ defined by the equation

$$f = \sum_{i=1}^{l} f_i U_i \tag{12}$$

satisfies all the requirements of the theorem.

Indeed, the function $f_i \in H_{pn}^{(r)}(M_i)$, and the function $U_1$ is bounded and continuous on $R_n$ along with its partial derivatives to order $\bar{r}+1$. Therefore the function $f_i U_i$ lies in $H_{pn}^{(r)}(M_i')$, where

$$M_i' < c_{1i} \| f_i \|_p^{(n)} + c_{2i} M_i, \tag{13}$$

$$\| f_i U_i \|_p^{(n)} < c_{1i} \| f_i \|_p^{(n)} + c_{2i} M_i \quad (i = 1, 2, \ldots, l). \tag{14}$$

This fact is proved in a way quite analogous to the way we proved in §2.2 that the function $\phi = fH$, defined using equation (4) of that subsection, lies in the class $H_p^{(r_1, \ldots, r_n)}$ (see inequality (5) of §2.2 and the remark to it). 1)
In view of (10), (11), (13) and (14), we obtain

$$M_i' < c_{2i} \sum_{(\lambda)k_i} \| F_{(\lambda)k_i \sigma_i} \|_{L_p(G_i)} + c_{3i} M, \tag{15}$$

$$\| f_i \|_p^{(n)} < c_{2i} \sum_{(\lambda)k_i} \| F_{(\lambda)k_i \sigma_i} \|_{L_p(G_i)} + c_{3i} M \quad (i = 1, 2, \ldots, l). \tag{16}$$

It will follow from (12) that the function $f \in H_{pn}^{(r)}(M^*)$, where

$$M^* \leqslant \sum_{i=1}^{l} M_i', \quad \| f \|_p^{(n)} \leqslant \sum_{i=1}^{l} \| f_i U_i \|_p^{(n)}. \tag{17}$$

Therefore, if we take into account the fact that each piece $\sigma_i$ lies in some piece $\sigma$, covering, according to the conditions of the theorem, our manifolds, and $G_i \subset G$, where $G_i$ and $G$ are the projections of $\sigma_i$ and $\sigma$ on the corresponding linear set of points $(x_1, \cdots, x_{m_i})$; and, moreover, if one notes that only a finite number of different pieces $\sigma_i$ belong to the same $\sigma$, then it is is obvious from (15)–(17) that there exist two positive constants $c'$ and $c''$ such that

1) In the case at hand the proof is somewhat simpler than it was in §2.2, since it is necessary to replace the region $G_{\mu/2}$ by $R_n$.

$$M^* < c' \sum_{\sigma} \sum_{(\lambda)_\sigma} \| F_{(\lambda)_\sigma \sigma} \|_{L_p(G)} + c'' M,$$

$$\| f \|_p^{(n)} < c' \sum_{\sigma} \sum_{(\lambda)_\sigma} \| F_{(\lambda)_\sigma \sigma} \|_{L_p(G)} + c'' M.$$

Here the second sum is extended over all possible admissible systems $\lambda_{m_k+1}, \cdots, \lambda_n$, corresponding to the piece $\sigma$ in correspondence with the number of its dimensions, and the first sum is extended over all possible pieces $\sigma$ covering (according to the theorem) our manifolds. As to the constants $c'$ and $c''$, they do not depend on the numbers

$$\sum_{\sigma} \sum_{(\lambda)_\sigma} \| F_{(\lambda)_\sigma \sigma} \|_{L_p(G)} \text{ and } M.$$

Thus the first assertion of our theorem, and along with it inequalities (4) and (5), are proved. We turn now to the proof of the second assertion.

We have already proved that $f \in H_{pn}^{(r)}$. Therefore on each piece $\sigma$ covering, according to the conditions of the theorem, one or another of the manifolds $S_k$, we may use Theorem 5.12 to assert, for admissible $(\lambda)_k$, the existence of a function

$$\left. \frac{\partial^\lambda f}{\partial N_{m_k+1}^{\lambda_{m_k+1}} \cdots \partial N_n^{\lambda_n}} \right|_\sigma = \left. \frac{\partial^\lambda F_{(h)}}{\partial h_{m_k+1}^{\lambda_{m_k+1}} \cdots \partial h_n^{\lambda_n}} \right|_0,$$

i.e. the fact that the partial derivative

$$\frac{\partial^\lambda F_{(h)}}{\partial h_{m_k+1}^{\lambda_{m_k+1}} \cdots \partial h_n^{\lambda_n}} \tag{18}$$

of the function $F_{(h)}$, defined by equation (1) of §5.11, tends, in the mean, on any region $G'$ whose closure lies in the projection of the piece $\sigma$ on the linear manifold of the variables $x_1, \cdots, x_{m_k}$, as $\sum_{\nu=m_k+1}^n h_\nu^2 \to 0$, to some function of these variables.

All that faces us now is the problem of proving that the derivative indeed tends to $F_{(\lambda)_k \sigma}$ for admissible systems $(\lambda)_k$.

The balls $V_i'$ $(i = 1, \cdots, l)$ covering our manifolds $S_k$ cut them into separate pieces. Select one such piece and denote it by $\Delta$. Thus it does not contain in its interior points of the boundaries of the spheres $V_i''$. Let

$$V_{i_1}'', \ V_{i_2}'', \ \ldots, \ V_{i_\tau}'' \tag{19}$$

be those of the spheres $V_i''$ which contain $\Delta$, i.e. let $\Delta = \Sigma_{s=1}^{\tau} \sigma_{i_s}''$. Our assertion will be proved if we show that the equation

$$\left.\frac{\partial^\lambda F_{(h)}}{\partial h_{m_k+1}^{\lambda_{m_k+1}} \ldots \partial h_n^{\lambda_n}}\right|_\Delta = F_{(\lambda)_k \Delta} \tag{20}$$

is valid for the admissible system $(\lambda)_k$ on any piece $\Delta$.

Suppose that $\Delta^*$ denotes any closed piece lying in $\Delta$. We note that the functions $U_i$ thus have

$$U_i = 0 \text{ on } R_n - V_i''. \tag{21}$$

Moreover, from the obvious equation

$$U_1 + U_2 + \ldots + U_l = 1 - u_1 u_2 \ldots u_l$$

it follows that

$$U_1 + U_2 + \ldots + U_l = 1 \tag{22}$$

on $V_i''$. Since the balls $V_i'$ cover $\Delta''$, we can find a region $W \supset \Delta^*$ on which equation (22) is satisfied.

Suppose that $E$ is the intersection of the spheres (19):

$$E = \prod_{s=1}^{\tau} V_{i_s}''.$$

On the basis of (21), if $i$ is not one of the indices $i_1, \cdots, i_\tau$, then the equation $H_i = 0$ holds for it on $E$, and thus

$$f = \sum_{s=1}^{\tau} f_{i_s} U_{i_s} \tag{23}$$

on $E$ and accordingly on $EW$. Moreover, it is obvious that

$$\sum_{s=1}^{\tau} U_{i_s} = 1 \text{ on } EW. \tag{24}$$

We further introduce the functions $F_{i(h)}$ and $T_{i(h)}$ of the variables $x_1, \cdots, x_{m_k}, h_{m_k+1}, \cdots, h_n$, constructed from the functions $f_i$ and $U_i$ in a way similar to that in which the function $F_{(h)}$ was constructed from $f$.

Now for the admissible system $(\lambda)_k$ we may write

$$\frac{\partial^\lambda F_{(h)}}{\partial h_{m_k+1}^{\lambda_{m_k+1}} \ldots \partial h_n^{\lambda_n}} = \sum_{s=1}^{\tau} \frac{\partial^\lambda F_{i_s(h)}}{\partial h_{m_k+1}^{\lambda_{m_k+1}} \ldots \partial h_n^{\lambda_n}} T_{i_s(h)} + Q, \tag{25}$$

where $Q$ is a sum each term of which is the product of some partial derivative of the functions $T_{i_s(h)}$ relative to the variables $h_j$ by the functions $F_{i_s(h)}$ or some of their partial derivatives. By the fourth property of the functions $u_i$ the partial derivatives of $T_{i_s(h)}$ relative to the variables $h_j$ tend uniformly to zero on $\Delta^*$ as $\Sigma_{j=m_k+1}^n h_j^2 \to 0$, and at the same time both the functions $F_{i_s(h)}$ and their partial derivatives entering into $Q$ converge in the mean as well. Hence as $\Sigma_{j=m_k+1}^n h_j^2 \to 0$ the function $Q$ (of the variables $x_1, \cdots, x_{m_k}$) converges in the mean on $\Delta^*$ to zero.

As to the partial derivatives

$$\frac{\partial^\lambda F_{i_s(h)}}{\partial h_{m_k+1}^{\lambda_{m_k+1}} \cdots \partial h_n^{\lambda_n}},$$

all of them, as functions of $x_1, \cdots, x_{m_k}$, tend as $\Sigma_{j=m_k+1}^n h_j^2 \to 0$ in the mean to one and the same function $F_{(\lambda)_k \Delta^*}$. This follows from equations (9) if one takes into account the fact that $\Delta^*$ lies in all the $\sigma''_{i_s}$ $(s = 1, 2, \cdots, \tau)$. Finally the functions $T_{i_s(h)}$ tend as well uniformly to $U_{i_s(0)}$, where $P$ is the corresponding point of $\Delta^*$.

From the observations just made, it follows on taking account of (24) that the partial derivative (25) tends as $\Sigma_{j=m_k+1}^n h_j^2 \to 0$ in the mean on $\Delta^*$ to the function

$$F_{(\lambda)_k \Delta^*} \sum_{s=1}^{\tau} U_{i_s}(P) = F_{(\lambda)_k \Delta^*}.$$

In view of the arbitrary choice of $\Delta^* \subset \Delta$, the second portion of our theorem is also proved.

6.21. Lemma. *Suppose that the piece $\sigma$ of the manifold $S$ is defined in the neighborhood of a point $P_0 \in S$ by the equations*

$$\begin{aligned} x_{m+1} &= \varphi_{m+1}(x_1, \ldots, x_m), \\ &\cdots\cdots\cdots\cdots \qquad (x_1, \ldots, x_m) \in G, \\ x_n &= \varphi_n(x_1, \ldots, x_m), \end{aligned}$$

*and that $V_0'$, $V_0''$ and $V_0$ are balls in $R_n$ with centers at $P_0$ and radii $r_0/3$, $r_0/2$ and $r_0$ respectively, such that the pieces $\sigma_0'$, $\sigma_0''$ and $\sigma_0$ which they excise in $S$ lie in $\sigma$. Then it is possible to construct in $R_n$ a function $u_0(P)$ having the following properties.*

1) *$u_0(P)$ is continuous along with its partial derivatives to order $\bar{r}+1$ inclusive.*

2) $u_0(P) = 0$ *for* $\overline{PP_0} \leq r_0/3$.

3) $u_0(P) = 1$ *for* $\overline{PP_0} \geq r_0/2$.

4) *Along the manifold S the function* $u_0(P)$ *has normal derivatives to order* $\bar{r}+1$ *inclusive (mixed and unmixed), equal to zero.*

Proof. Denote by $G_0'$, $G_0''$ and $G_0$ respectively the projections of $\sigma_0'$, $\sigma_0''$ and $\sigma_0$ on the linear manifold $R_m$ of variables $x_1, \cdots, x_m$. Suppose that $W_1$ and $W_2$ are two pairwise nonintersecting regions of $R_m$ such that the closure of $G_0'$ lies in $W_1$ and the closure of $R_m - G_0''$ lies in $W_2$.

We shall suppose that on the region $W_1 + W_2$ there is given a function $\psi$ defined by the equations

$$\psi = \begin{cases} 0 & \text{on } W_1, \\ 1 & \text{on } W_2. \end{cases} \tag{1}$$

On the basis of Theorem 2.3 we may extend it beyond the limits of $W_1$ and $W_2$ to $R_m$, in such a way that the extended function, which we shall denote by $\phi$, will have on $R_m$ continuous partial derivatives to order $\bar{r}+1$ inclusive. [1)]

Now we introduce the transformation

$$z_i = x_i + \sum_{j=m+1}^{n} \alpha_i^{(j)} h_j \quad (i = 1, 2, \ldots, m),$$

$$z_k = \varphi_k + \sum_{j=m+1}^{n} \alpha_i^{(j)} h_j \quad (k = m+1, m+2, \ldots. n)$$

of the points $(x_1, \cdots, x_m, h_{m+1}, \cdots, h_n)$ constituting the region $W^{(\epsilon)}$, defined by the relations

$$(x_1, \ldots, x_m) \in G_0, \quad \sum_{j=m+1}^{n} h_j^2 < \varepsilon.$$

From §4.11 we know that for sufficiently small $\epsilon$ the region $W^{(\epsilon)}$ goes over into some region $\Omega^{(\epsilon)}$ of points $(z_1, \cdots, z_n)$ in a one-to-one way, the inverse functions of $z_1, \cdots, z_n$ having continuous partial derivatives of orders $\bar{r}+1$.

Define a function $u_1(P)$ as follows:

---

1) In order to apply Theorem 2.3, we need to suppose $\psi = 0$ on $W_1'$ and $\psi = 1$ on $W_2'$, where $W_1' W_2' = 0$, and $W_1$ and $W_2$ lie strictly in $W_1'$ and $W_2'$.

$$u_1(P) = u_1(z_1, \ldots, z_n) = \begin{cases} 0 & \text{on } E'_{\varepsilon_1}\left\{\overline{PP_0} \leqslant \frac{r_0}{3} + \varepsilon_1\right\}, \\ 1 & \text{on } E''_{\varepsilon_1}\left\{\overline{PP_0} \geqslant \frac{r_0}{2} - \varepsilon_1\right\}, \\ \psi(x_1 \ldots, x_m) & \text{on } \Omega^{(\varepsilon)}. \end{cases} \quad (2)$$

On each of the sets $E'_{\epsilon_1}$, $E''_{\epsilon_1}$, $\Omega^{(\epsilon)}$ the function $u_1$ has continuous partial derivatives of order $\bar{r} + 1$. But for sufficiently small $\epsilon_1$ this same fact holds also on the set-theoretic sum

$$E = E'_{\varepsilon_1} + E''_{\varepsilon_1} + \Omega^{(\varepsilon)}$$

of these sets, in spite of the fact that $\Omega^{(\epsilon)}$ intersects, on the one hand, $E'_{\epsilon_1}$, and on the other hand with $E''_{\epsilon_1}$. Indeed, since the region $W_1$ of variables $x_1, \cdots, x_m$ strictly contains the region $G_0$, and since the function

$$\psi = 0 \text{ on } W_1, \quad (3)$$

we see from the third equation (2) that the function $u_1 = 0$ for all the points $P \in \Omega^{(\epsilon)}$ for which $r_0/3 \leq \overline{PP}_0 \leq r_0/2 - \eta$, where $\eta$ is sufficiently small. Thus, if one takes $\epsilon_1 < \eta$, one will obviously achieve continuity of the partial derivatives of order $\bar{r} + 1$ of the function $u_1$ on $E'_{\epsilon_1} + \Omega^{(\epsilon)}$. Similar arguments may be carried out for the set $E''_{\epsilon_1} + \Omega^{(\epsilon)}$.

Now we may again use Theorem 2.3 and extend $u_1(P)$ beyond the limits of the set $E'_{\epsilon_1/2} + E''_{\epsilon_1/2} + \Omega^{(\epsilon/2)}$ to the entire space $R_n$ in such a way that the resulting function, which we shall denote by $u(P)$, has on $R_n$ continuous partial derivatives of order $\bar{r} + 1$. Obviously $u(P)$ satisfies all four requirements of the lemma. In particular, that the fourth requirement is satisfied follows from the fact that, because of the third equation (2), the function $u_1$ in the neighborhood of points of $\sigma$ is constant along the normals to $\sigma$, and thus does not depend on $h_{m+1}, \cdots, h_n$.

## BIBLIOGRAPHY

[1] S. M. Nikol'skiĭ, *Inequalities for entire functions of finite degree and their application in the theory of differentiable functions of several variables*, Trudy Mat. Inst. Steklov. 38 (1951), 244–278; English transl., Amer. Math. Soc. Transl. (2) 80 (1968), 1–38. MR 14, 32.

[2] ———, *On the continuation of differentiable functions of several variables*, Dokl. Akad. Nauk SSSR 82 (1952), 521–524. (Russian) MR 13, 635.

[3] ———, *Second note on the continuation of differentiable functions of several variables*, Dokl. Akad. Nauk SSSR 88 (1953), 17–19. (Russian) MR 15, 425.

[4] ———, *Properties of differentiable functions of several variables on closed smooth manifolds*, Dokl. Akad. Nauk SSSR 88 (1953), 213–216. (Russian) MR 15, 425.

[5] S. L. Sobolev, *On a theorem of functional analysis*, Mat. Sb. 4 (46) (1938), 471–497; English transl., Amer. Math. Soc. Transl. (2) 34 (1963), 39–68.

[6] S. L. Sobolev, *Applications of functional analysis in mathematical physics*, Izdat. Leningrad. Gos. Univ., Leningrad, 1950; English transl., Transl. Math. Monographs, vol. 7, Amer. Math. Soc., Providence, R. I., 1963. MR 14, 565; MR 29 #2624.

[7] V. I. Kondrašov, *On some properties of functions from $L_p$-spaces*, Dokl. Akad. Nauk 48 (1945), 563–566. (Russian)

[8] T. I. Amanov, *On the embedding theorem for differentiable functions of several variables*, Dokl. Akad. Nauk SSSR 83 (1953), 5–8. (Russian) MR 15, 205.

[9] ———, *On the solution of a biharmonic problem*, Dokl. Akad. Nauk SSSR 88 (1953), 389–392. (Russian) MR 15, 124.

[10] G. M. Fihtengol'c, *Course of differential and integral calculus*. Vol. 1, Fizmatgiz, Moscow, 1958; German transl., Hochschulbücher für Math., Band 61, 1966, VEB Deutscher Verlag, Berlin, 1966.

[11] S. N. Bernšteĭn, *Collected works*. Vol. 1: *Constructive theory of functions*, Izdat. Akad. Nauk SSSR, Moscow, 1952. MR 14, 2.

[12] S. M. Nikol'skiĭ, *On the Dirichlet problem*, Dokl. Akad. Nauk SSSR 83 (1952), 23–25; erratum 84 (1952), 652. (Russian) MR 13, 943.

[13] ———, *On the solution of the polyharmonic equation by a variational method*, Dokl. Akad. Nauk SSSR 88 (1953), 409–411. (Russian) MR 15, 425.

Translated by:
J. M. Danskin

# GENERALIZATION OF A THEOREM OF HAYMAN ON SUBHARMONIC FUNCTIONS IN AN $m$-DIMENSIONAL CONE

UDC 517.53

V. S. AZARIN

## §1

Suppose that $u(z)$ is a subharmonic function in the right halfplane $\Pi$ ($\operatorname{Re} z > 0$), bounded by zero on the imaginary axis, i.e. $u(z)$ satisfies the condition

$$\overline{\lim_{x\to 0}}\, u(x+iy) \leqslant 0. \tag{1.1}$$

Suppose that $u(z)$ has order 1 in $\Pi$ and is of normal type. This means that

$$\overline{\lim_{r\to\infty}} \frac{M_u(r)}{r} = \sigma < \infty,$$

where

$$M_u(r) = \max_{|\theta|\leqslant\frac{\pi}{2}} u^+(re^{i\theta}), \qquad u^+(z) = \max\{u(z), 0\}.$$

Denote by $h_u(\theta)$ the indicator of the function $u(z)$:

$$h_u(\theta) = \overline{\lim_{r\to\infty}} \frac{u(re^{i\theta})}{r}.$$

For functions of the indicated class the following representation holds, as may be obtained from the theorem of Riesz [1]:

$$u(z) - \alpha x = v(z) - \int_{\zeta\in\Pi} g(z,\zeta)\, d_\xi\mu, \quad z = x+iy, \quad \xi = \zeta + i\eta, \tag{1.2}$$

where $\alpha = h_u(0)$, $g(z,\zeta) = \ln|(\zeta+\bar z)/(\zeta - z)|$ is the Green's function for $\Pi$, $v(z)$ is the best harmonic majorant of the function $u(x) - \alpha x$ in $\Pi$, and $d_\zeta u$ is a nonnegative Radon measure in $\Pi$.

The function $v(z)$ in its turn is representable in the form $v(z) = -\int_{-\infty}^{\infty}(\partial g(z, i\eta)/\partial\xi)\, d_\eta v$, where the quantity $d_\eta v$, defined by the equation $d_\eta v = -\overline{\lim}_{x\to 0}\, u(x+i\eta)\, d\eta$, is a nonegative measure on the imaginary axis.

Using this representation and his estimates for the potential of a nonnegative distribution of mass, Hayman proved in [2] a theorem characterizing the regularity of growth of the functions in question in the halfplane $\Pi$. He showed that for functions of the indicated class there exists a uniform limit

$$\lim_{r\to\infty}\frac{u(re^{i\theta})}{r}=\alpha\cos\theta,\quad \theta\leqslant\frac{\pi}{2}, \tag{1.3}$$

if $r$ on tending to infinity does not take on values lying in a certain set $\Delta_0\subset(0,\infty)$ of finite logarithmic length.[1)] In other words, the limit (1.3) exists in the case that the point $re^{i\theta}$ tends to infinity outside some set of half-rings whose intersection with the positive ray gives the set $\Delta_0$.

The result of Hayman is stronger than the corresponding result of Ahlfors and Heins [3], who showed earlier the existence of the limit (1.3) for $|\theta|\leq\pi/2-\delta$, where $\delta$ is arbitrarily small.

I. V. Ušakova showed in [4] that the limit (1.3) exists if $|\theta|\leq\pi/2-\delta$, and $z=re^{i\theta}\to\infty$ in such a way that $z$ does not lie in some set $C^0$ of circles for which the radii $\delta_j$ and centers $z_j$ satisfy the condition $\sum_j\delta_j/|z_j|<\infty$. This theorem strengthens the Heins-Ahlfors result, but however does not overlap the result of Hayman.

The object of the present paper is to give a generalization of the theorem of Riesz and Hayman. We introduce the following notation. Let $\Omega$ be a region on the unit sphere $S_1$ in $m$-dimensional Euclidean space $E_m$, whose boundary is a smooth contour $l$ of bounded curvature.

Suppose further that $L$ is a spherical operator,

$$L=\sum_{i=0}^{m-2}\frac{1}{\Pi}\frac{\partial}{\partial\theta_i}\frac{\Pi}{\Pi_i}\frac{\partial}{\partial\theta_i},$$

where $\theta_i$ are spherical coordinates: $(0\leq\theta_0<2\pi,\ 0\leq\theta_i\leq\pi,\ i=1,\cdots,m-2)$, and $\Pi$ and $\Pi_i$ are defined by the equations

$$\Pi=\prod_{0}^{m-2}\sin^j\theta_j,\quad \Pi_i=\prod_{i+1}^{m-2}\sin^2\theta_j,\quad \Pi_{m-2}=1.$$

It is clear that $L$ is the spherical part of a Laplace operator. Denote by $\lambda_1$ the first eigenvalue of the boundary problem

$$LY+\lambda Y=0,\quad Y(\bar{x})|_{\bar{x}\in l}=0, \tag{1.4}$$

and by $\rho_1$ the nonnegative root of the equation

$$\rho(\rho+m-2)=\lambda_1. \tag{1.5}$$

---

1) This means that $\int_{\Delta_0}r^{-1}dr<\infty$.

The value of the function $u(Q)$ at the point $Q \in E_m$ will be denoted by $u(r\bar{x})$, where $r\bar{x}$ is the radius vector of the point $Q$, drawn from the origin of coordinates $O$, $r = r_{OQ}$, $\bar{x} \in S_1$.

The subharmonic function $u(Q)$ has (by definition) *order* $\rho$ in the cone $K^\Omega\{r\bar{x}: 0 < r < \infty, \bar{x} \in \Omega\}$, given by the region $\Omega \subset S_1$, if

$$\overline{\lim_{r\to\infty}} \frac{\ln M_u(r)}{\ln r} = \rho,$$

where

$$M_u(r) = \max u^+(r\bar{x}), \quad \bar{x} \in \Omega, \quad u^+(Q) = \max\{u(Q), 0\},$$

and *normal type* $\sigma$ (for order $\rho$), if

$$\overline{\lim_{r\to\infty}} \frac{M_u(r)}{r^\rho} = \sigma < \infty.$$

In what follows we will put $\rho = \rho_1$, where $\rho_1$ is found according to (1.5).

We shall say that the function $u(Q)$ is bounded by zero on the boundary $K^l$ of the cone $K^\Omega$, if

$$\overline{\lim_{Q\to Q'}} u(Q) \leqslant 0, \quad Q' \in K^l, \quad Q \in K^\Omega. \tag{1.6}$$

Denote by $\gamma(x, \bar{y})$ the angle between the vectors $\bar{x}$ and $\bar{y}$ ($\bar{x}, \bar{y} \in S_1$), and by $l_\delta(\bar{x}_0)$ the contour[1)] on the unit sphere given by the condition

$$\gamma(\bar{x}, \bar{x}_0) = \delta, \quad \bar{x} \in S_1. \tag{1.7}$$

We introduce the notation

$$k_\delta(\bar{x}_0) = \overline{\lim_{r\to\infty}} \max_{\bar{x}\in l_\delta(\bar{x}_0)} \frac{u(r\bar{x})}{r^{\rho_1}}, \tag{1.8}$$

$$k(\bar{x}_0) = \underline{\lim_{\delta\to 0}} K_\delta(\bar{x}_0). \tag{1.9}$$

The generalization of (1.2) is given by the following theorem.

**Theorem 1.** *Suppose that $u(Q)$ is a function subharmonic in the cone $K^\Omega$, having order $\rho_1$ there and normal type, and suppose that $u(Q)$ is bounded by zero on $K^l$. Then for $u(Q)$ one has the representation*

---

1) For $m = 3$ this is a circumference.

$$u(r\bar{x}) - \frac{k(\bar{x}_0)}{Y_1(\bar{x}_0)} Y_1(\bar{x}) r^{\rho_1} = v(r\bar{x}) - \int_{Q \in K^\Omega} G^\Omega(r\bar{x}, Q)\, d_Q\mu, \quad (1.10)$$

*where* $\bar{x}_0$ *is any vector of* $\Omega$, $k(\bar{x}_0)$ *is defined by* (1.9), $Y_1(x)$ *is the first eigenfunction of the boundary problem* (1.4), $d_Q\mu$ *is the nonnegative Radon measure which corresponds to* $u(Q)$ *relative to the Riesz representation in a finite region (see* [1]), $G^\Omega$ *is the Green's function of the Laplace operator for the cone* $K^\Omega$, *and* $v(r\bar{x})$ *is the best harmonic majorant of the function*

$$u(r\bar{x}) - \frac{k(\bar{x}_0)}{Y_1(\bar{x}_0)} Y_1(\bar{x}) r^{\rho_1}.$$

*Here, for the function* $v(r\bar{x})$, *the following representation holds:*

$$v(r\bar{x}) = -\int_{K^l} \frac{\partial}{\partial n_Q} G^\Omega(r\bar{x}, Q)\, d_Q\nu,$$

*where* $d_Q\nu$ *is a nonnegative Radon measure on* $K^l$, *related to* $u(Q)$ *by the relation*

$$d_Q\nu = -\overline{\lim_{Q\to Q'}}\, u(Q)\, d\sigma \quad (Q \in K^\Omega,\ Q' \in K^l),$$

*in which* $d\sigma$ *is the element of area on the surface* $K^l$, *and* $\partial/\partial n_Q$ *is the derivative along the normal to* $K^l$.

We note that from the definition of the function $k(\bar{x}_0)$ it follows that it is bounded below ($<\sigma$). From the representation (1.10) it is clear that $k(\bar{x}_0) \neq -\infty$ for any point $\bar{x}_0$, since otherwise one would have $u(r\bar{x}) \equiv -\infty$. Moreover, it follows from (1.10) that the quantity $k = k(\bar{x}_0)/Y_1(\bar{x}_0)$ does not depend on $\bar{x}_0$.

The following result generalizes the theorem of Hayman in [2].

**Theorem 2.** *If the function* $u(r\bar{x})$ *satisfies the conditions of Theorem* 1, *then the function* $H(r\bar{x}) = u(r\bar{x})r^{-\rho_1}$ *tends uniformly to the function* $h(\bar{x}) = kY_1(\bar{x})$, *when* $r\bar{x} \to \infty$ *in* $K^\Omega$, *but outside a certain set* $C^0$ *of spheres such that the radii* $\delta_j$ *of these spheres and the distances* $R_j$ *of their centers from the origin of coordinates satisfies the condition*

$$\sum_j \left(\frac{\delta_j}{R_j}\right)^{m-1} < \infty.$$

If $m = 2$ and $\Omega$ is the circular are $\{re^{i\theta}: r = 1, |\theta| < \pi/2\}$, then we obtain an assertion which strengthens the Hayman theorem and the result of I. V. Ušakova.

For simplicity of exposition we shall suppose in what follows that $m \geq 3$.

## §2

For the proof of Theorems 1 and 2 we need a number of auxiliary propositions.

We shall present below two lemmas which give an estimate of the Green's function $G^{\Omega}(r\bar{x}, t\bar{y})$ for the cone $K^{\Omega}\{r\bar{x}: 0 < r < \infty, \bar{x} \in \Omega\}$ and the Green's function $G_R^{\Omega}(r\bar{x}, t\bar{y})$ for the conical sector $K_R^{\Omega}\{r\bar{x}: 0 < r < R, \bar{x} \in \Omega\}$ in terms of the first eigenfunction $Y_1(\bar{x})$ of the boundary problem (1.4) and the order $\rho_1$ defined by equation (1.5).

Lemma 1. *For the function* $G^{\Omega}(r\bar{x}, t\bar{y})$ *the following inequalities hold:*

$$k_1 \cdot \frac{Y_1(\bar{x}) Y_1(\bar{y})}{t^{m-2}} \left(\frac{r}{t}\right)^{\rho_1} \leqslant G^{\Omega}(r\bar{x}, t\bar{y}) \leqslant k_2 \cdot \frac{Y_1(\bar{x}) Y_1(\bar{y})}{t^{m-2}} \left(\frac{r}{t}\right)^{\rho_1} \quad (\bar{x}, \bar{y} \in \Omega),$$

$$k_1 \cdot \frac{Y_1(\bar{x}) \frac{\partial}{\partial n} Y_1(\bar{y})}{t^{m-1}} \left(\frac{r}{t}\right)^{\rho_1}$$

$$\leqslant \frac{\partial}{\partial n_t} G^{\Omega}(r\bar{x}, t\bar{y}) \leqslant k_2 \cdot \frac{Y_1(\bar{x}) \frac{\partial}{\partial n} Y_1(\bar{y})}{t^{m-1}} \cdot \left(\frac{r}{t}\right)^{\rho_1} (\bar{x} \in \Omega, \bar{y} \in l),$$

*if* $0 < r/t \leq 4/5$, *and*

$$k_1 \cdot \frac{Y_1(\bar{x}) Y_1(\bar{y})}{r^{m-2}} \left(\frac{t}{r}\right)^{\rho_1} \leqslant G^{\Omega}(r\bar{x}, t\bar{y}) \leqslant k_2 \cdot \frac{Y_1(\bar{x}) Y_1(\bar{y})}{r^{m-2}} \left(\frac{t}{r}\right)^{\rho_1} \quad (\bar{x}, \bar{y} \in \Omega),$$

$$k_1 \cdot \frac{Y_1(\bar{x}) \frac{\partial}{\partial n} Y_1(\bar{y})}{r^{m-2} t} \left(\frac{t}{r}\right)^{\rho_1}$$

$$\leqslant \frac{\partial}{\partial n_t} G^{\Omega}(r\bar{x}, t\bar{y}) \leqslant k_2 \cdot \frac{Y_1(\bar{x}) \frac{\partial}{\partial n} Y_1(\bar{y})}{r^{m-2} t} \left(\frac{t}{r}\right)^{\rho} (\bar{x} \in \Omega, \bar{y} \in l),$$

*if* $0 < t/r \leq 4/5$, *where* $k_1$ *and* $k_2$ *are constants depending only on the region* $\Omega$, *and* $\partial/\partial n$ *is the derivative along the normal to* $K^l$.

Lemma 2. *For the function* $G_R^{\Omega}(r\bar{x}, t\bar{y},)$ *we have*

$$c_1\left(\frac{r}{R}\right)^{\rho_1}\frac{Y_1(\bar{x})Y_1(\bar{y})}{R^{m-1}}\leqslant\frac{\partial}{\partial t}G_R^{\Omega}(r\bar{x},t\bar{y})|_{t=R}\leqslant c_2\left(\frac{r}{R}\right)^{\rho_1}\frac{Y_1(\bar{x})Y_1(\bar{y})}{R^{m-1}},$$

*where $c_1$ and $c_2$ depend only on the region $\Omega$.*

The proof of these lemmas is omitted from this exposition for reasons of conciseness.[1)]

Further on we shall need the following analog of the Phragmén-Lindelöf theorem, established by Deny and Lelong [5].

(A) Suppose that $K^{\Omega}$ is a cone with with lateral surface $K^l$ ($\Omega$ is the region of the unit sphere giving the cone, and $l$ its boundary). Suppose that $u(P)$ is a function subharmonic in $K^{\Omega}$ and satisfying the following conditions:

a) $\overline{\lim\limits_{P\to P'}}\, u(P)\leqslant 0\quad (P\in K^{\Omega},\ P'\in K^l)$,

b) $\lim\limits_{r\to\infty}\dfrac{M_u(r)}{r^{\rho_1}}=0$,

where $M_u(r)=\sup u^+(r\bar{x}), \bar{x}\in\Omega$ and $\rho_1$ is found from equation (1.5).

Then $u(P)\leq 0$ everywhere inside $K^{\Omega}$, while equality at even one point $P\in K^{\Omega}$ is possible only in the case $u(P)\equiv 0$.

Lemma 3. *Suppose that $u(Q)$ is a function subharmonic in the cone $K^{\Omega}$ and satisfying the conditions of Theorem 1. Then the inequality $u(r\bar{x})-(k(\bar{x}_0)/Y_1(\bar{x}_0))\,Y_1(\bar{x})\,r^{\rho_1}\leq 0$, is satisfied in the entire cone $K^{\Omega}$, where $k(\bar{x}_0)$ is defined by equation (1.9) and $\bar{x}_0$ is any vector of $\Omega$.*

Proof. Suppose that $l_\delta(\bar{x}_0)$ is the contour defined by equation (1.7), and $C_\delta(\bar{x}_0)$ the region on $S_1$ bounded by this contour. Introduce the notation

$$\varepsilon(\delta,\bar{x}_0)=\max_{x\in C_\delta(\bar{x}_0)}|Y_1(\bar{x})-Y_1(\bar{x}_0)|\,.$$

We choose $\delta$ so small that the region $C_\delta(\bar{x}_0)$ lies inside $\Omega$ and satisfies the condition $Y_1(\bar{x}_0)-\epsilon(\delta,\bar{x}_0)>0$.

Further, we denote by $v(r\bar{x},\delta,\epsilon)$ the subharmonic function

$$v(r\bar{x},\delta,\varepsilon)=u(r\bar{x})-\frac{k_\delta(\bar{x}_0)+\varepsilon}{Y_1(\bar{x}_0)-\varepsilon(\delta,\bar{x}_0)}Y_1(\bar{x})\,r^{\rho_1},$$

---

[1)]Furthermore, a proof of analogous inequalities was given in the paper [9].

where $k_\delta(\bar{x}_0)$ is defined by equation (1.8). The function $v(r\bar{x}, \delta, \epsilon)$ is bounded by zero on $K^l$, and on the conical surface $K^{l\,\delta}(r\bar{x}\colon 0 < r < \infty, \bar{x} \in l_\delta)$ the inequality $v(r\bar{x}, \delta, \epsilon) \leq 0$ is satisfied for $r > R_\epsilon$. If we introduce the notation

$$m(\delta, \varepsilon) = \Big|\max_{0<r<R_\varepsilon,\ \bar{x}\in l_\delta(\bar{x}_0)} v(r\bar{x}, \delta, \varepsilon)\Big|,$$

then the function $v(r\bar{x}, \delta, \epsilon) - m(\delta, \epsilon)$ will be bounded by zero both on $K^l$ and on $K^{l\,\delta}$.

The first eigenvalue of the boundary problem (1.4) for the region $C_\delta(\bar{x}_0)$ will be denoted by $\lambda^*$, and for the region $\tilde{C}_\delta = \Omega \backslash C_\delta \cup l_\delta$ by $\tilde{\lambda}$. Since $\lambda_1$, an eigenvalue for the region $\Omega$, satisfies the conditions $\lambda_1 < \lambda^*$, $\lambda_1 < \tilde{\lambda}$, the same inequalities are satisfied as well for the quantities $\rho^*, \tilde{\rho}, \rho_1$ related to $\lambda^*, \tilde{\lambda}, \lambda_1$ by (1.5).

Accordingly, for the function $v(r\bar{x}, \delta, \epsilon) - m(\delta, \epsilon)$ subharmonic in the cone $K^{C\,\delta(\bar{x}_0)}$, the conditions of Theorem (A) are satisfied, which means that for $r\bar{x} \in K^{C\,\delta(x_0)}$ we have

$$v(r\bar{x}, \delta, \varepsilon) - m(\delta, \varepsilon) \leqslant 0. \tag{2.1}$$

This inequality holds in the cone $K^{\tilde{C}\,\delta}$ as well, and therefore it holds in the entire cone $K^\Omega$. Since the function $v(r\bar{x}, \delta, \epsilon)$ is bounded by zero on $K^l$ and the number $m(\epsilon, \delta)$ in $\Omega$, applying Theorem (A) to it once again we find that

$$u(r\bar{x}) \leqslant \frac{k_\delta(\bar{x}_0) + \varepsilon}{Y_1(\bar{x}_0) - \varepsilon(\delta, \bar{x}_0)} \cdot Y_1(\bar{x})\, r^{\rho_1}$$

in $K^\Omega$. Since the left side of this inequality does not depend on $\epsilon$ and $\delta$, on taking the limit on the right as $\epsilon \to 0$, and the $\underline{\lim}$ as $\delta \to 0$ we obtain the assertion of the lemma.

We note that there exists a sequence of points $r_j\bar{x}_j$, where $\bar{x}_j \to \bar{x}_0$, such that

$$\frac{u(r_j\bar{x}_j)}{r_j^{\rho_1}} - \frac{k(\bar{x}_0)}{Y_1(\bar{x}_0)} \cdot Y_1(\bar{x}_j) \to 0 \quad \text{as} \quad j \to \infty.$$

It is also clear that $k(\bar{x}_0) \neq -\infty$. Otherwise we would have $u \equiv -\infty$.

Now we turn to the proof of Theorem 1. We introduce the notation

$$v(r\bar{x}) = u(r\bar{x}) - \frac{k(\bar{x}_0)}{Y_1(\bar{x}_0)} Y(\bar{x}) r^{\rho_1}.$$

The function $v(r\bar{x})$ is nonpositive and subharmonic in $K^\Omega$, while on some sequence of points $r_j\bar{x}_j$

$$\lim_{j\to\infty} \frac{v(r_j \bar{x}_j)}{r_j^{\rho_1}} = 0, \quad \bar{x}_j \to \bar{x}_0.$$

Let us write down for $v(r, \bar{x})$ the Riesz representation ([1]; see also [6]) in the sector $K_R^\Omega\{r\bar{x}: 0 < r < R, \bar{x} \in \Omega\}$:

$$v(r\bar{x}) = -\int\limits_{K_R^l} \frac{\partial}{\partial n} G_R^\Omega (r\bar{x}, Q)\, d_Q\nu - \int\limits_{K^\Omega \cap S_R} \frac{\partial}{\partial R} G_R^\Omega\, d\nu_1 - \int\limits_{K_R^\Omega} G_R^\Omega\, d_Q\mu, \tag{2.2}$$

where $G_R^\Omega$ is the Green's function of the $m$-dimensional Laplace operator for $K_R^\Omega$, $K_R^l = \{r\bar{x}: 0 < r < R, \bar{x} \in l\}$ is the lateral surface of the sector $K_R^\Omega$, $S_R$ is a sphere of radius $R$ with center at the origin, $d\mu$ is the mass distribution of the function $v(Q)$, and $d\nu$, $d\nu_1$ are nonnegative Radon measures on $K^l$ and $S_R$ respectively, connected with $v(Q)$ by the relations

$$d\nu = -\overline{\lim_{P\to Q'}} v(P)\, d\sigma, \quad P \in K_R^\Omega,\ Q' \in K_R^l;$$

$$d\nu_1 = -\overline{\lim_{P\to Q''}} v(P)\, d\sigma_1, \quad P \in K_R^\Omega,\ Q'' \in S_R;$$

here $d\sigma$ and $d\sigma_1$ are the area elements of the surfaces $K^l$ and $S_R$ respectively,

Putting $v^-(r\bar{x}) = -v(r\bar{x})$, we obtain

$$v^-(r\bar{x}) = \int\limits_{K_R^l} \frac{\partial}{\partial n} G_R^\Omega (r\bar{x}, Q)\, d\nu + \int\limits_{K^\Omega \cap S_R} \frac{\partial}{\partial n} G_R^\Omega (r\bar{x}, Q)\, d\nu_1 + \int\limits_{K_R^\Omega} G_R^\Omega (r\bar{x}, Q)\, d\mu. \tag{2.3}$$

Suppose that $v^-(r\bar{x}) < \infty$. Since $G_R^\Omega \uparrow G^\Omega$ as $R \to \infty$, formula (2.3) gives us

$$\int\limits_{K_R^\Omega} G_R^\Omega\, d\mu \to \int\limits_{K^\Omega} G^\Omega d\mu < \infty, \quad \int\limits_{K_R^l} \frac{\partial}{\partial n} G_R^\Omega\, d\nu \to \int\limits_{K^l} \frac{\partial}{\partial n} G^\Omega\, d\nu < \infty.$$

We shall show that as $R \to \infty$

$$\int_{K^\Omega \cap S_R} \frac{\partial}{\partial R} G_R^\Omega (r\bar{x}, R\bar{y})\, d\nu_1 \to 0.$$

On the sequence $r_j \bar{x}_j$ we have

$$v^-(r_j \bar{x}_j) \leqslant \varepsilon_j r_j^{\rho_1}, \quad \varepsilon_j \to 0.$$

Since on the sequence $R_j = 2r_j$, in view of Lemma 2,

$$v^-(r_j\bar{x}_j) \geqslant \int_{S_{R_j} \cap K^\Omega} \frac{\partial}{\partial R} G_R^\Omega\, d\nu_1 \geqslant c_1 Y_1(\bar{x}_j) \left(\frac{r_j}{R_j}\right)^{\rho_1} \int_{K^\Omega \cap S_{R_j}} Y_1\, d\nu_1,$$

it follows, putting $I(R_j) = \int_{K^\Omega \cap S_{R_j}} Y_1\, d\nu_1$, that

$$\frac{I(R_j)}{R_j^{\rho_1}} c_1 Y_1(\bar{x}_j) \leqslant \varepsilon_j \to 0, \quad Y_1(\bar{x}_j) \to Y_1(\bar{x}_0) > 0.$$

Then for any $r\bar{x}$ we have, according to Lemma 2,

$$\int_{S_{R_j} \cap K^\Omega} \frac{\partial}{\partial R} G_R^\Omega (r\bar{x}, R_j \bar{y})\, d\nu_1 \leqslant c_2 Y_1(\bar{x}) \frac{I(R_j)}{R_j^{\rho_1}} \leqslant c_2 Y_1(\bar{x})\, r^{\rho_1} \frac{\varepsilon_j}{c_1 Y_1(\bar{x}_j)} \to 0. \tag{2.4}$$

Passing to the limit as $j \to \infty$, we obtain the following representation for $v(r\bar{x})$:

$$v(r\bar{x}) = -\int_{K^l} \frac{\partial}{\partial n} G^\Omega (r\bar{x}, Q)\, d_Q\nu - \int_{K^\Omega} G^\Omega d\mu \tag{2.5}$$

for all points $r\bar{x}$, where $v^-(r\bar{x}) < \infty$.

Now suppose that $r\bar{x}$ is such that $v^-(r\bar{x}) = \infty$. It follows from (2.4) that the second term in the right side of formula (2.5) is bounded. Accordingly the sum of the first and third terms is infinite. As $R$ increases the sum does not decrease and therefore it remains infinite.

Thus relation (2.5) is proved also under the condition $v^-(r\bar{x}) = \infty$. It is easy to see that the quantity $d\nu$ for the functions $v(Q)$ and $u(Q)$ is one and the same. The same is true also for $d\mu$. Therefore the assertion of Theorem 1 follows from the representation (2.5).

## §3

In the proof of Theorem 2 we shall use basically the method of Hayman, modified for our case. However, the application of this method for a space of three or more dimensions requires that some auxiliary inequalities be established.

Denote by $K^L(M, \lambda)$ the set of points $r\bar{x} \in E_m$ depending on $L$, the latter being a set on $S_1$, and on the numbers $M$ and $\lambda$, and given by the relation

$$K^L(M, \lambda) = \left\{ r\bar{x}\colon \bar{x} \in L \subset S_1; \frac{1}{\lambda} < \frac{r}{M} < \lambda \right\}. \tag{3.1}$$

$K^L(M, \lambda)$ is the piece of the cone defined by the set $L$. An estimate of the Green's function $G^\Omega(P, Q)$ in the middle part of the cone $K^\Omega$ is given by the following lemma.

Lemma 4. *Suppose that $P \in K^\Omega$, and $Q \in K^\Omega(r_P, 5/4)$, where $K^\Omega(r_P, 5/4)$ is defined by* (3.1). *Then for the Green's function $G^\Omega(P, Q)$ of the cone $K^\Omega$ the inequality*

$$G^\Omega(P, Q) \leqslant \frac{Y_1(\bar{x})\, Y_1(\bar{y})}{t^{m-2}} \left( A + B \cdot \frac{rt^{m-1}}{r_{PQ}^m} \right) \tag{3.2}$$

*holds, where $Y_1$ is the first eigenfunction of the boundary problem* (1.4), *the vectors $r\bar{x}$ and $t\bar{y}$ correspond to the points $P$ and $Q$, and $A$ and $B$ are constants not depending on $P$ and $Q$.*

Proof. We note that we shall frequently use, instead of a point, its radius vector. We observe that

$$G^\Omega(r\bar{x}, t\bar{y}) = \frac{1}{t^{m-2}} G^\Omega(\bar{x}, t'\bar{y}), \tag{3.3}$$

where $t' = t/r, \bar{x}, \bar{y} \in \Omega$. Suppose that $\tilde{K}$ is any region satisfying the conditions

$$K^\Omega\left(1, \frac{3}{2}\right) \subset \tilde{K} \subset {}'K^\Omega(1, 2) \tag{3.4}$$

and such that its boundary has bounded curvature. Denote by $G_1(P, Q)$ the Green's function for the region $\tilde{K}$ and consider the function

$$v(r\bar{x}) = -\int_{K^l} \frac{\partial}{\partial n} G^\Omega(r\bar{x}, Q)\, d_Q\nu - \int_{K^\Omega} G^\Omega d\mu.$$

The function $\Phi_1(Q')$ is harmonic in the region $K^\Omega(1, 5/4)$. By Lemma 1, for $Q' = 4\bar{y}/5$, we have

$$\Phi_1\left(\frac{4}{5}\bar{y}\right) \leqslant k_2\left(\frac{4}{5}\right)^{\rho_1} Y_1(\bar{x})\, Y_1(\bar{y}), \tag{3.5}$$

and, for $Q' = 5\bar{y}/4$,

$$\Phi_1\left(\frac{5}{4}\bar{y}\right) \leqslant k_2\left(\frac{4}{5}\right)^{\rho_1+m-2} Y_1(\bar{x})\, Y_1(\bar{y}). \tag{3.6}$$

We apply the maximum principle in the region $K^\Omega(1, 5/4)$ to the harmonic function

$$\Phi_2(Q') = \Phi_1(Q') - k_2 Y_1(\bar{x})\,[Y_1(\bar{y}')(t')^{\rho_1} + Y_1(\bar{y}')(t')^{-\rho_1-m+2}],$$

where $Q' = \bar{y}'t'$. The function $\Phi_1(Q')$ is equal to zero on the lateral surface $K^\Omega(1, 5/4)$.

From (3.5) and (3.6) it follows that on the remaining portion of the boundary

$$\Phi_2(Q') \leqslant 0, \tag{3.7}$$

so that inequality (3.7) is satisfied also in the entire region $K^\Omega(1, 5/4)$; that is

$$\Phi_1(Q') \leqslant k_2 Y_1(\bar{x})\, Y_1(\bar{y}')\,[(t')^{\rho_1} + (t')^{-\rho_1-m+2}]. \tag{3.8}$$

From (3.8) the following estimate holds for $G^\Omega(\bar{x}, Q')$:

$$G^\Omega(\bar{x}, Q') \leqslant k_2 Y_1(\bar{x})\, Y_1(\bar{y}')\,[(t')^{\rho_1} + (t')^{-\rho_1-m+2}] + G_1(\bar{x}, Q'). \tag{3.9}$$

For the estimation of the function $G_1(\bar{x}, Q')$ we need the following inequality, obtained by M. V. Keldyš and M. A. Lavrent′ev [7].

**Lemma** (Keldyš-Lavrent′ev).[1)] *Suppose that $D$ is a region in $E_m$, whose boundary $D'$ is a Ljapunov surface.*[2)] *Then the following estimate holds for its Green's function $G^D(P, Q)$:*

---

[1)] The possibility of applying this lemma to obtain the needed estimates was kindly pointed out to the author by N. S. Landkof.

[2)] This means that $D'$ is a smooth surface and that there exists a number $\alpha$, $0 < \alpha \leq 1$, and a constant $k$ such that $\gamma(P, Q)$, the angle between the normals at the points $P$ and $Q$, satisfies the condition

$$\gamma(P, Q) < r_{PQ}^{\alpha} k.$$

If $\alpha = 1$, then the surface has bounded curvature.

$$G^D(P, Q) \leqslant M \frac{d(P)\, d(Q)}{r_{PQ}^m},$$

*where $d(P)$ is the distance from the point $P \in D$ to the closest point $P' \in D'$, and $M$ is a constant not depending on $P$ and $Q$.*

It is easy to see that for the distances $d(P)$, $d(Q)$ from the points $P, Q \in K^{\Omega}(1, 4/5)$ to the boundary of a region $\widetilde{K}$ satisfying the conditions of this lemma and relations (3.4), the inequalities $d(P) \leq M_1 Y_1(\bar{x})$, $d(Q') \leq M_2 Y_1(\bar{y}')\, t'$ are satisfied, where $P = \bar{x}$, $Q' = t'\bar{y}'$, and $M_1$ and $M_2$ do not depend on $P$ and $Q$.

Therefore for $G_1(\bar{x}, Q')$ we have the estimate

$$G_1(\bar{x}, Q') \leqslant M_1 M_2 M \frac{Y_1(\bar{x}) Y_1(\bar{y}')\, t'}{r_{PQ'}^m}, \tag{3.10}$$

where $P = \bar{x}$, $Q' = t'\bar{y}'$.

From this relation and (3.9) we find that

$$G^{\Omega}(\bar{x}, Q') \leqslant Y_1(\bar{x}) Y_1(\bar{y}') \left[A + B_1 \frac{t'}{r_{PQ'}^m}\right], \tag{3.11}$$

where $A = \max_{4/5<t<5/4} k_2(t^{\rho_1} + t^{-\rho_1 - m + 2})$, $B_1 = M_1 M_2 M$, $P = \bar{x}$, $Q' = t'\bar{y}'$.

By changing the constant $B_1$ somewhat and using the fact that the condition $4/5 < t' < 5/4$ is satisfied by $t'$, we obtain the inequality

$$G^{\Omega}(\bar{x}, Q') \leqslant Y_1(\bar{x}) Y_1(\bar{y}') \left[A + B \frac{(t')^{m-1}}{r_{PQ'}^m}\right],$$

where $B$ is that same changed constant $B_1$.

Putting $t' = t/r$, $P = r\bar{x}$, $Q = t\bar{y}$ and using (3.3), we obtain the assertion of the lemma.

**Remark.** For $G_1$ we obviously have

$$G_1(\bar{x}, Q') \leqslant \frac{1}{r_{PQ'}^{m-2}}.$$

Hence for $G^{\Omega}(P, Q)$ we obtain the relation

$$G^{\Omega}(P, Q) \leqslant \frac{Y_1(\bar{x}) Y_1(\bar{y})}{t^{m-2}} A + \frac{1}{r_{PQ}^{m-2}}, \tag{3.12}$$

where $P = r\bar{x}$, $Q = t\bar{y}$, $Q \in K^{\Omega}(r, 5/4)$.

## §4

We turn to the proof of Theorem 2. Suppose that $u(Q)$ satisfies the conditions of Theorem 2. Using the nonnegativity of the function $d\nu$ on the boundary $K^l$ of the cone $K^\Omega$, the representation (1.10) may be rewritten in

$$u(r\bar{x}) = kY_1(\bar{x})r^{\rho_1} - \int_{K^\Omega \cup K^l} K(r\bar{x}, Q)\, d_Q m, \quad P = r\bar{x}, \quad Q = t\bar{y},$$

where $K(P, Q)$ and $d_Q m$ are defined by

$$K(r\bar{x}, t\bar{y}) = \begin{cases} \dfrac{G^\Omega(r\bar{x}, t\bar{y})}{Y_1(\bar{y})} t^{\rho_1+m-2} & \text{for } \bar{y} \in \Omega,\ t \geqslant 1, \\ \dfrac{G^\Omega(r\bar{x}, t\bar{y})}{Y_1(\bar{y})} t^{-\rho_1} & \text{for } \bar{y} \in \Omega,\ t < 1, \\ \dfrac{\frac{\partial}{\partial n_Q} G^\Omega(r\bar{x}, t\bar{y})}{\frac{\partial}{\partial n} Y_1(\bar{y})} t^{\rho_1+m-1} & \text{for } \bar{y} \in l,\ t \geqslant 1, \\ \dfrac{\frac{\partial}{\partial n_Q} G^\Omega(r\bar{x}, t\bar{y})}{\frac{\partial}{\partial n} Y_1(\bar{y})} t^{-(\rho_1+1)} & \text{for } \bar{y} \in l,\ t < 1; \end{cases}$$

$$d_Q m = \begin{cases} Y_1(\bar{y}) t^{-\rho_1-m+2}\, d\mu & \text{for } \bar{y} \in \Omega,\ t \geqslant 1, \\ Y_1(\bar{y}) t^{\rho_1} d\mu & \text{for } \bar{y} \in \Omega,\ t < 1, \\ \frac{\partial}{\partial n} Y_1(\bar{y}) t^{-\rho_1-m+1}\, d\nu & \text{for } \bar{y} \in l,\ t \geqslant 1, \\ \frac{\partial}{\partial n} Y_1(\bar{y}) t^{\rho_1+1} d\nu & \text{for } \bar{y} \in l,\ t < 1 \end{cases}$$

(here $P = rx$, $Q = ty$ and $\partial/\partial n$ is the derivative along the normal to $K^l$).

Without loss of generality we may suppose that $k = 0$, i.e. we consider the function $u_1 = u(r\bar{x}) - kY_1(\bar{x})r^{\rho_1}$. Then

$$u_1(r\bar{x}) = -\int_{K^\Omega \cup K^l} K(r\bar{x}, Q)\, d_Q m. \tag{4.1}$$

Thus it is sufficient to prove Theorem 2 for the functions defined by equation (4.1).

We first observe that, from Lemma 1, the following inequalities hold for $K(P, Q)$:

$$k_1 \leqslant \frac{K(P,Q)}{Y_1(x) r^{\rho_1}} \leqslant k_2 \quad \text{for } \frac{r}{t} \leqslant \frac{4}{5}, \quad t \geqslant 1; \tag{4.2}$$

$$k_1 \leqslant \frac{K(P,Q)\, t^{2\rho_1+m-2}}{Y_1(\bar{x})\, r^{\rho_1}} \leqslant k_2 \quad \text{for } \frac{r}{t} < \frac{4}{5}, \quad t < 1; \tag{4.3}$$

$$k_1 \leqslant \frac{K(P,Q)\, r^{\rho_1+m-2}}{Y_1(x)\, t^{2\rho_1+m-2}} \leqslant k_2 \quad \text{for } \frac{t}{r} < \frac{4}{5}, \quad t \geqslant 1; \tag{4.4}$$

$$k_1 \leqslant \frac{K(P,Q)\, r^{\rho_1+m-2}}{Y(x)} \leqslant k_2 \quad \text{for } \frac{t}{r} < \frac{4}{5}, \quad t < 1. \tag{4.5}$$

Introduce the notation $\bar{\Omega} = \Omega \cup l$. We shall prove that

$$m(\bar{\Omega}) = \int_{K^{\bar{\Omega}}} dm < \infty. \tag{4.6}$$

Putting $r = 1/2$, we find from (4.1) and (4.2) that

$$-u(r\bar{x}) > \int_{K^{\bar{\Omega}}, t \geqslant 1} K(r\bar{x}, t\bar{y})\, dm > k_1 Y_1(\bar{x}) r^{\rho_1} \int_{K^{\bar{\Omega}}, t \geqslant 1} dm,$$

and hence $\int_{t \geq 1} dm < \infty$. Analogously, putting $r > 2$, we find that $\int_{t<1} dm < \infty$, from which formula (4.6) follows.

For a given point $P = r\bar{x}$ we shall denote the subsets

$$\left\{ t\bar{y}:\ 0 < \frac{t}{r} < \frac{4}{5},\ \bar{y} \in \bar{\Omega} \right\},\ \left\{ t\bar{y}:\ \frac{4}{5} < \frac{t}{r} < \frac{5}{4},\ \bar{y} \in \bar{\Omega} \right\},$$

$$\left\{ t\bar{y}:\ \frac{5}{4} < \frac{t}{r},\ \bar{y} \in \bar{\Omega} \right\}$$

by $K_1, K_2, K_3$ respectively.

Using inequalities (4.2)–(4.5) and the relation (4.6), it can be shown that

$$\int_{K_1} K(P,Q)\, d_Q m = o(1)\, r^{\rho_1} Y_1(\bar{x}), \quad \int_{K_3} K(P,Q)\, d_Q m = o(1)\, r^{\rho_1} Y_1(\bar{x}) \tag{4.7}$$

as $r \to \infty$, uniformly relative to $\bar{x} \in \bar{\Omega}$. The following is a consequence.

Lemma 5. *The function* $u(Q)$ *given by* (4.1) *satisfies the equation*

$$u(r\bar{x}) = -\int_{K_2} K(P,Q)\, d_Q m + o(1)\, r^{\rho_1} Y_1(\bar{x})$$

*uniformly for* $r \to \infty, \bar{x} \in \bar{\Omega}$.

In order to prove Theorem 2 we estimate the integral in Lemma 5. To this end we use two lemmas.

**Definition.** Suppose that $\epsilon > 0$ is a fixed number. We shall say that the point $Q$ is $\epsilon$-normal,[1] if for all $0 < \lambda < r_Q/2$ the inequality

$$\int_{K_\lambda(Q)} dm < \frac{\varepsilon \lambda^{m-1}}{r_Q^{m-1}}$$

holds, where $K_\lambda(Q)$ is the sphere of radius $\lambda$ with center at $Q$.

**Lemma 6.** *If* $r_P > 2$ *and the point* $P$ *is* $\epsilon$*-normal, then*

$$\int_{K_2} K(P, Q)\, d_Q m < A\, \varepsilon r_P^{\rho_1}, \tag{4.8}$$

*where* $A$ *is a constant not depending on* $P$ *and* $\epsilon$.

**Lemma 7.** *The set* $\Delta(\epsilon)$ *of points not* $\epsilon$*-normal may be covered by a system* $C(\epsilon)$ *of spheres* $K_j$ *whose radii* $\delta_j$ *and distances* $R_j$ *from their centers to the origin of coordinates satisfy the condition*

$$\sum_j \left(\frac{\delta_j}{R_j}\right)^{m-1} < \infty. \tag{4.9}$$

Suppose for the moment that the lemmas have been proved; we shall present the proofs below. According to Lemma 7, relation (4.9) holds for $K_j \in C(\epsilon)$. It is possible to choose a sequence of positive numbers $r_n \to \infty$ in such a way that for the spheres $K_j \in C(1/n)$ the relation

$$\sum_{R_j > r_n} \left(\frac{\delta_j}{R_j}\right)^{m-1} < 2^{-n} \tag{4.10}$$

is satisfied. Denote by $C(r_n, 1/n)$ the set of spheres whose radii appear in the sum (4.10), and by $C^0$ the following set of spheres: $C^0 = \bigcup_n C(r_n, 1/n)$. It is easily seen that for the spheres of $C^0$ the relation $\Sigma_j (\delta_j/R_j)^{m-1} < 1$ holds. On the other hand, if the point $P$ lies outside $C^0$ and $r_p > r_m$, then from Lemma 6 we have

$$\int_{K_2} K(P, Q)\, d_Q m < \frac{A}{n} r_P^{\rho_1},$$

from which Theorem 2 follows.

---

[1] This definition is a modification of the corresponding definition of Hayman.

Now we shall prove Lemmas 6 and 7.

Proof of Lemma 6. We note that from relations (3.2) and (3.12) the estimate

$$K(P,Q) \leqslant Y_1(\bar{x})\, t^{\rho_1}\left[A + B\frac{rt^{m-1}}{r_{PQ}^m}\right] \tag{4.11}$$

for $K(P,Q)$ follows if $P = r\bar{x}$, $Q = t\bar{y}$, $r > 2$, $Q \in K_2$ and the estimate

$$K(P,Q) < AY(\bar{x})\, t^{\rho_1} + \frac{t^{\rho_1+m-2}}{Y_1(y)\, r_{PQ}^{m-2}} \tag{4.12}$$

follows if $Q$ is an interior point of the subset $K_2$.

Using these estimates, we get

$$\int_{K_2} K(P,Q)\, d_Q m \leqslant AY_1(\bar{x}) \int_{K_2} t^{\rho_1} d_Q m + \int_{K_2} \bar{K}(P,Q)\, d_Q m, \tag{4.13}$$

where

$$\bar{K}(P,Q) = \min\left\{B\frac{Y_1(\bar{x})\, rt^{\rho_1+m-1}}{r_{PQ}^m},\ \frac{t^{\rho_1+m-2}}{Y_1(\bar{y})\, r_{PQ}^{m-2}}\right\}. \tag{4.14}$$

It is easy to see that as $r \to \infty$ the first integral in (4.13) is $o(1)\, r^{\rho_1}$. Therefore it remains to estimate the second integral.

We denote by $K_3(P)$ the set of points $Q = t\bar{y}$:

$$K_3 = \left\{t\bar{y} : \sin\,(\widehat{\bar{x}, \bar{y}}) < \frac{3}{8}\right\} \cap K_2,$$

and put $K_4 = K_2 \setminus K_3$. For $Q \in K_4$ we have $r_{PQ} > 3t/8\ (t\bar{y} = Q)$. Hence, using (4.14), we obtain

$$\bar{K}(P,Q) \leqslant B_1 Y_1(\bar{x}) r^{\rho_1}, \tag{4.15}$$

where $B_1$ does not depend on $P$ and $Q$. From (4.15) it follows that

$$\int_{K_4} \bar{K}(P,Q)\, d_Q m < A\varepsilon r^{\rho_1} \tag{4.16}$$

for sufficiently large $A$, not depending on $r$ and $\epsilon$, since $\int_{K_4} d_Q m \to 0$ as $r \to \infty$.

Now we turn to the estimate of the integral over the set $K_3$. We note that $K_3$ lies entirely in a sphere of radius $r_P/2$ with center at the point $P$. We decompose the set $K_3$ into the sum of spherical layers $C_n$:

$$2^{n-1} r_{PP'} \leqslant r_{PQ} < 2^n r_{PP'} \quad (-\infty < n < +\infty),$$

where $r_{PP'}$ is the distance from the point $P$ to the lateral surface of the cone $K^{\Omega}$. (For $n > 0$ only a finite number of the layers $C_n$ are nonempty.) Since the point $P$ is $\epsilon$-normal we have $m(P) = 0$, so that

$$\int_{K_1} \overline{K}(P, Q)\, d_Q m = \sum_{-\infty}^{\infty} \int_{C_n} \overline{K}(P, Q)\, d_Q m. \tag{4.17}$$

We shall estimate each term in the right side of this equation.

For $n < 0$ we have

$$r_{PQ}^{m-2} > 2^{(n-1)(m-2)} r_{PP'}^{m-2}, \quad t < \frac{5}{4} r \quad (P = r\bar{x}, Q = t\bar{y}).$$

Moreover, since for $n < 0$ the inequalities $r_{QQ'} > \frac{1}{2} r_{PP'}$ and $r_{QQ'} < CY_1(\bar{y})t$ hold, we have

$$rY_1(\bar{y}) > \frac{1}{2} \cdot \frac{4}{5C} r_{PP'} = \frac{4}{10 \cdot C} r_{PP'}.$$

Thus for $\overline{K}(P, Q)$ we get the estimate

$$\overline{K}(P, Q) \leqslant \frac{10 \cdot C}{4} \left(\frac{5}{4}\right)^{\rho_1 + m - 2} r_{PP'}^{-(m-1)} 2^{-(n-1)(m-2)} r^{\rho_1 + m - 1}$$

Using the $\epsilon$-normality of the point $P$, we get

$$m(C_n) \leqslant \frac{\varepsilon \cdot 2^{n(m-1)} r_{PP'}^{m-1}}{r^{m-1}},$$

so that for the integrals over $C_n$ we have the estimate

$$\int_{C_n} \bar{K}(P, Q)\, d_Q m \leqslant \varepsilon B_1 2^n r^{\rho_1},$$

where

$$B_1 = \frac{10 \cdot C}{4} \left(\frac{5}{4}\right)^{\rho_1 + m - 1} 2^{m-2}.$$

Thus for negative $n$ we have

$$\sum_{-\infty}^{-1} \int_{C_n} \overline{K}(P, Q)\, d_Q m \leqslant \varepsilon B_1 r^{\rho_1}. \tag{4.18}$$

Consider the case $n \geq 0$. Using relation (4.14), we get

$$\overline{K}(P, Q) \leqslant \frac{B r_{PP'} t^{\rho_1+m-1}}{r_{PQ}^m} \leqslant B_2 r_{PP'}^{1-m} 2^{-(n-1)m} r^{\rho_1+m-1} \quad (P=r\overline{x},\ Q=t\overline{y}),$$

where $B_2$ is a constant not depending on $P$ and $\epsilon$. As a consequence of the $\epsilon$-normality of the point $P$, we have

$$m(C_n) \leqslant \frac{\varepsilon \cdot 2^{n(m-1)} r_{PP'}^{m-1}}{r^{m-1}},$$

and hence

$$\int_{C_n} \overline{K}(P, Q)\, d_Q m \leqslant \varepsilon B_2 2^m r^{\rho_1} 2^{-n}.$$

Thus for positive $n$ we have

$$\sum_0^\infty \int_{C_n} \overline{K}(P, Q)\, d_Q m \leqslant \varepsilon B_2 2^m r^{\rho_1}. \tag{4.19}$$

The assertion of Lemma 6 follows from (4.18), (4.19) and (4.13).

Proof of Lemma 7. Suppose that $\Delta(\epsilon)$ is the set of points not $\epsilon$-normal. For each point $P \in \Delta(\epsilon)$ there exists a sphere $K_\lambda(P)$ with center at that point, such that

$$\int_{K_\lambda(P)} d_Q m > \varepsilon \frac{\lambda^{m-1}}{r_P^{m-1}}, \tag{4.20}$$

where $\lambda \leq r_P/2$.

Denote by $\Delta_n$ the set of those points of $\Delta(\epsilon)$ which lie in the layer $\{t\overline{y}\colon \overline{y} \in \overline{\Omega},\ 4^n < t \leq 4^{n+1}\}$.

Obviously the spheres with centers at the points $\Delta_{2k}$ $(k = 0, 1, \cdots)$ do not intersect one another if the $k$ are distinct. The same is true relative to the spheres with centers at the points $\Delta_{2k+1}$ $(k = 0, 1, \cdots)$.

We shall use the following assertion, proved recently by N. S. Landkof [8].

If the set $E \subset E_m$ is covered by spheres of bounded radii such that each point of the set is the center of a sphere, then from this covering one may select a subsystem of spheres which covers that set, each point of $E_m$ being covered not more than $\nu$ times by the spheres of this subsystem, where $\nu$ is a number depending only on the dimension of the space.

In view of this assertion, from the system of spheres whose centers lie at points of $\Delta_n$ one may select a subsystem $C_n$ covering the set $\Delta_n$, while the covering of each point of $E_m$ is not more than $\nu$-multiple.

We shall denote by $\Delta_{\text{even}}$ the set $\bigcup_0^\infty \Delta_{2k}$, and by $C_{\text{even}}$ the system of spheres covering the set $\Delta_{\text{even}}$:

$$C_{\text{even}} = \bigcup_0^\infty C_{2k}.$$

As was noted above, the spheres of $C_{2k}$ do not intersect, so that the covering of $\Delta_{\text{even}}$ by the system $C_{\text{even}}$ remains not more than $\nu$-multiple.

If we denote by $\delta_j$ the radii of the spheres $K_j \in C_{\text{even}}$, and by $R_j$ the distances from their centers to the origin of coordinates, then we find from (4.20) that

$$\nu \int_{K\overline{\Omega}} dm \geqslant \varepsilon \sum_j \left(\frac{\delta_j}{R_j}\right)^{m-1},$$

which implies that the sum on the right is finite.

Analogously we find that the corresponding series converges for the spheres of the system $C_{\text{odd}} = \bigcup_0^\infty C_{2k+1}$ covering the set $\Delta_{\text{odd}} = \bigcup_0^{2k+1} \Delta_{2k+1}$. Accordingly, the system $C(\epsilon) = C_{\text{even}} \cup C_{\text{odd}}$, covering the set $\Delta(\epsilon) = \Delta_{\text{even}} \cup \Delta_{\text{odd}}$, satisfies the condition $\Sigma_{K_j \in C(\epsilon)} (\delta_j/R_j)^{m-1} < \infty$; this proves Lemma 7.

The proof of Theorem 2 is thus completed.

The author thanks B. Ja. Levin for posing the questions considered in this paper, and N. S. Landkof for a series of valuable observations.

## BIBLIOGRAPHY

[1] F. Riesz, *Sur les fonctions subharmoniques et leur rapport à la théorie du potentiel*, Acta Math. 48 (1926), 329–343.

[2] W. K. Hayman, *Questions of regularity connected with the Phragmén-Lindelöf principle*, J. Math. Pures Appl. (9) 35 (1956), 115–126. MR 17, 1073.

[3] L. Ahlfors and M. Heins, *Questions of regularity, connected with the Phragmén-Lindelöf principle*, Ann. of Math. (2) 50 (1949), 341–346. MR **10**, 522.

[4] I. V. Ušakova, *Some estimates of subharmonic functions in the circle*, Zap. Meh.-Mat. Fak. *i* Har'kov. Mat. Obšč. 29 (1963), 53–66. (Russian)

[5] J. Deny and P. Lelong, *Sur une généralisation de l'indicatrice de Phragmén-Lindelöf*, C. R. Acad. Sci. Paris 224 (1947), 1046–1048. MR 8, 514.

[6] I. I. Privalov, *Subharmonic functions,* GITTL, Moscow, 1937. (Russian)

[7] M. V. Keldyš and M. A. Lavrent'ev, *Sur une évaluation pour la fonction de Green,* Dokl. Akad. Nauk SSSR 24 (1939), 102–103. MR 2, 57.

[8] N. S. Landkof, *Capacity and Hausdorff measures. Estimates of potentials,* Seventh All-Union Conference on the Theory of Functions of a Complex Variable, Abstracts of reports, Rostov-on-Don, 1964, p. 100.

[9] V. S. Azarin, *On the indicatrix of a function which is subharmonic in a higher-dimensional space,* Mat. Sb. 58 (100) (1962), 87–94. (Russian) MR 26 #343.

Translated by:
J. M. Danskin

# ON THE REPRESENTATION OF CONTINUOUS POSITIVE-DEFINITE KERNELS

N. N. ČAUS

Necessary and sufficient conditions for existence and uniqueness of an integral representation of an arbitrary positive-definite (p. d.) kernel $K$ in terms of solutions of certain equations were given by Ju. M. Berezanskiĭ ([1], Theorem 1). As was shown there, in the case of continuous p. d. kernels $K = K(x, y)$ $(x, y \in E_n)$ and differential equations with constant coefficients sufficient conditions may be formulated in a much more concrete form when an essential role is played by the growth of the kernel $K(x, y)$ at $\infty$ (Theorem 4).

In the present article we consider continuous p. d. kernels $K(x, y) = K(x_1, \cdots, x_n; y_1, \cdots, y_n)(x, y \in G_{mn} \subset E_n, m \leq n)$, where $x \in G_{mn}$ denotes that $x_1, \cdots, x_m$ run through the values from $-\infty$ to $+\infty$, and $x_{m+1}, \cdots, x_n$ run from $0$ to $+\infty$. We investigate the possibility of representing such kernels in terms of solutions of the equations $L^{(i)}u_i = \lambda_i u_i$, where $L^{(1)}, \cdots$ $\cdots, L^{(n)}$ are differential expressions with constant coefficients (it is additionally assumed that $L^{(m+1)}, \cdots, L^{(n)}$ are expressions of the second order). We prove Theorem 1, which in the case $m = n$ differs from Theorem 4 [1] by its less stringent limitations on the growth of the kernel at $\infty$. The conditions of Theorem 1 are sufficient for the existence and uniqueness of integral representations of p. d. kernels. With regard to uniqueness these conditions turn out to be in a certain sense also necessary. This is discovered in an example of representing a kernel in terms of cosines by comparing the result with a theorem of E. B. Vul [2].

The practical side of the proof of Theorem 1, according to results obtained in [1], reduces to establishing the uniqueness of the weak solution of the Cauchy problem $du_t/dt \pm iL^* u_t$, where $u_t$ lies in the narrowest possible Hilbert space.

## §1. Formulation of the basic theorem. General facts on representations of p. d. kernels

Let us introduce the following definition.

We will say that $l(x)$ $(x \geq 0)$ is a slowly increasing function if:

1) $l(x)$ is nondecreasing positive and $\lim_{x \to \infty} l(x) = \infty$;

2) $l(x) < C_\epsilon x^\epsilon$ for an arbitrary $\epsilon > 0$ with some constants $C_\epsilon$;

3) the limit $\lim_{x\to\infty} xl''(x)/l'(x) = \gamma$ exists ($\gamma$ is finite or infinite);

4) the function $xl'(x)/l(x)$ is nonincreasing.

Denote by $\chi_1^j(x_j, \lambda_j), \cdots, \chi_{r_j}^j(x_j, \lambda_j)$ the fundamental system of solutions of the equation $L^{(j)}u = \lambda_j u$ ($r_j$ is the order of the expression $L^{(j)}$) and form the product

$$X_j(x, \lambda) = \chi_{j_1}^1(x_1, \lambda_1) \ldots \chi_{j_n}^n(x_n, \lambda_n)$$

$(x = (x_1, \cdots, x_n) \in G_{mn};\ \lambda = (\lambda_1, \cdots, \lambda_n) \in E_n)$. Here $j = (j_1, \cdots, j_n)$ is the combined index, which varies over the set $N$ of points with coordinates $j_l = 1, \cdots, r_l$ $(l = 1, \cdots, n)$.

We recall that a kernel $K(x, y)$ $(x, y \in G_{mn})$ is called p. d. if for an arbitrary finite on $G_{mn}$ function $\phi(x)$

$$\int_{G_{mn}} \int_{G_{mn}} K(x, y)\, \varphi(x)\, \overline{\varphi(y)}\, dx dy \geqslant 0.$$

We have the following theorem on representations of p. d. kernels.

Theorem 1. *In the domain $G_{mn}$ let there be defined a continuous p. d. kernel $K(x, y)$ $(x, y \in G_{mn})$, satisfying in the sense of generalized functions the relations*

$$L_{x_j}^{(j)} K(x, y) = \overline{L_{y_j}^{(j)}} K(x, y) \qquad (j = 1, \ldots n),$$

*where $L^{(1)}, \cdots, L^{(n)}$ are differential expressions with real* constant coefficients with respect to the variables $x_1, \cdots, x_n$, respectively, of orders $r_1, \cdots, r_m > 1$, $r_{m+1} = \cdots = r_n = 2$. If*

$$|K(x, y)| \leqslant C \exp[|x_1|^{r_1'} l_1(|x_1|) + |y_1|^{r_1'} l_1(|y_1|) + \ldots + |x_m|^{r_m'} l_m(|x_m|) \\ + |y_m|^{r_m'} l_m(|y_m|) + a_{m+1} x_{m+1}^{2-\varepsilon} + a_{m+1} y_{m+1}^{2-\varepsilon} + \ldots + a_n x_n^{2-\varepsilon} + a_n y_n^{2-\varepsilon}] \tag{1}$$

*with some $C, a_i, \epsilon > 0$, $1/r_k' + 1/r_k = 1$ and $l_j(s) \in I(r_j)$, then we have the unique representation*

$$K(x, y) = \int_{E_n} \sum_{j,k \in N} X_j(x, \lambda)\, \overline{X_k(y, \lambda)}\, d\varrho_{jk}(\lambda), \tag{2}$$

*where the matrix $\|d\rho_{jk}(\lambda)\|_{j,\, k \in N}$ is p. d. in the sense that*

$$\sum_{j,k \in N} \varrho_{jk}(\Delta)\, \xi_j \bar{\xi}_k \geqslant 0 \qquad (\varrho_{jk}(\Delta) = \varrho_{jk}(\lambda'') - \varrho_{jk}(\lambda'),\ \ \Delta = [\lambda'', \lambda'))$$

*for any $\Delta$ and any collection of numbers $\xi_j$. The integral is absolutely convergent. The inclusion $l(s) \in I(q)$ means here that $l(s)$ is a slowly increas-*

* In the course of the proof of the theorem it will be seen that this requirement may be changed to the requirement of equality of the defect indices at each of the operators $A_j$ in $H_k$, which are defined later. For this reason we leave the bar over $L_{y_j}^{(j)}$ in the theorem.

*ing function with the condition* $\int_1^\infty (1/sl(s)^{q-1})ds = \infty$.

We will now, making use of [1], describe briefly that situation for which one has a representation of a p. d. kernel in terms of solutions of certain equations. We shall require this for the proof of Theorem 1.

In the sequel we will denote by $L_2(\tau(x)\,dx)$ $(\tau(x) \geq 1)$ a Hilbert space of functions $\phi(x)$ with the scalar product $(\phi, \psi) = \int_{-\infty}^{\infty} \phi(x)\,\overline{\psi(x)}\,\tau(x)dx$ ($\|\cdot\|$ is the norm), and by $L_2(\rho(x)dx)$ $(\rho(x) \geq 1)$ a Hilbert space of functions $\phi(x)$ with the scalar product $(\phi, \psi)_1 = \int_0^\infty \phi(x)\,\overline{\psi(x)}\rho(x)dx$ ($\|\cdot\|_1$ is the norm).

Let us now consider $n$ Hilbert spaces $L_2(\tau_1(x_1)dx_1), \cdots, L_2(\tau_m(x_m)dx_m)$, $L_2(\rho_{m+1}(x_{m+1})dx_{m+1}), \cdots, L_2(\rho_n(x_n)dx_n)$ and define in them, respectively, the operators $B_1, \cdots, B_n$ in the following manner: $B_j$ is the closure in $L_2(\cdot\, dx_j)$ of the operator $\phi(x) \to (\overline{L_j^+ \phi})(x)$ on finite infinitely differentiable functions $\phi(x)$ from $L_2(\cdot\, dx_j)$ (here $L_j^+$ is the expression formally adjoint to $L^{(j)}$, and the coefficients of $L^{(j)}$ are not necessarily real).

Let us denote by $H_+$ the tensor product of all the spaces introduced:

$$H_+ = L_2(\tau_1(x_1)\,dx_1) \otimes \ldots \otimes L_2(\varrho_n(x_n)\,dx_n)$$

and by $H_-$:

$$H_- = L_2\left(\frac{1}{\tau_1(x_1)}dx_1\right) \otimes \ldots \otimes L_2\left(\frac{1}{\varrho_n(x_n)}dx_n\right).$$

We will now make one assumption.

a) We will assume that the kernel $K(x, y)$ p. d. on $G_{mn}$ and the weight functions $\tau_i(x_i)$ and $\rho_k(x_k)$ are such that $K(x, y) \in H_- \otimes H_-$. For this it is sufficient, for example, that

$$\int_{G_{mn}} \int_{G_{mn}} \frac{|K(x, y)|^2}{\sigma(x)\,\sigma(y)}\,dxdy < \infty \;(\sigma(x) = \tau_1(x_1) \cdots \varrho_n(x_n)).$$

Let us introduce in the space $H_+$ the scalar product

$$\langle \varphi, \psi \rangle = \int_{G_{mn}} \int_{G_{mn}} K(x, y)\,\varphi(y)\,\overline{\psi(x)}dxdy \qquad (\varphi, \psi \in H_+)$$

and denote by $H_K$ the Hilbert space obtained after completion with respect to this scalar product. Since each $L_2(\cdot\, dx_j)$ has its own operator $B_j$, it follows that in $H_K$ there are defined $n$ operators $A_j = E \otimes \cdots \otimes E \otimes B_j \otimes E \otimes \cdots \otimes E$ ($B_j$ in the $j$th place) with the obvious domains of definition. It is obvious that the operators $A_j$ commute in $H_K$.

We now make two more assumptions.

b) We will assume that the operators $A_j$ are Hermitian in $H_k$ and each of them has in $H_k$ the same defect indices. The Hermitian property is here equivalent to

$$L^{j}_{x_j} K(x, y) = \overline{L^{(j)}_{y_j}} K(x, y),$$

where the equality is understood in the sense of generalized functions, and $L^{(j)}_{x_j}$ denotes the application of the expression $L^{(j)}$ with respect to the variable $x_j$; the bar denotes the change of coefficients of $L^{(j)}$ to their complex conjugates. The above condition is obtained from the equality $\langle A_j \phi, \psi\rangle = \langle\phi, A_j \psi\rangle$ $(j = 1, \cdots, n)$ for infinitely differentiable functions $\phi(x)$ and $\psi(x)$ finite on $G_{mn}$.

c) We will assume that the operators $A_j$ after closure in $H_k$ become selfadjoint and also commute (in the sense of partitions of unity).

According to [1], in the presence of conditions a), b) and c) there holds the unique representation (2). Thus for the proof of our theorem it is sufficient to verify that the conditions of the theorem imply the satisfaction of assumptions a) and c). For this purpose we let $\tau_i(x_i) = \exp[|x_i|^{r'_i} l_i(|x_i|)]$ and $\rho_k(x_k) = \exp[a_k x_k^{2-\epsilon/2}]$, where $l_i(x_i)$, $a_k$ and $\epsilon$ are the same as the ones in the formulation of the theorem. It is obvious that in this case condition a) is satisfied. Condition c) will also be satisfied. We will show this making use of the following theorem of Berezanskiĭ [1], whose formulation we will simplify in application to our conditions.

Theorem 2 [1]. *For each of the equations*

$$\frac{du_t}{dt} \pm iB_j^* u_t = 0, \qquad t \in [0, T],$$

*considered, respectively, in* $L_2(\cdot\, dx_j)$ $(j = 1, \cdots, n)$, *let the uniqueness of weak solutions of the Cauchy problem hold. Then the closures of the operators* $A_j$ *are selfadjoint in* $H_k$ *and their partitions of unity commute.*

By a weak solution of the Cauchy problem for the equation $du_t/dt - C^* u_t = 0$ $(t \in [0, T])$, where $C$ is an operator in some Hilbert space $H$ with a dense domain of definition $D(C)$, we mean a vector function $u_t$ with values in $H$, which assumes the given value $u_0$ at $t = 0$, is weakly differentiable on $[0, T]$, and satisfies the relation $d(u_t, \phi)/dt - (u_t, C\phi) = 0$ $(\phi \in D(C))$. Uniqueness of the weak solution means that $u_0 = 0$ implies $u_t = 0$ for $0 \le t \le T$.

Thus we see that Theorem 1 will have been proved if we prove the following statement, which for convenience is formulated as an independent theorem.

Theorem 3. *Let* $L$ *be an ordinary differential expression of order* $r > 1$ *with constant coefficients, and* $B$ *an operator in* $L_2(\tau(x)dx)$ $(\tau(x) =$

$\exp [|x|^{r'} l(|x|)]$; $1/r + 1/r' = 1$; $l(x) \in I(r)$), *constructed as the closure of the operator* $\phi \to L\phi$ *on finite infinitely differentiable* (*on* $(-\infty, \infty)$) *functions* $\phi = \phi(x)$. *Then the weak solution of the Cauchy problem for the equation*

$$\frac{du_t}{dt} - B^* u_t = 0, \qquad t \in [0, T]$$

*is unique in* $L_2(r(x)dx)$. *An analogous statement holds when* $L = P$ *is an expression of the second order and* $B = R$ *is its corresponding operator in* $L_2(\rho(x)dx)$ ($\rho(x) = \exp[ax^{2-\epsilon}]$; $a, \epsilon > 0$ *are constants*).

The next section is devoted to the proof of this theorem.

## §2. Establishment of uniqueness of weak solutions

1. In this subsection we will do some preparatory work, introducing entire analytic functions of a special kind. The sets of functions under consideration will be denoted by $W_{M,a}^{\Omega,b}$ and $M_\theta(P)$.

We set $M(x) = \Omega(x) = |x|^{r'} l(|x|)$, where $r' > 1$ and $l(x)$ is a slowly increasing function. It may be shown that the space $W_{M,a}^{\Omega,b}$ can be correctly defined in terms of such $M(x)$ and $\Omega(y)$. As is known [4], it consists of entire analytic functions $\phi(z) = \phi(x + iy)$ for which for arbitrary $\tilde{a} = a - \epsilon$, $\tilde{b} = b + \delta$, $\epsilon, \delta > 0$ one has

$$|\varphi(z)| \leqslant C_{\varphi,\varepsilon,\delta} e^{-M(\tilde{a}x) + \Omega(\tilde{b}y)}$$

with some constants $C_{\phi,\epsilon,\delta}$. We note that in the spaces $W_{M,a}^{\Omega,b}$ there is defined the operation of differentiation ($\phi(z) \to \phi'(z)$), which does not lead out of the space. In addition, we have the following lemma [7].

LEMMA 1. *For a sufficiently large* $b$ *the space* $W_{M,1}^{\Omega,b}$ *is sufficiently rich in functions on the entire axis. The consecutive derivatives of* $\phi(x) \in W_{M,1}^{\Omega,b}$ *satisfy the inequalities*

$$|\varphi^{(q)}(x)| \leqslant C_{\varphi,\varepsilon,\delta} K_\varepsilon^q q^{\frac{q}{r}} l(q)^{\frac{q}{r'}} e^{-l(|\tilde{a}x|)|\tilde{a}x|^{r'}},$$

*where* $\tilde{a} = 1 - \epsilon$, $\epsilon > 0$, $K_\epsilon$ *is independent of* $q$, $q = 0, 1, \cdots$; $1/r + 1/r' = 1$.

Here and below we will say that a set $\Phi$ is sufficiently rich in functions $\phi(x)$ on $G$, if for an arbitrary locally integrable function $u(x)$ the absolute convergence of the integral

$$\int_G u(x) \varphi(x)\, dx$$

and its vanishing for all $\phi(x) \in \Phi$ implies that $u(x) = 0$ almost everywhere on $G$.

LEMMA 2. *Let* $P(d/dx)$ *be a differential expression of the second order with constant coefficients, whose coefficient of the highest derivative is* *different from zero, and let $\theta$ be a fixed number subject to the condition $1/2 < \theta < 1$. There exists a set $M_\theta(P)$ of entire analytic functions such that*

$$1)\ \left[P^n\left(\frac{d}{dx}\right)\varphi(x)\right]_{x=0} = 0 \qquad (n = 0, 1, \ldots; \varphi(z) \in M_\theta(p)); \tag{3}$$

$$2)\ |\varphi^{(q)}(x)| \leqslant CN^q q^{q\left(1-\frac{1}{2\theta}\right)} e^{-a|x|^{2\theta}} \qquad (q = 0, 1, \ldots; -\infty < x < \infty) \tag{4}$$

*with some constants $C$, $N$, $a > 0$ which depend on $\phi(z)$; and*

*r) $M_\theta(P)$ is sufficiently rich in functions on the right semiaxis $(x \geq 0)$.*

The proof of this lemma will be conducted in several stages.

Denote by $V_\theta$ the set of entire analytic functions $f(z)$ for which

$$|f(z)| \leqslant Ce^{b|z|^\theta}, \tag{5}$$

$$|f(x)| \leqslant C'e^{-ax^\theta} \qquad \text{for } x \geqslant 0, \tag{6}$$

where $C$, $C'$, $a$ and $b$ are positive and depend on $f(z)$. An example of such a nontrivial function may be found in [3], §8, Chapter 4.

Lemma 3. *The space $V_\theta$ is sufficiently rich in functions on the right semiaxis $(x \geq 0)$.*

Proof. It is easy to show that if $f(z) \in V_\theta$, then for the function $f_1(z) = f(x + h + iy)$ $(h \gtrless 0)$ estimates (5) and (6) are satisfied with some constants $C_1$, $C_1'$, $a_1$, $b_1 > 0$, i.e. in the space $V_\theta$ the shift operator along the real axis is defined.

We next consider the entire analytic function $w = \cos\sqrt{\lambda z}$, where $\lambda$ varies in some interval $[\lambda'', \lambda']$. Since

$$|\cos\sqrt{\lambda z}| \leqslant C_0 e^{k|z|^{\frac{1}{2}}}$$

with some $C_0$, $k > 0$, in the space $V_\theta$ the operation of multiplication by $\cos\sqrt{\lambda z}$ is defined (since $1/2 < \theta$), i.e. $\cos\sqrt{\lambda z}\, f(z) \in V_\theta$ $(\lambda \in [\lambda'', \lambda'])$, if $f(z) \in V_\theta$.

We now assume that for some $\omega(x)$ we have

$$\int_0^\infty |\omega(x) f(x)|\, dx < \infty \qquad \text{and} \qquad \int_0^\infty \omega(x) f(x)\, dx = 0$$

for all $f(x) \in V_\theta$. It remains to show that $\omega(x) = 0$ almost everywhere for $x > 0$. For this purpose we consider the equality

$$\int_0^\infty \omega(x) f_0(x) \cos\sqrt{\lambda x}\, dx = 0,$$

where $f_0(x) \in V_\theta$ and is not identically equal to zero. Upon the substitution $x = t^2$ this yields

$$\int_0^{\infty} t\omega(t^2) f_0(t^2) \cos \sqrt{\lambda} t\, dt = 0,$$

from which, in view of the familiar uniqueness property of the Fourier transformations, we obtain

$$t\omega(t^2) f_0(t^2) = 0, \qquad \text{or} \qquad \sqrt{x}\omega(x) f_0(x) = 0$$

almost everywhere for $x > 0$. Taking into account the fact that in $V_\theta$ the operation of shift is defined, and shifting, if necessary, the function $f_0(x)$ from the very beginning by $h \gtrless 0$, we easily obtain that $\omega(x) = 0$ almost everywhere for $x > 0$.

We will also require the following lemma.

Lemma 4. *Let* $\alpha(x)$ *be an arbitrary infinitely differentiable function and* $P(d/dx) = (d/dx - \lambda_1)(d/dx - \lambda_2)$. *Then the function*

$$\varphi(x) = \alpha(x) e^{\lambda_1 x} - \alpha(-x) e^{\lambda_2 x}$$

*satisfies the condition*

$$\left[P^n\left(\frac{d}{dx}\right)\varphi(x)\right]_{x=0} = 0 \qquad (n = 0, 1, \ldots).$$

Proof. Differentiating $\phi(x)$, we obtain

$$P\left(\frac{d}{dx}\right)\varphi(x) = (\alpha''(x) + \lambda_1\alpha'(x) - \lambda_2\alpha'(x)) e^{\lambda_1 x}$$

$$+(-\alpha''(-x) + \lambda_2\alpha'(-x) - \lambda_1\alpha'(-x)) e^{\lambda_2 x}.$$

Denoting $\alpha''(x) + (\lambda_1 - \lambda_2)\alpha'(x) = \alpha_1(x)$, we thus obtain

$$P\left(\frac{d}{dx}\right)[\alpha(x) e^{\lambda_1 x} - \alpha(-x) e^{\lambda_2 x}] = \alpha_1(x) e^{\lambda_1 x} - \alpha_1(-x) e^{\lambda_2 x}.$$

Therefore

$$P^n\left(\frac{d}{dx}\right)[\alpha(x) e^{\lambda_1 x} - \alpha(-x) e^{\lambda_2 x}] = \alpha_n(x) e^{\lambda_1 x} - \alpha_n(-x) e^{\lambda_2 x},$$

which implies that

$$\left[P^n\left(\frac{d}{dx}\right)\varphi(x)\right]_{x=0} = \alpha_n(0) - \alpha_n(0) = 0,$$

which is what we set out to prove.

We will now go over directly to the proof of Lemma 2. For the set $M_\theta(P)$ we take the set of functions $\phi(z)$, defining them in the following manner:

$$\varphi(z) = \alpha(z) e^{\lambda_1 z} - \alpha(-z) e^{\lambda_2 z}, \qquad \alpha(z) = zf(z^2), \qquad f(z) \in V_\theta$$

(here $\lambda_1, \lambda_2$ are the roots of the polynomial $P(s)$).

Let us verify that this $M_\theta(P)$ satisfies all the conditions of the lemma. The fact that $M_\theta(P)$ satisfies condition (3) is proved in Lemma 4. Verification of the fact that $M_\theta(P)$ is sufficiently rich in functions on the right semiaxis is conducted very simply by utilizing Lemma 3. It thus remains to prove the estimates (4).

First of all, it follows from the definition of the space $V_\theta$ that for $f(z) \in V_\theta$ for the function $g(z) = f(z^2)$ we have

$$|g(z)| \leqslant Ce^{b|z|^{2\theta}}, \tag{7}$$

$$|g(x)| \leqslant C_1 e^{-a|x|^{2\theta}} \qquad (-\infty < x < \infty). \tag{8}$$

If we make use of Theorems 1 and 2 of [3], Chapter IV, §7, then from conditions (7) and (8) it is easy to obtain the following estimates for $g(z)$ with some $C'$, $a_1$, $b_1 > 0$:

$$|g(z)| \leqslant C' e^{-a_1|x|^{2\theta}+b_1|y|^{2\theta}}. \tag{9}$$

An entire function $g(z)$ which is subject to condition (9) is said [3] to belong to the space $S_\alpha^\beta$ with $\alpha = 1/2\theta$, $\beta = 1 - 1/2\theta$. The spaces $S_\alpha^\beta$ are a particular case of the spaces $W_M^\Omega$; they possess, in particular, the property that in them the operations of multiplication by certain entire functions are defined. In our case it is not difficult to show that $\phi(z) = f(z^2)z(e^{\lambda_1 z} + e^{\lambda_2 z})$ belongs to the space $S_\alpha^\beta$ together with $f(z^2)$, i.e.

$$|\varphi(z)| \leqslant C'' e^{-a_2|x|^{2\theta}+b_2|y|^{2\theta}}$$

with some $C''$, $a_2$, $b_2 > 0$. But now it is seen that estimates (4) are a well-known fact in the theory of the spaces $S_\alpha^\beta$ ([3], §7).

Thus Lemma 2 is completely proved.

2. In this subsection we will prove Theorem 3. We recall that the operators $B$ and $R$ which appear there were defined through ordinary differential expressions $L$ and $P$, respectively, in the spaces $L_2(\tau(x)dx)$ and $L_2(\rho(x)dx)$, where $\tau(x) = \exp[|x|^{r'} l(|x|)]$ and $\rho(x) = \exp[ax^{2-\epsilon}]$. We will now show that $W_{M,1}^{\Omega,b}$ with $M(x) = \Omega(x) = \tau(x)$ and $M_\theta(P)$ with $\theta = 1 - \epsilon/4$ are in the domains of definition of the operators $B$ and $R$, respectively, (the fact that $W_{M,1}^{\Omega,b} \subset L_2(\tau(x)dx)$ and $M_\theta(P) \subset L_2(\rho(x)dx)$ is obvious). Moreover, we will show that $B\phi = L\phi$ ($\phi \in W_{M,1}^{\Omega,b}$) and $R\phi = P\phi$ ($\phi \in M_\theta(P)$).

For this purpose we choose some infinitely differentiable function $\omega(x)$ which is equal to 1 on the interval $[-1, 1]$ and which vanishes outside the interval $[-2, 2]$. Further, for each $\phi(x)$ from $W_{M,1}^{\Omega,b}$ we construct a sequence of finite functions $\phi_\nu(x) = \omega(x/\nu)\phi(x)$. In this connection, for an arbitrary fixed $q \geq 0$ we have

$$\|\varphi^{(q)}(x) - \varphi_\nu^{(q)}(x)\|^2 = \int_{|x|>\nu} |\varphi^{(q)}(x) - \varphi_\nu^{(q)}(x)|^2 \tau(x)\,dx \leqslant$$

$$\leqslant \int_{|x|>\nu} \left\{ |\varphi^{(q)}(x)|^2 + \left| \left[ \omega\left(\frac{x}{\nu}\right) \varphi(x) \right]^{(q)} \right|^2 \right\} \tau(x)\, dx \to 0,$$

since $\phi^{(i)}(x) \in L_2(\tau(x)dx)$ and $|[\omega(x/\nu)]^{(i)}| \leq C_q$ $(0 \leq i \leq q)$. This immediately implies that for the sequence $\phi_\nu(x)$

$$\|\varphi(x) - \varphi_\nu(x)\| \to 0 \text{ and } \|L\varphi - L\varphi_\nu\| \to 0 \qquad (\nu \to \infty).$$

But this means that $W^{\Omega, b}_{M, 1} \subset D(B)$ and $B\phi = L\phi$ $(\phi \in W^{\Omega, b}_{M, 1})$.

We will now do the same thing for the case of the operator $R$ and the space $L_2(\rho(x)dx)$. The sequence of finite on $[0, \infty)$ infinitely differentiable functions $\tilde{\phi}_\nu(x)$ for the function $\phi(x) \in M_\theta(P)$ will be chosen in such a way that $\tilde{\phi}_\nu(x)$ is equal to zero on $|0, 1/2\nu|$, and on the semiaxis $x > \nu$ it coincides with $\phi_\nu(x)$ which was constructed before. It is obvious that $\|\phi(x) - \tilde{\phi}_\nu(x)\|_1 \longrightarrow 0$ as $\nu \longrightarrow \infty$. In addition,

$$\|P\varphi - P\tilde{\varphi}_\nu\|_1^2 = \int_0^{\frac{1}{\nu}} |P\varphi(x) - P\tilde{\varphi}_\nu(x)|^2 \varrho(x)\, dx + \int_\nu^\infty |P\varphi(x) - P\tilde{\varphi}_\nu(x)|^2 \varrho(x)\, dx$$

$$\leqslant \frac{1}{\nu} \max_{0 \leqslant x \leqslant \frac{1}{\nu}} (|P\varphi(x) - P\tilde{\varphi}_\nu(x)|^2 \varrho(x)) + \int_\nu^\infty |P\varphi(x) - P\tilde{\varphi}_\nu(x)|^2 \varrho(x)\, dx.$$

It is clear that the last integral tends to zero as $\nu \longrightarrow \infty$. Further, from the continuity of the functions $P\phi(x)$ and $P\tilde{\phi}_\nu(x)$ and from the conditions $[P\phi(x)]_{x=0} = [P\tilde{\phi}_\nu(x)]_{x=0} = 0$ it follows that

$$\frac{1}{\nu} \max_{0 \leqslant x \leqslant \frac{1}{\nu}} (|P\varphi(x) - P\tilde{\varphi}_\nu(x)|^2 \varrho(x)) \to 0 \qquad (\nu \to \infty),$$

which implies that $\|P\phi - P\tilde{\phi}_\nu\|_1 \longrightarrow 0$ as $\nu \longrightarrow \infty$. Q. E. D.

We note that from the properties of $W^{\Omega, b}_{M, 1}$, from Lemma 2, and from the statement just proved it follows that $W^{\Omega, b}_{M, 1}$ and $M_\theta(P)$ are invariant subspaces of the operators $B$ and $R$ respectively.

We will now assume that $u_t$ belongs to $L_2(\tau(x)dx)$ for each $t \in [0, T]$ and is a weak solution of the Cauchy problem

$$\frac{du_t}{dt} - B^* u_t = 0, \qquad u_0 = 0.$$

Since for $u_t$

$$\frac{d}{dt}(u_t, \varphi) = (u_t, B\varphi) \qquad (\varphi \in W^{\Omega, b}_{M, 1}),$$

and $W^{\Omega, b}_{M, 1}$ is an invariant subspace of the operator $B$, the function $F_\phi(t) = (u_t, \phi)$ $(\phi \in W^{\Omega, b}_{M, 1})$ is infinitely differentiable with respect to $t$, and

$$\frac{d^N}{dt^N} F_\varphi(t) = (u_t, B^N\varphi) = (u_t, L^N\varphi),$$

$$\left|\frac{d^N}{dt^N} F_\varphi(t)\right| = |(u_t, L^N\varphi)| \leqslant \|u_t\|\cdot\|L^N\varphi\| = \|u_t\|\cdot\sqrt{\int_{-\infty}^{\infty} |L^N\varphi|^2 \tau(x)\,dx}\,.$$

Since $L^n$ is an expression of order $Nr$, it follows from Lemma 1 that

$$|L^N\varphi(x)| \leqslant A_\varepsilon^N N^N l(N)^{N\frac{r}{r'}} e^{-|\tilde{a}x|^{r'} l(|\tilde{a}x|)},$$

where $A_\epsilon > 0$ is a quantity independent of $N$, $\tilde{a} = 1 - \epsilon$, $\epsilon \to 0$. This estimate implies

$$\left|\frac{d^N}{dt^N} F_\varphi(t)\right| \leqslant \|u_t\| A_\varepsilon^N N^N l(N)^{N\frac{r}{r'}} \sqrt{\int_{-\infty}^{\infty} e^{-2|\tilde{a}x|^{r'} l(|\tilde{a}x|)} \tau(x)dx}$$

$$\leqslant C_T A_\varepsilon^N N^N l(N)^{N\frac{r}{r'}} C_1 = m_N,$$

where $C_1^2$ is the value of the integral (it is clear that it is finite) and $C_T$ is a constant which bounds the aggregate of $\|u_t\|$, $t \in [0, T]$. Existence of such a $C_T$ follows from the weak continuity of $u_t$, which in turn follows from the assumption of weak differentiability of $u_t$.

It is easy to verify that the sequence $m_N$ satisfies the condition $\Sigma\, 1/\sqrt[N]{m_N} = \infty$. In addition, it is seen that all derivatives $F_\phi^{(N)}(t)$ vanish at $t = 0$. From the theory of quasianalytic functions it is known [6] that a function $F_\phi(t)$ with such properties is identically equal to zero on $[0, T]$; i.e. we have obtained that

$$F_\varphi(t) = \int_{-\infty}^{\infty} u_t(x)\overline{\varphi(x)}\tau(x)\,dx = 0, \quad t \in [0, T]$$

for functions $\phi(x) \in W_{M,1}^{\Omega,b}$. But the space $W_{M,1}^{\Omega,b}$ is sufficiently rich in functions on $(-\infty, \infty)$, and therefore $u_t(x)\tau(x) = 0$ almost everywhere for each $t \in [0, T]$, and consequently $u_t = 0$ in $L_2(\tau(x)dx)$ $(t \in [0, T])$.

It remains to show that the weak solution $u_t$ of the Cauchy problem

$$\frac{du_t}{dt} - R^*u_t = 0, \qquad u_0 = 0, \qquad (t \in [0, T])$$

is also equal to zero in $L_2(\rho(x)\,dx)$. However, we omit this argument, since it reduces to obvious modifications of the preceding proof.

Thus the proof of Theorem 3 is complete.

## §3. Nonuniqueness of the representation of one p. d. kernel

On the basis of a concrete p. d. kernel and a concrete expression $L$ we will show here that the estimates (1), given by Theorem 1 for uniqueness of its integral representation, cannot be improved.

Thus, let $k(t)$ $(-\infty < t < \infty)$ be a continuous, real, even function whose kernel $K(x, y) = \frac{1}{2}[k(x+y) + k(y-x)]$ is p. d., and let $L = -d^2/dx^2 = L^+$. The eigenfunctions of $L$ are $\cos\sqrt{\lambda}x$ and $\sin\sqrt{\lambda}x$.

If we assume that

$$|K(x, y)| \leqslant Ce^{x^2l(|x|)+y^2l(|y|)} \qquad (C > 0,\ l(s) \in I(2)),$$

then according to Theorem 1 we have the unique representation (2), or, since $k(t)$ is even,

$$K(x, y) = \int_{-\infty}^{\infty} \cos\sqrt{\lambda}x \cos\sqrt{\lambda}y\, d\sigma(\lambda),$$

where $d\sigma(\lambda)$ is a nonnegative measure finite on $(-\infty, \infty)$. From this it is easy to see that with the same $d\sigma(\lambda)$ we have

$$k(t) = \int_{-\infty}^{\infty} \cos\sqrt{\lambda}t\, d\sigma(\lambda), \qquad (10)$$

which, in turn, implies the preceding equality. In particular, in the last representation $d\sigma(\lambda)$ will be defined uniquely if $|k(t)| \leq C_1 \exp[x^2 l_1(|x|)]$ $(C_1 > 0,$ $l_1(x) \in I(2))$, since this condition easily implies the necessary condition for the kernel $K(x, y)$. Moreover, this last condition, which guarantees the uniqueness of representation (10), cannot be improved. To be more exact, we have the following theorem.

Theorem 4. *Consider the set of all functions $k(t)$ having the representation* (10) *with a nonnegative measure $d\sigma(\lambda)$ and such that $|k(t)| \leq$ $C\exp[t^2h(t)]$, where $h(t)$ is a slowly increasing function which does not belong to $I(2)$, i.e. $\int_1^\infty (1/th(t))dt = C_0 < \infty$. Then in this set there exists a function $k_0(t)$ such that the representation* (10) *holds for it with different $d\sigma_i(\lambda)$.*

We will prove this theorem starting from an analogous theorem by E. B. Vul [2].

Theorem 5 [2]. *Let us consider the class of functions $f(x)$ which are representable in the form*

$$f(x) = \int_{-\infty}^{\infty} \cos\sqrt{\lambda}\, x\, d\omega(\lambda), \qquad (11)$$

*where* $\omega(\lambda)$ *is a complex-valued function of bounded variation such that*

$$\int_{-\infty}^{0} \exp[\sqrt{|\lambda|}x]\,|d\omega(\lambda)| \leqslant C\exp[\varrho(x)], \quad x > 0. \tag{12}$$

*If* $\rho(x)$ *is differentiable, if* $\rho'(x)$ *is monotonically increasing, if*

$$\lim_{x\to\infty}\frac{x\varrho'(x)}{\varrho(x)} = \gamma > 1 \ \ \textit{and} \ \int_1^\infty t^{-2}\tau(t)\,dt < \infty,$$

*where* $\tau(x)$ *is the inverse of* $\rho'(x)$, *then there exists a function* $f_0(t)$ *whose representation in the form* (11) *is not unique in the class of functions* $\omega(\lambda)$ *satisfying* (12).

Proof of Theorem 4. First of all, on the basis of Theorem 5, we construct for an arbitrary function $\rho(x)$, satisfying the conditions of Theorem 5, a function $k_0(t)$ such that $|k_0(t)| \le C_0 \exp[\rho(x)]$ and it possesses a nonunique representation (10). After this it remains to show that for $\int_1^\infty s^{-1}h^{-1}(s)ds = C_0 < \infty$ the function $\rho(x) = x^2h(x)$ satisfies the conditions of Theorem 5.

Let $f_0(t)$ be one of the functions of Theorem 5, i. e.

$$f_0(t) = \int_{-\infty}^{\infty}\cos\sqrt{\lambda}t\,d\omega'(\lambda) = \int_{-\infty}^{\infty}\cos\sqrt{\lambda}t\,d\omega''(\lambda) \qquad (d\omega'(\lambda) \neq d\omega''(\lambda)).$$

Then at least one of the functions

$$F_1(t) = f_0(t) + \overline{f_0(t)}, \qquad F_2(t) = -i(f_0(t) - \overline{f_0(t)})$$

(for definiteness let it be $F_1(t)$) possesses two representations:

$$F_1(t) = \int_{-\infty}^{\infty}\cos\sqrt{\lambda}t\,d\sigma_1(\lambda) = \int_{-\infty}^{\infty}\cos\sqrt{\lambda}t\,d\sigma_2(\lambda)$$

with different real functions $\sigma_1(\lambda) = 2\omega_1'(\lambda)$, $\sigma_2(\lambda) = 2\omega_1''(\lambda)$ of bounded variation (we denoted here $\omega'(\lambda) = \omega_1'(\lambda) + i\omega_2'(\lambda)$, $\omega''(\lambda) = \omega_1''(\lambda) + i\omega_2''(\lambda)$). Further, we set $\sigma_i(\lambda) = \sigma_i^+(\lambda) - \sigma_i^-(\lambda)$, where $\sigma_i^+(\lambda)$ and $\sigma_i^-(\lambda)$ are nondecreasing bounded functions, and for the function $k_0(t)$ we take

$$k_0(t) = F_1(t) + \int_{-\infty}^{\infty}\cos\sqrt{\lambda}t\,d[\sigma_1^-(\lambda) + \sigma_2^-(\lambda)].$$

Indeed, $k_0(t)$ possesses two different representations (10) with nonnegative measures $d[\sigma_1^+(\lambda) + \sigma_2^-(\lambda)]$ and $d[\sigma_2^+(\lambda) + \sigma_1^-(\lambda)]$. In addition,

$$|k_0(t)| \leqslant \int_{-\infty}^{\infty}|\cos\sqrt{\lambda}t|\,|d(\sigma_1^+(\lambda) + \sigma_2^-(\lambda))| \leqslant \int_{-\infty}^{\infty}|\cos\sqrt{\lambda}t|\,|d\omega'(\lambda)|$$

$$+ \int_{-\infty}^{\infty}|\cos\sqrt{\lambda}t|\,|d\omega''(\lambda)| \leqslant \int_{-\infty}^{0}\exp[\sqrt{|\lambda|}|t|]\,|d\omega'(\lambda)| + \int_{-\infty}^{0}\exp[\sqrt{|\lambda|}|t|]\,|d\omega''(\lambda)|$$

$$+ \operatorname{var}\omega'(\lambda) + \operatorname{var}\omega''(\lambda) \leqslant C\exp[\varrho(x)].$$

Thus the necessary function $k_0(t)$ has been constructed from $\rho(x)$.

We will now show that the function $\rho(x) = x^2h(x)$ satisfies the conditions of Theorem 5. It is clear that $\rho(x)$ is differentiable and that $\rho'(x)$ increases (since $x^2h(x)$ is convex). In addition,

$$\lim_{x\to\infty} \frac{x\varrho'(x)}{\varrho(x)} = \lim_{x\to\infty} \frac{2x^2h(x) + x^2h'(x)}{x^2h(x)} \geqslant 2.$$

It thus remains to verify that $\rho(x) = x^2h(x)$ satisfies the condition $\int_1^\infty (\tau(x)/x^2)dx < \infty$, where $\tau(x)$ is the inverse of $[x^2h(x)]'$. For this purpose we note that $\rho(x)$ and $\int_0^x \tau(y)dy = \nu(x)$ are dual functions in the sense of Jung, and consequently for them we have the inequalities [5] a) $\nu(s) \leq s\nu'(s) \leq \nu(2s)$ $(s > 0)$, and b) $s < \nu^{-1}(s)\rho^{-1}(s) \leq 2s$ $(s > 0)$, where $\nu^{-1}(s)$ and $\rho^{-1}(s)$ are the inverses of $\nu(s)$ and $\rho(s)$ respectively.

Let us consider the chain of integrals

$$\int_{a_1}^\infty \frac{\nu'(s)}{s^2}\,ds = \int_{a_1}^\infty \frac{\tau(s)}{s^2}\,ds, \quad \int_{a_2}^\infty \frac{dz}{[\nu^{-1}(z)]^2}, \quad \int_{a_3}^\infty \frac{[\varrho^{-1}(z)]^2}{z^2}\,dz,$$

$$\int_{a_4}^\infty \frac{t^2}{\varrho(t)^2}\varrho'(t)\,dt, \quad \int_{a_5}^\infty \frac{sds}{\varrho(s)} = \int_{a_6}^\infty \frac{ds}{sh(s)}.$$

It is easy to see from inequalities a) and b) that each of the above integrals is equivalent to the preceding one in the sense of convergence. But since $\int_1^\infty (1/sh(s))\,ds < \infty$ we also have $\int_1^\infty (\tau(s)/s^2)\,ds < \infty$. This completes the proof of Theorem 4.

In conclusion the author thanks Ju. M. Berezanskiĭ for his help in this work.

## BIBLIOGRAPHY


[1] Ju. M. Berezanskiĭ, *A generalization of a multidimensional theorem of Bochner*, Dokl. Akad. Nauk SSSR 136 (1961), 1011–1014 = Soviet Math. Dokl. 2 (1961), 143–147. MR 28 #489.

[2] E. B. Vul, *Uniqueness theorems for a certain class of functions represented by integrals*, Dokl. Akad. Nauk SSSR 129 (1959), 722–725. (Russian) MR 22 #8093.

[3] I. M. Gel'fand and G. E. Šilov, *Generalized functions*. Vol. 2: *Functions and generalized function spaces*, GITTL, Moscow, 1958; English transl., Academic Press and Gordon and Breach, New York, (to appear). MR 21 #5142a.

[4] ———, *Generalized functions*. Vol. 3: *Some questions in the theory of differential equations*, Fizmatgiz, Moscow, 1958; English transl., Academic Press, New York, 1967. MR 21 #5142b; MR 36 #506.

[5] M. A. Krasnosel'skiĭ and Ja. B. Rutickiĭ, *Convex functions and Orlicz spaces*, GITTL, Moscow, 1958 and Fizmatgiz, Moscow, 1959; English transl., Noordhoff, Groningen, 1961. MR 21 #5144; MR 23 #A4016.

[6] S. Mandelbrojt, *Quasianalytic classes of functions*, ONTI, Moscow, 1937. (Russian)

[7] N. N. Čaus, *Uniqueness of solution of the Cauchy problem for a differential equation with constant coefficients*, Ukrain. Mat. Ž. 16 (1964), no. 3. (Russian)

Translated by:
V. I. Filippenko

UDC 517.22+519.5

# ON TOTAL DIFFERENTIALS

G. H. SINDALOVSKIĬ

## §1. Introduction

Let $f(x)$ be a function of several variables $x = (x_1, \cdots, x_n)$ defined in a region $G$ of $n$-dimensional Euclidean space $R^{(n)}$, and let $Q$ be some totally disconnected set in $R^{(n)}$ having the origin of the coordinates as a limit point.

We propose the following question. What necessary and sufficient conditions must $Q$ satisfy so that the fulfillment of the condition

$$\overline{\lim_{h\to 0,\ h\in Q}} \frac{|f(x+h)-f(x)|}{|h|} < +\infty, \text{ }^{1)} \tag{1}$$

at each point $x$ of some set $E$ necessarily implies the existence of the total differential $df$ at almost every point of the set $E$? Here it is assumed that the set $Q$ does not depend on the point $x$, i.e. the differential property (1) is considered relative to congruent sets.

This question can be posed in connection with the class of arbitrary functions $f(x)$, the class of measurable functions, or the class of continuous functions. It will be shown below that the solution of the problem is exactly the same in the classes of arbitrary functions and of measurable functions.

For the case of functions of one variable in the classes of measurable and of continuous functions this problem (in a somewhat weaker formulation than in the present paper) is already solved (cf. [1], [2]). In the class of measurable functions the solution is given by the collection of linear sets $Q$ satisfying the condition $\underline{\lim}_{\delta\to 0+} \operatorname{mes} Q(-\delta, \delta)/2\delta > 0$; in [1] and [2] this collection of sets is called class $(B)$. In the class of continuous functions the analogous role is played by the sets $Q$ with the property $\underline{\lim}_{\delta\to 0+} \operatorname{mes} \overline{Q}(-\delta, \delta)/2\delta > 0$ (class $(C)$); here $\overline{Q}$ is the closure of $Q$.

Some deepening of these results for functions of one variable in the question of comparing the ordinary derivatives and the derivatives relative to congruent sets is given in [3]. The problem of the connection between asymptotic and congruent differentiability for functions of one variable is solved in [4] and [5].

The question naturally arises of whether it is possible to generalize the re-

1) Here $|h|$ is the distance from the point $h \in Q$ to the origin of the coordinates; $x+h$ is the point appearing at the end of the radius vector obtained by the addition of the radius vectors of the points $x$ and $h$.

sults of [1] and [2] to functions of several variables.

For any natural number $n$ ($n$ is the dimension of the space), we shall define classes of sets $Q$ generalizing the concept of classes $(B)$ and $(C)$.

**Definition 1.** We shall put a set $Q \subset R^{(n)}$ in class $(B^{(n)})$ if $\underline{\lim}_{\delta\to 0+} \text{mes}\, QS_\delta/\text{mes}\, S_\delta > 0$; here $S_\delta$ is the $n$-dimensional ball of radius $\delta$ with center at the origin.

**Definition 2.** We shall put a set $Q \subset R^{(n)}$ in class $(C^{(n)})$ if its closure $\overline{Q}$ belongs to class $(B^{(n)})$, i.e. if

$$\underline{\lim}_{\delta\to 0+} \frac{\text{mes}\overline{Q}S_\delta}{\text{mes}S_\delta} > 0.$$

It is clear that $(B)$ and $(B^{(1)})$ (and also $(C)$ and $(C^{(1)})$ are identical.

In these definitions, instead of balls one may take cubes with center at the point $O$, with edge $\delta$ and with faces parallel to the coordinate planes or having arbitrary direction. Here the inequalities holding for the limits indicated below are not violated.

In §3 below the following theorems will be proved.

**Theorem 1.** *Let $f(x)$, $x = (x_1, \cdots, x_n)$ be a function defined in the region $G \subset R^{(n)}$, and let the set $Q$ belong to $(B^{(n)})$. If the inequality $\overline{\lim}_{h\to 0, h\in Q} |F(x+h) - f(x)|/|h| < +\infty$ is valid at each point $x$ of some set $E \subset G$, then at almost every point of $E$ the function $f(x)$ has a total differential.*

*If the set $Q$ does not belong to $(B^{(n)})$, then there exists a (measurable) function $\Phi(x)$ on a cube $K \subset R^{(n)}$ which satisfies the condition $\lim_{h\to 0, h\in Q} |\Phi(x+h) - \Phi(x)|/|h| = 0$ almost everywhere in $K$ and which, nevertheless, does not have a total differential at any point of this cube.*[1)]

Thus the solution of the problem turns out to be the same for both arbitrary functions and measurable functions (for any $n \geq 1$); namely, the desired class of sets $Q$ is the class $(B^{(n)})$. Therefore in [1] and [2], where the case $n = 1$ is considered, the requirement of measurability of the function $f(x)$ in the first part of the theorem may be removed.

**Theorem 2.** *Let $f(x)$, $x = (x_1, x_2, \cdots, x_n)$, be a continuous function in the region $G \subset R^{(n)}$, and let the set $Q$ belong to the class $(C^{(n)})$. If the inequality $\overline{\lim}_{h\to 0, h\in Q} |f(x+h) - f(x)|/|h| < +\infty$ is valid at each point $x$ of some set $E \subset G$, then at almost every point of $E$ the function $f(x)$ has a total differential.*

*If the set $Q$ does not belong to $(C^{(n)})$ and $G$ is an arbitrary region of $R^{(n)}$, then there exists a function $\Phi(x)$ continuous in $\overline{G}$ which satisfies the condition*

1) It is obvious that if this can be done in any cube $K$, then an analogous function $\Phi(x)$ can be constructed in all of the space $R^{(n)}$ as well.

$\lim_{h\to 0, h\in Q} |\Phi(x+h) - \Phi(x)|/|h| = 0$, *but which does not have a total differential almost everywhere in the region* $G$. *Here* $\Phi(x)$ *may be assumed to take any continuous values specified in advance on the boundary of* $G$.

In the simplest case, when the set $Q$ is a whole neighborhood of the origin of the coordinates, the first parts of Theorem 1 and 2 become the well-known theorems of Stepanov [6] (for measurable functions) and Rademacher [7]. If the set $Q$ is a solid ($n$-dimensional) angle with vertex at the origin, then the assertion of the first part of Theorem 1 actually coincides with a result of Haslam-Jones (see [8]). It is easy to see that in this case the noncongruency of the sets $Q$ for different points is not a generalizing assumption.

In the last section of the present paper one of the author's theorems about $Q$-continuity of functions of several variables will be generalized as an auxiliary result; this result is used in the proof of the fundamental results of the paper.

It is proved that the condition that $\text{mes}\, Q > 0$ in any neighborhood of the origin of the coordinates is necessary and sufficient for ordinary continuity almost everywhere to follow from $Q$-continuity. Earlier [9] this result was obtained under the assumption that $f(x)$ is measurable. By somewhat modifying the proof we free ourselves from this limitation.

## §2

Two auxiliary propositions are proved below.

In Lemma 1 properties of a set $Q$ from class $(B^{(n)})$ relating to its intersection with the set $Q^h$ are established; $Q^h$ designates the set obtained from $Q$ by translation by the vector $\overrightarrow{Oh}$; here $h$ is a point of $R^{(n)}$.

Lemma 1. *Let* $Q$ *be a set in the space* $R^{(n)}$ *belonging to class* $(B^{(n)})$, *and let* $\overrightarrow{O\theta}$ *be a ray from the origin of the coordinates. Then on* $\overrightarrow{O\theta}$ *there exists a linear set* $H$ *satisfying the following conditions:*

1) $\underline{\lim}_{l\to 0+} \text{mes}\, H(0, l)/l > 0$, *i.e.*, $H \in (B^{(1)})$ *on the axis* $\overrightarrow{O\theta}$;

2) *for each* $h \in H$ *there exists a cube* $K_r$ *with center at the point* $O$, *edge* $r$, *and boundaries parallel to the coordinate planes, such that* $\text{mes}\, QQ^hK_r > C_2 \text{mes}\, K_r$ *where* $h/r > C_1$. *The positive constants* $C_1$ *and* $C_2$ *depend on* $Q$ *and* $n$.

Proof. Since $Q$ belongs to $(B^{(n)})$, there exist numbers $\alpha > 0$ and $\delta_0 > 0$ such that

$$\text{mes}\, QK_\delta > \alpha\, \text{mes}\, K_\delta \quad (\delta \leqslant \delta_0) \tag{2}$$

($K_\delta$ is a cube with center at $O$ and side $\delta$).

We shall consider the semiaxis $\overrightarrow{O\theta}$ to be the real semiaxis. Let $l$ be an arbitrary point (positive number) of this semiaxis satisfying the condition $2l < \delta_0$.

We fix the integer $m = [2^n/\alpha] + 1$ ($n$ is the dimension of the space). We define the interval $\Delta = (l/m - l/m^2, l/m)$; $\operatorname{mes}\Delta = l/m^2$. If $x \in \Delta$, then the points $x, 2x, \cdots, mx$ lie in the interval $(l/m - l/m^2, l)$.

For any fixed $x \in \Delta$ we examine the $m$ sets in $R^{(n)}$: $QQ^x, \cdots, QQ^{mx}$. Here two cases are possible.

1) At least one of the sets $QQ^{kx}$ ($k = 1, \cdots, m$) has the following property: $\operatorname{mes} QQ^{kx}K_{4l} > \overline{C}_2 \operatorname{mes} K_{4l}$, where $\overline{C}_2$ is a fixed constant which, as will be shown below, depends only on $n$ and $\alpha$; it will be determined below.

2) For all values of $k$ $(1 \le k \le m)$ the inequality $\operatorname{mes} QQ^{kx}K_{4l} \le \overline{C}_2 \operatorname{mes} K_{4l}$ is fulfilled.

In case 2 we shall examine the following two variants:

2a)
$$\operatorname{mes} Q^{k_1x}Q^{k_2x}K_{4l} \le \overline{C}_2 \operatorname{mes} K_{4l} \tag{3}$$

for any pair of integers $k_1, k_2$ $(1 \le k_1 < k_2 \le m)$;

2b) $\operatorname{mes} Q^{k_1x}Q^{k_2x}K_{4l} > \overline{C}_2 \operatorname{mes} K_{4l}$ for some pair of numbers $k_1, k_2$ $(1 \le k_1 < k_2 \le m)$.

Variant 2a). It is easy to see that for any $kx < l$ the inequality

$$\operatorname{mes} Q^{kx}K_{4l} \geqslant \operatorname{mes} QK_{2l} \tag{4}$$

is valid (since $\operatorname{mes} QK_{2l} = \operatorname{mes} Q^{kx}K_{2l}^{kx} \le \operatorname{mes} Q^{kx}K_{4l}$; we take into consideration that $K_{2l}^{kx} \subset K_{4l}$).

From inequalities (2)–(4) and the equation $2^n \operatorname{mes} K_{2l} = \operatorname{mes} K_{4l}$ we get

$$\operatorname{mes}\left(\sum_{k=1}^{m} Q^{kx}K_{4l}\right) \geqslant \sum_{k=1}^{m} \operatorname{mes} Q^{kx}K_{4l} - \sum_{\substack{k\neq l\\ k<l}} \operatorname{mes} Q^{kx}Q^{lx}K_{4l}$$

$$\geqslant \sum_{k=1}^{m} \operatorname{mes} QK_{2l} - C_m^2\overline{C}_2 \operatorname{mes} K_{4l} = m \cdot \operatorname{mes} QK_{2l} - C_m^2\overline{C}_2 \operatorname{mes} K_{4l}$$

$$> m\alpha \operatorname{mes} K_{2l} - C_m^2\overline{C}_2 \operatorname{mes} K_{4l} = \operatorname{mes} K_{4l}\left(\frac{m\alpha}{2^n} - C_m^2\overline{C}_2\right)$$

$$> \operatorname{mes} K_{4l}\left\{\left(\left[\frac{2^n}{\alpha}\right]+1\right)\frac{\alpha}{2^n} - C_m^2\overline{C}_2\right\} > \operatorname{mes} K_{4l}.$$

The last inequality will be valid if $\overline{C}_2$ is chosen sufficiently small so that $\{([2^n/\alpha] + 1)\alpha/2^n - C_m^2\overline{C}_2\} > 1$; $\overline{C}_2$ depends only on $\alpha$ and $n$, which are fixed.

Since $\sum_{k=1}^m Q^{kx}K_{4l} \subset K_{4l}$, the inequality $\operatorname{mes}(\sum_{k=1}^m Q^{kx}K_{4l}) > \operatorname{mes} K_{4l}$ makes no sense. This shows that variant 2a) is impossible.

Variant 2b). We suppose that $k_2 > k_1$. The relative location of the sets $Q^{k_1x}$ and $Q^{k_2x}$ is the same as the relative location of the sets $Q$ and $Q^{(k_2-k_1)x}$

($k_2 - k_1$ is an integer). Since $\operatorname{mes} Q^{k_1 x} Q^{k_2 x} K_{4l} > \overline{C}_2 \operatorname{mes} K_{4l}$, we have $\operatorname{mes} QQ^{(k_1-k_2)x} K_{4l}^{-k_1 x} > \overline{C}_2 \operatorname{mes} K_{4l}$. Taking into account that $K_{4l}^{-k_1 x} \subset K_{6l}$ $(-l < -k_1 x < 0)$, we come to the conclusion that

$$\operatorname{mes} QQ^{(k_2-k_1)x} K_{6l} > \overline{C}_2 \operatorname{mes} K_{4l} = \frac{\overline{C}_2}{(1{,}5)^n} \operatorname{mes} K_{6l} = C_2 \operatorname{mes} K_{6l};$$

the constant $C_2 = \overline{C}_2/(1.5)^n$ depends only on $\alpha$ and $n$ (i.e. depends on $Q$ and $n$).

We recall that in case 1 the inequality $\operatorname{mes} QQ^{kx} K_{4l} > \overline{C}_2 \operatorname{mes} K_{4l}$ holds for some value of $k$ $(1 \leq k \leq m)$, whence it follows that

$$\operatorname{mes} QQ^{kx} K_{6l} > \overline{C}_2 \operatorname{mes} K_{4l} = C_2 \operatorname{mes} K_{6l}.$$

As was shown above, the same inequality holds in variant 2b) as well (with $k = k_2 - k_1$).

Thus for any $x \in \Delta$ there will be at least one integral value of $k = k(x)$ $(1 \leq k \leq m)$ for which

$$\operatorname{mes} QQ^{kx} K_{6l} > C_2 \operatorname{mes} K_{6l}. \tag{5}$$

We define the sets $E_k$ $(k = 1, \cdots, m)$ as follows:

$$E_1 = \{x \in \Delta, \operatorname{mes} QQ^{1x} K_{6l} > C_2 \operatorname{mes} K_{6l}\} \quad (k = 1),$$

$$E_k = \left\{x \in \Delta, \ x \notin \sum_{i=1}^{k-1} E_i, \ \operatorname{mes} QQ^{kx} K_{6l} > C_2 \operatorname{mes} K_{6l}\right\} \quad (k > 1)$$

$$E_k \cdot E_l = 0 \quad \text{for} \quad k \neq l.$$

The following result was obtained above (see (5)): every $x \in \Delta$ belongs to some set $E_k$, i.e. $\Delta = \sum_{k=1}^m E_k$.

The set $E_1$ is open, since the condition $\operatorname{mes} QQ^x K_{6l} > C_2 \operatorname{mes} K_{6l}$ implies that $\operatorname{mes} QQ^{x'} K_{6l} > C_2 \operatorname{mes} K_{6l}$ for all $x'$ sufficiently close to $x$. This is a consequence of the fact that the positive measure of the common part of two measurable sets depends continuously on translation of one set relative to the other.

The set $E_2$ is the difference of two open sets: the set $E_k$ (for any $k$) is the difference of an open set and a measurable set, i.e. is measurable. From the equation $\Delta = \sum_{k=1}^m E_k$ it follows that $\operatorname{mes} E_{k_0} \geq (1/m) \operatorname{mes} \Delta$ for at least one value $k = k_0$.

If the point $x$ runs through the set $E_{k_0}$, then the point $k_0 x$ runs through some set $R_{k_0}$,

$$\operatorname{mes} R_{k_0} = k_0 \operatorname{mes} E_{k_0} \geqslant \frac{k_0}{m} \operatorname{mes} \Delta \geqslant \frac{1}{m} \operatorname{mes} \Delta = \frac{1}{m} \cdot \frac{l}{m^2} = \frac{l}{m^3}.$$

Taking into account that $1 \leq k_0 \leq m$, we obtain $R_{k_0} \subset (l/m - l/m^2, l)$.

By the definition of the set $E_{k_0}$, for every point $k_0 x \in R$ we have $\operatorname{mes} QQ^{k_0 x} K_{6l} > C_2 \operatorname{mes} K_{6l}$. In addition, for every point $k_0 x \in R_{k_0}$ we get

$$\frac{k_0 x}{6l} > \frac{1}{6l}\left(\frac{l}{m} - \frac{l}{m^2}\right) = \frac{1}{6}\,\frac{m-1}{m^2} = \frac{1}{6}\left(\left[\frac{2^n}{\alpha}\right]\right)\left(\left[\frac{2^n}{\alpha}\right]+1\right)^{-2} = C_1 > 0. \qquad (6)$$

Here $C_1$ is the constant which (as we shall see below) is required by the lemma ($C_1$ does not depend on $l$, but only on $Q$ and $n$).

We designate by $H_1$ the set of all points $h$ on the semiaxis $\overrightarrow{O\theta}$ for which there exists a cube $K_r$ such that $\operatorname{mes} QQ^h K_r > C_2 \operatorname{mes} K_r$ and $h/r > C_1$ ($C_1$ and $C_2$ were determined above). Inequality (6) says that $R_{k_0} \subset H_1$.

Thus on each interval $(0, l)$ ($l$ sufficiently small) there exists a measurable set of type $R_{k_0}$ (depending on $l$) which lies in $H_1$ and for which the inequality $\operatorname{mes} R_{k_0}/\operatorname{mes}(0, l) \geq lm^{-2}/l = 1/m^3 = ([2^n/\alpha] + 1)^{-3} > 0$ is valid (the constant on the right does not depend on $l$). Hence it is possible to select a measurable subset $H$ from $H_1$ having the property

$$\overline{\lim_{l \to 0+}}\, \frac{1}{l} \operatorname{mes} H(0, l) > 0.$$

Lemma 1 is proved.

Lemma 2. *Let the function $f(x)$, $x = (x_1, \cdots, x_n)$, be defined in the region $G \subset R^{(n)}$, and let the set $Q$ belong to class $(B^{(n)})$. If $f(x)$ satisfies the condition $\overline{\lim}_{h \to 0, h \in Q} |f(x+h) - f(x)|/|h| < +\infty$ at each point $x$ of the set $E \subset G$, then it has an asymptotic differential at almost every point of $E$.*

Proof. We naturally assume that $\overline{\operatorname{mes}}\, E > 0$. We may suppose that all points of the set $Q$ are density points; for this it suffices to pass from the set $Q$ to a subset of the same measure. The conditions $Q \in (B^{(n)})$ and (1) are not violated by this. Since $Q \in (B^{(n)})$ we have $\operatorname{mes} Q > 0$ in any neighborhood of the origin. Since condition (1) implies that $\lim_{h \to 0, h \in Q} f(x+h) = f(x)$, according to the theorem [1]) of §4 we get that the function $f(x)$ is continuous in the usual sense at almost every point of $E$, i.e. at each point of some set $E_1' \subset E$; $\overline{\operatorname{mes}}\, E_1' = \overline{\operatorname{mes}}\, E > 0$. For the proof of the lemma it suffices to show that there exists an asymptotic differential at almost every point of $E_1'$.

We designate by $D'$ the set of all points in the region $G$ at which $f(x)$ is

1) The proof of this theorem does not depend on the rest of the present paper.

continuous in the usual sense. The set $D'$ is of type $(G_\delta)$, i.e. measurable. It is clear that $D' \supset E_1'$. Since $\overline{\text{mes}}\, E_1' > 0$ we have $\text{mes}\, D' > 0$. We examine the set $D$ consisting only of density points of the set $D'$; $\text{mes}\, D = \text{mes}\, D'$. All points of $D$ are density points. It is sufficient to prove asymptotic differentiability of $f(x)$ at almost every point of the set $E_1 = E_1' D$.

For every natural number $m$ we define the set

$$F_m = \left\{x \in D,\ \frac{|f(x+h)-f(x)|}{|h|} \leqslant m,\ h \in QK_{\frac{1}{m}},\ x+h \in D\right\}. \tag{7}$$

Hence $K_{1/m}$ is the open cube with center at the origin of the coordinates, edge $1/m$ and faces parallel to the coordinate planes.

For every point $x \in E_1$ inequality (1) is valid. Hence there exists a natural number $m = m(x)$ for which $x \in F_m$. Thus $E_1 \subset \sum_{m=1}^{\infty} F_m$. Therefore the lemma will be proved if the asymptotic differentiability of the function $f(x)$ is proved at almost every point of $F_m$ (for each value of $m$).

Let the set $F_m$ be fixed. We shall show that it is measurable. To do this it is sufficient to show that $F_m F$ is closed for any closed subset $F$ of $D\,(F_m \subset D)$. We suppose that $x^{(k)} \in F_m F$ is a sequence of points, $x^{(k)} \to \bar{x}$, $\bar{x} \in F$. We shall show that $\bar{x} \in F_m$. Let $h \in K_{1/m}$ and $\bar{x} + h \in D$; it must be proved that $|f(\bar{x}+h) - f(\bar{x})|/|h| \leq m$.

We fix a neighborhood $R$ of the point $h$ ($h$ is a density point of the set $Q$ and belongs to the open cube $K_{1/m}$) small enough so that $QR \subset QK_{1/m}$. We designate by $L_k$ the set of points of the form $x^{(k)} + \eta$, where $\eta \in QR$. All the $L_k$ are congruent to each other. It is clear that the point $x^{(k)} + h \in L_k$ and is one of its density points. If $x^{(k)} \to \bar{x}$ (as $k \to \infty$), then $x^{(k)} + h \to \bar{x} + h \in D$; $\bar{x} + h$ is a density point of the set $D$. The sets $L_k$ and $D$ intersect in a neighborhood of the point $\bar{x} + h$ for all sufficiently large $k$; moreover, taking $k$ sufficiently large, one can guarantee the existence of common points of the sets $L_k$ and $D$ arbitrarily close to the point $\bar{x} \in h$. [1] For each sufficiently large value $k$ we choose the point $x^{(k)} + h_k \in L_k D$ in such a way that $x^{(k)} + h_k \to \bar{x} + h$ $(k \to \infty)$, $h_k \in QK_{1/m}$.

Since $x^{(k)} \in F_m$, $x^{(k)} + h_k \in D$ and $h_k \in QK_{1/m}$, we have $|f(x^{(k)} + h_k) - f(x^{(k)})|/|h_k| \leq m$. Passing to the limit gives $|f(\bar{x} + h) - f(\bar{x})|/|h| \leq m$, since $f(x)$ is continuous in the usual sense at the points $\bar{x} + h \in D$ and $\bar{x} \in D$. The measurability of $F_m$ is proved.

We assume that $\text{mes}\, F_m > 0$ and prove the asymptotic differentiability of the function $f(x)$ almost everywhere on $F_m$. For this it suffices to prove this fact for any closed subset of the set $F_m$. Therefore we shall henceforth consider $F_m$

1) This follows from the fact that the density points of $L_k$ and $D$ come arbitrarily close together, while the sets $L_k$ are congruent to each other.

to be a closed set.

Almost all points of $F_m$ are points of linear density with respect to the variable $x_i$ and points of ordinary density (see [8], Russian pp. 196 and 431). Let $x_0$ be any of these points. For simplicity we shall take $x_0$ as the origin $(x_0 = O)$.

Applying Lemma 1 to the coordinate semiaxis $Ox_i$ (in place of the semiaxis $\overrightarrow{O\theta}$), we get on $Ox_i$ a corresponding set $H_i$ (in place of $H$) and constants $C_1^{(i)}$ and $C_2^{(i)}$ (in place of $C_1$ and $C_2$).

Since $H_i$ has positive lower linear density at the origin of the coordinates, the set $H_i F_m$ has the same property.

Let $h \in H_i F_m$ and satisfy the inequality $|h| < 1/2m$; then by Lemma 1 $\operatorname{mes} QQ^h K_r > C_2^{(i)} \operatorname{mes} K_r$ for some cube $K_r$, $|h|/r > C_1^{(i)}$. If $|h|$ is taken sufficiently small $r$ will be sufficiently small $(r < 1/m\sqrt{n})$ and the set $QQ^h K_r F_m$ will be nonempty, since the origin of the coordinates is a density point of $F_m$. Let $\tilde{x} \in QQ^h K_r F_m$; by the definition (7) we have

$$|f(\tilde{x}) - f(O)| \leqslant m\,|O\tilde{x}|, \quad \text{since} \quad O \in F_m,\ \tilde{x} \in F_m \subset D,\ \tilde{x} - O = \tilde{x} \in QK_r \subset QK_{\frac{1}{m}}, \tag{8}$$

$$|f(\tilde{x}) - f(h)| \leqslant m\,|h\tilde{x}|, \quad \text{since} \quad h \in F_m,\ \tilde{x} \in F_m \subset D,\ \tilde{x} - h \in QK_{\frac{1}{m}}. \tag{9}$$

The inclusion $\tilde{x} - h \in K_{1/m}$ is easily established:

$$|\tilde{x} - h| \leqslant |O\tilde{x}| + |Oh| = |\tilde{x}| + |h| < \frac{r\sqrt{n}}{2} + \frac{1}{2m} < \frac{1}{2m} + \frac{1}{2m} = \frac{1}{m}$$

(here we took into account that $\tilde{x}$ belongs to a cube with center $O$ and edge $r$, as well as the inequalities $|h| < 1/2m$ and $r < 1/m\sqrt{n}$).

From (8) and (9) it follows that

$$|f(h) - f(O)| = |[f(\tilde{x}) - f(O)] + [-f(\tilde{x}) + f(h)]|$$
$$\leq m|O\tilde{x}| + m|h\tilde{x}| \leq m|O\tilde{x}| + m(|Oh| + |O\tilde{x}|) = 2m|O\tilde{x}| + m|Oh|$$
$$\leq mr\sqrt{n} + m|h| \leq m\sqrt{n} \cdot \frac{|h|}{C_1^{(i)}} + mh = \gamma h$$

($\gamma$ does not depend on $h$). Thus $|f(h) - f(0)|/|h| \leq \gamma$. This inequality is valid for a set of points $h$ (on the semiaxis $Ox_i$) which has positive lower density at the origin. The analogous fact holds at almost every point $x$ of $F_m$:

$$\frac{|f(x+h) - f(x)|}{|h|} \leqslant \gamma \tag{10}$$

for a set of points $x + h$ (on the straight line parallel to the axis $Ox_i$ and passing through the point $x$) which has positive lower linear density at the point $x$; $\gamma$ can depend on $x$.

We take the cross-section $(F_m)_i$ of the set $F_m$ by any straight line parallel to the axis $Ox_i$ on which $F_m$ has positive linear measure. At almost every point of almost every such straight line condition (10) holds. The indicated cross-section is a closed linear set, and the function $f(x)$ is continuous with respect to this set, since $f(x)$ is continuous with respect to $F_m \subset D$. By a theorem of d'Anjois-Hinčin [8] the function $f(x)$ has a finite asymptotic derivative with respect to the variable $x_i$ at almost every point of the nonspecial cross-section $(F_m)_i$. Hence follows the existence of the asymptotic derivative with respect to the variable $x_i$ almost everywhere on $F_m$.[1] Since the analogous fact holds for all $x_i$ $(i = 1, \cdots, n)$, by a theorem of V. V. Stepanov [8] the function $f(x)$ has an asymptotic differential almost everywhere on $F_m$. Lemma 2 is proved.

## §3

We shall prove the two fundamental theorems formulated in the introduction.

Proof of Theorem 1. We prove the first part of the theorem.

We assume that $\overline{\text{mes}}\, E > 0$. According to Lemma 2, the function $f(x)$ has an asymptotic differential almost everywhere on the set $E$. We assume that $E_1 \subset E$ is the set of points where $f(x)$ has an asymptotic differential; $\overline{\text{mes}}\, E_1 = \overline{\text{mes}}\, E > 0$.

We designate by $K(x, \rho)$ the open cube with center at the point $x$ whose edge equals $\rho$ and whose faces are parallel to the coordinate planes (the point $x = O$ is considered the origin of the coordinates). For each natural number $l$ we define the set

$$F_l = \left\{x \in E_1, \frac{|f(x+h) - f(x)|}{|h|} < l,\ h \in QK\left(O, \frac{1}{l}\right)\right\}.$$

Since the inequality $\overline{\lim}_{h\to 0, h \in Q} |f(x + h) - f(x)|/|h| < +\infty$ is valid at every point $x \in E_1$, we have $E_1 = \sum_{l=1}^{\infty} F_l$.

It suffices to prove that the total differential of the function $f(x)$ exists almost everywhere on each of the sets $F_l$.

We fix a value $l = l_0$ for which $\overline{\text{mes}}\, F_{l_0} > 0$. Let $x_0 \in F_{l_0}$ be any point of outer density of the set $F_{l_0}$. We shall find a constant $C > 0$ and a measurable set $R$ of points $x$, with density 1 at the point $x_0$, such that $|f(x) - f(x_0)|/|xx_0| < C$. This can be done, since at the point $x_0$ the asymptotic differential exists.

---

1) The last conclusion is obtained by considering the fact that the set of all points of $F_m$ where there exists an asymptotic derivative with respect to $x_i$ is a measurable set (cf. [8], p. 432).

As the set $Q$ belongs to $(B^{(n)})$, there exist numbers $\alpha > 0$ and $\delta_0 > 0$ such that

$$\operatorname{mes} QK(O, \delta) > \alpha \operatorname{mes} K(O, \delta) \quad \text{for all} \quad \delta \leqslant \delta_0. \tag{11}$$

We shall find a number $\epsilon_0 > 0$ $(\epsilon_0 < \delta_0)$ so small that for every $\epsilon < \epsilon_0$ the inequality

$$\overline{\operatorname{mes}}\, K(x_0, \varepsilon) F_{l_0} R > \operatorname{mes} K(x_0, \varepsilon) \cdot \left(1 - \frac{\alpha}{4^n \cdot 3}\right) \tag{12}$$

will be valid, where $n$ is the dimension of $R^{(n)}$.

We fix a number $\rho_0 > 0$ satisfying the inequalities $\rho_0 < \epsilon_0/2 < \delta_0$, $\rho_0 < 2/l_0$ and a point $x \neq x_0$ of the cube $K(x_0, \rho_0)$. A number $\rho \leq \rho_0$ can be chosen so that $x \in K(x_0, \rho) \setminus K(x_0, \rho/2)$. It is easy to see that the inclusion $K(x, \rho/2) \subset K(x_0, 2\rho)$ is valid, since any point of the cube $K(x, \rho/2)$ is distant from $x_0$ by no more than $|x_0 x| + \rho\sqrt{n}/2 \cdot 2 < \rho\sqrt{n}/2 + \rho\sqrt{n}/4 < \rho\sqrt{n}$, while $\rho\sqrt{n}$ is the half-diagonal [1] of the cube $K(x_0, 2\rho)$.

We define the set $M = \{y \in K(x, \rho/2),\ y + h = x,\ h \in Q\}$. It is clear that $M$ is congruent to the set obtained from $Q$ by reflection about the origin of the coordinates in the cube $K(0, \rho/2)$. Therefore, taking into account that $\rho/2 < \delta_0$, we have (according to (11))

$$\operatorname{mes} M > \alpha \operatorname{mes} K\left(O, \frac{\rho}{2}\right) = \frac{\alpha \rho^n}{2^n}; \quad M \subset K\left(x, \frac{\rho}{2}\right) \subset K(x_0, 2\rho). \tag{13}$$

Since $2\rho < \epsilon_0$ by (12) we have

$$\overline{\operatorname{mes}}\, K(x_0, 2\rho) F_{l_0} R > \operatorname{mes} K(x_0, 2\rho) \cdot \left(1 - \frac{\alpha}{4^n \cdot 3}\right)$$

$$= (2\rho)^n \cdot \left(1 - \frac{\alpha}{4^n \cdot 3}\right) = \operatorname{mes} K(x_0, 2\rho) - \frac{\alpha \rho^n}{2^n \cdot 3}. \tag{14}$$

Comparing (13) and (14), we conclude that the sets $K(x_0, 2\rho) F_{l_0} R$ and $M$ intersect. Let $y_0$ be a common point; $y_0 \in K(x_0, 2\rho) F_{l_0} RM$. We define $h_0$ by the equation $y_0 + h_0 = x$. Taking into account the definition of the set $M$ and the inequality $\rho/2 < 1/l_0$, we have $h_0 \in QK(0, \rho/2) \subset QK(0, 1/l_0)$; since $y_0 \in F_{l_0}$, we have $|f(y_0 + h_0) - f(y_0)| < l_0 |h_0|$ (see the definition of $F_l$). Since $y_0 \in R$, $|f(y_0) - f(x_0)| < C|y_0 x_0|$. Using the last two inequalities, we bound the quantity $|f(x) - f(x_0)| = |f(y_0 + h_0) - f(x_0)|$ from above:

1) The diagonal of an $n$-dimensional cube with edge $r$ equals $r\sqrt{n}$.

$$|f(x)-f(x_0)|\leqslant|f(x)-f(y_0)|+|f(y_0)-f(x_0)|$$
$$\leqslant l_0|h_0|+C|y_0x_0|\leqslant C_1(|h_0|+|y_0x_0|) \tag{15}$$

(we note that here $C_1$ does not depend on $\rho$). Taking into account that $|h_0|\leq \rho\sqrt{n}/4$, $|y_0x_0|\leq\rho\sqrt{n}$, $|xx_0|\geq\rho/4$, we will have

$$\frac{|h_0|+|y_0x_0|}{|xx_0|}\leqslant\frac{\frac{\rho}{4}\sqrt{n}+\rho\sqrt{n}}{\frac{\rho}{4}}=C_2 \tag{16}$$

(here $C_2$ also does not depend on $\rho$). Further, considering (15) and (16), we obtain $|f(x)-f(x_0)|\leq C_1C_2|xx_0|$, i.e.

$$\frac{|f(x)-f(x_0)|}{|xx_0|}\leqslant C_3. \tag{17}$$

The inequality (17), with one and the same constant $C_3$, is fulfilled for any point $x\neq x_0$ of the cube $K(x_0,\rho_0)$.[1] In other words, $\overline{\lim}_{x\to x_0}|f(x)-f(x_0)|/|xx_0|<+\infty$. The last inequality is valid at every point $x_0$ of outer density of the set $F_{l_0}$, i.e. almost everywhere on $F_{l_0}$.

By a theorem of V. V. Stepanov and of Haslam-Jones (see [8]) the total differential of the function $f(x)$ exists at almost every point of $F_{l_0}$. The first part of Theorem 1 is proved.

We pass to the second part of the theorem. Let the set $Q$ not belong to the class $(B^{(n)})$; then there exists a sequence of concentric balls $S_{\rho_k}$ with center at the origin and radius $\rho_k$ ($\rho_k\to 0$ as $k\to\infty$), for which

$$\lim_{k\to\infty}\frac{\operatorname{mes} QS_{\rho_k}}{\operatorname{mes} S_{\rho_k}}=0. \tag{18}$$

We construct an arbitrary number sequence $\{\epsilon_l\}$, $\epsilon_l>0$, for which

$$\sum_{l=1}^{\infty}\varepsilon_l^{\frac{1}{3}}<+\infty. \tag{19}$$

Using the fact that $\operatorname{mes} S_{\rho_k}=C\rho_k^n$ (the constant $C$ depends only on the dimension $n$), and also equation (18), we select from the sequence $\{S_{\rho_k}\}$ a subsequence $\{S_{a_l}\}$ so sparse that

---

1) The constant $C_3$ in general changes depending on $x_0$.

$$\text{mes}\, QS_{a_l} < \varepsilon_l a_l^n. \tag{20}$$

Further, we introduce the series $\{r_l\}$ and $\{\tau_l\}$ by means of the equations

$$r_l = a_l \varepsilon_l^{\frac{1}{3n}}, \quad \tau_l = r_l \varepsilon_l^{\frac{1}{3n}}. \tag{21}$$

We fix any natural number $l$. We construct points $d_l^k$ $(k = 1, \cdots, i_l)$ inside the given cube $K$ such that they are vertices of a rectilinear cubical network; let the diameter of the cubes composing the network be equal to $\tau_l$, and let the faces of the cubes be parallel to the faces of the cube $K$.

We shall assume that $\tau_l < \sqrt[n]{\text{mes}\, K}$; then for any $l$ there must be vertices $d_l^k$ inside $K$. The last inequality can be easily satisfied, replacing, if necessary, the sequences $\{\epsilon_l\}$, $\{a_l\}$, $\{r_l\}$, $\{\tau_l\}$ by subsequences with a new indexing. The length of an edge of an $n$-dimensional cube with diameter $\tau_l$ equals $\tau_l/\sqrt{n}$.

The number of vertices $d_l^k$ located in the cube $K$ will be designated by $i_l$;[1) ] it is clear that $i_l \le (\sqrt{n}\sqrt[n]{\text{mes}\, K}/\tau_l + 1)^n$. Since $(\sqrt{n}\sqrt[n]{\text{mes}\, K}/\tau_l + 1)^n$: $\text{mes}\, K/\tau_l^n < C_n$ ($C_n$ can be chosen so that it does not depend on $l$), we have

$$i_l \leqslant C_n \frac{\text{mes}\, K}{\tau_l^n}. \tag{22}$$

We define the sets $M_l^k = \{x \in K,\ x = d_l^k - h,\ h \in QS_{a_l}\}$ $(l = 1, 2, \cdots;\ k = 1, 2, \cdots, i_l)$, and the sets $L_l = \sum_{k=1}^{i_l} M_l^k$ and $R = \overline{\lim}_{l\to\infty} L_l$. The set $M_l^k$ is congruent to the set $QS_{a_l}$ reflected about the origin, or to a part of it; therefore $\text{mes}\, M_l^k \le \text{mes}\, QS_{a_l}$.

Further, taking into account the definition of $L_l$ and conditions (20)–(22), we obtain

$$\text{mes}\, L_l \leqslant i_l\, \text{mes}\, M_l^k \leqslant i_l\, \text{mes}\, QS_{a_l} \leqslant i_l \varepsilon_l a_l^n$$

$$\leqslant C_n \frac{\text{mes}\, K}{\tau_l^n} \varepsilon_l a_l^n = C_n \varepsilon_l \frac{a_l^n}{r_l^n} \cdot \frac{r_l^n}{\tau_l^n} \text{mes}\, K = C_n \varepsilon_l (\varepsilon_l^{-\frac{1}{3}})(\varepsilon_l^{-\frac{1}{3}})\, \text{mes}\, K$$

$$= C_n \varepsilon_l^{\frac{1}{3}}\, \text{mes}\, K.$$

Since $R = \prod_{t=1}^{\infty} \sum_{l=t}^{\infty} L_l$, while $\text{mes} \sum_{l=t}^{\infty} L_l \le C_n \text{mes}\, K \sum_{l=t}^{\infty} \epsilon_l^{1/3}$ and consequently tends to zero as $t \to \infty$ (cf. (19)), we have $\text{mes}\, R = 0$.

We designate the set of all points $d_l^k$ by $S$:

1) The points $d_l^k$ can be constructed so that for $n_1 \ne n_2$ and for any $k_1$ and $k_2$ the inequality $d_{n_1}^{k_1} \ne d_{n_2}^{k_2}$ holds.

$$S=\{d_l^k\}\ (l=1,2,\ldots;\ k=1,2,\ldots,i_l).$$

We define the desired set $\mathfrak{S}=K\setminus S\setminus R$ of total measure: $\operatorname{mes}\mathfrak{S}=\operatorname{mes}K$.

We construct the desired function $\Phi(x)$ in the following way: $\Phi(x)=r_l$ for $x=d_l^k$ and $\Phi(x)=0$ for $x\in K\setminus S$. Let $x\in\mathfrak{S}$. We shall show that $\lim_{h\to 0, h\in Q}(\Phi(x+h)-\Phi(x))/|h|=0$. Since $x\notin R$, $x\notin M_l^k$ for all $l>l(x)$. We consider only values of $h$ for which the magnitude $|h|$ is sufficiently small; namely $x+h\neq d_l^k$ for all $l\leq l(x)$ (there are only a finite number of points $d_l^k$ satisfying the condition $l\leq l(x)$).

Since $\Phi(x)=0$ we have $(\Phi(x+h)-\Phi(x))/|h|=\Phi(x+h)/|h|$. If $\Phi(x+h)\neq 0$ for some value $h\in Q$ this means that $x+h=d_l^k$ $(x=d_l^k-h)$ for some values of $l$ and $k$, $l>l(x)$. Since $x\notin M_l^k$ for $l>l(x)$, we have $|h|\geq a_l$ by the definition of $M_l^k$. Therefore, considering (21), we obtain

$$\frac{\Phi(x+h)-\Phi(x)}{|h|}=\frac{\Phi(x+h)}{|h|}=\frac{\Phi(d_l^k)}{|h|}=\frac{r_l}{|h|}\leqslant\frac{r_l}{a_l}=\varepsilon_l^{\frac{1}{3n}}.$$

If $h\to 0$ then the point $x+h$ can fall on points $d_l^k$, but if this happens $l$ grows without bound. Therefore in the last inequality $\epsilon_l^{1/3n}\to 0$, and consequently

$$\lim_{h\to 0,\, h\in Q}\frac{\Phi(x+h)-\Phi(x)}{|h|}=0.$$

It remains to show that $\Phi(x)$ does not have a total differential anywhere in $K$. If $x\in S$ this is obvious, since $\Phi(x)$ is discontinuous at points of the set $S$. Let $x\notin S$. For any natural number $l$ we find a point $d_l^k$ such that $|d_l^k-x|\leq\tau_l$ (recall that the diameter of the network whose vertices are the points $d_l^k$ is equal to $\tau_l$). Then

$$\frac{|\Phi(d_l^k)-\Phi(x)|}{|d_l^k-x|}=\frac{\Phi(d_l^k)}{|d_l^k-x|}\geqslant\frac{r_l}{\tau_l}=\frac{1}{\varepsilon_l^{\frac{1}{3n}}}.$$

Since such points $d_l^k$ can be found arbitrarily close to $x$ (since $\tau_l>0$), while $\epsilon_l\to 0$ as $l\to\infty$, we have the quantity $(\Phi(d_l^k)-\Phi(x))/|d_l^k-x|$ unbounded in a neighborhood of the point $x$, which is not compatible with the differentiability of the function $\Phi(x)$ at the point $x$.

The second part of Theorem 1 is proved.

Proof of Theorem 2. We prove the first part of the theorem. Since the set $Q$ belongs to $(C^{(n)})$ its closure $\overline{Q}$ belongs to $(B^{(n)})$.

Since $f(x)$ is continuous the inequality

$$\overline{\lim_{h\to 0,\, h\in Q}} \frac{|f(x+h)-f(x)|}{|h|} < +\infty$$

implies the inequality

$$\overline{\lim_{h\to 0,\, h\in \overline{Q}}} \frac{|f(x+h)-f(x)|}{|h|} < +\infty.$$

Now applying the first part of Theorem 1 to the set $\overline{Q} \in (B^{(n)})$, we get that the function $f(x)$ has a total differential almost everywhere on $E$.

We prove the second part of the theorem. Suppose that the set $Q$ does not belong to class $(C^{(n)})$ and $K \subset G$ is an arbitrary cube. We shall first construct a continuous function $\phi(x)$ which does not have a total differential and which satisfies the inequality $\lim_{h\to 0, h\in Q} (\phi(x+h) - \phi(x))/|h| = 0$ at every point of some set $E \subset K$ of positive neasure, but not almost everywhere.

To prove the existence of the continuous function $\Phi(x)$ having these same properties almost everywhere in $G$, it will be sufficient to refer to a very general theorem in [10].

Since $Q \notin (C^{(n)})$ there exists a sequence of closed spheres $S_{\rho_k}$ with center at the origin and radius $\rho_k \to 0$ for which

$$\lim_{k\to\infty} \frac{\operatorname{mes} \overline{Q} S_{\rho_k}}{\operatorname{mes} S_{\rho_k}} = \lim_{k\to\infty} \frac{\operatorname{mes} \overline{Q} S_{\rho_k}}{\rho_k^n} = 0.$$

We construct a number sequence $\{\alpha_l\}$ $(\alpha_l > 0)$ for which

$$\sum_{l=1}^{\infty} \alpha_l^{\frac{1}{3}} < \frac{1}{2C_n} \tag{23}$$

$(C_n$ will be defined below to depend only on the dimension $n$ of the space). We construct a number sequence $\{a_l\}$ $(a_l > 0;\ a_l \to 0)$, for which $\operatorname{mes} \overline{Q} S_{a_l} < \alpha_l a_l^n$; for this it suffices to take a sufficiently sparse subsequence of the sequence $\{\rho_k\}$ to be $\{a_l\}$. We also define the sequences

$$r_l = a_l \alpha_l^{\frac{1}{3n}}, \quad \tau_l = r_l \alpha_l^{\frac{1}{3n}}. \tag{24}$$

We cover the closed set $\overline{Q} S_{a_l}$ by an open set $T_l$ $(T_l \supset \overline{Q} S_{a_l})$ in such a way that

$$\operatorname{mes} T_l \leqslant \frac{9}{8} \operatorname{mes} \overline{Q} S_{a_l} < \frac{9}{8} \alpha_l a_l^n \tag{25}$$

(if mes $\overline{Q}S_{a_l} = 0$, we shall assume that mes $T_l < 9\alpha_l a_l^n/8$).

The distance $\mu_l$ between the boundary of $T_l$, which is a closed set, and the closed set $\overline{Q}S_{a_l}$ is positive: $\mu_l > 0$.

For every natural number $l$ we divide the whole space into equal cubes by drawing $n$ systems of parallel planes situated at equal distances from each other. Let the diameters of the cubes equal the numbers $\tau_l$ determined earlier. The faces of the cubes will be taken parallel to the faces of the cube $K$. The set of vertices of these cubes will be denoted by $F_l$.

Let the number $l_1$ be large enough for the network $F_{l_1}$ to have points inside the cube $K$ (we are using $\lim_{l\to\infty}\tau_l = 0$). Let $d_{l_1}^k$ $(k = 1, \cdots, i_1)$ be all the vertices of the network $F_{l_1}$ lying strictly inside $K$. We construct sufficiently small closed cubes $\sigma_{l_1}^k$ $(k = 1, \cdots, i_1)$ with centers $d_{l_1}^k$ which do not intersect each other, which lie strictly inside the cube $K$, and which satisfy the conditions: $\operatorname{diam}\sigma_{l_1}^k < \mu_{l_1}$, $\sum_{k=1}^{i_1}\operatorname{mes}\sigma_{l_1}^k < \operatorname{mes} K/4 \cdot 2^1$ (otherwise these cubes are arbitrary).

We find a number $l_2 > l_1$ so large that the network $F_{l_2}$ has points inside the cube $K$ but outside the set $\sum_{k=1}^{i_1}\sigma_{l_1}^k$; let these be the points $d_{l_2}^k$ $(k = 1, \cdots, i_2)$. Taking $l_2$ sufficiently large (i.e. $\tau_{l_2}$ sufficiently small), we can obtain that for every $x \in K\setminus\sum_{k=1}^{i_1}\sigma_{l_1}^k$ a point of $\{d_{l_2}^k\}$ can be found for which $|x - d_{l_2}^k| < \tau_{l_2}$. We construct sufficiently small closed cubes $\sigma_{l_2}^k$ (with centers $d_{l_2}^k$) which do not intersect each other or the cubes $\sigma_{l_1}^k$ and which satisfy the conditions: $\operatorname{dim}\sigma_{l_2}^k < \mu_{l_2}$, $\sum_{k=1}^{i_2}\operatorname{mes}\sigma_{l_2}^k < \operatorname{mes} K/4 \cdot 2^2$.

Suppose the cubes $\sigma_{l_{m-1}}^k$ $(k = 1, 2, \cdots, i_{m-1})$ are already constructed. We select a number $l_m > l_{m-1}$ so large that the network $F_{l_m}$ has points inside $K$ but outside the set $\sum_{t=1}^{m-1}\sum_{k=1}^{i_t}\sigma_{l_t}^k$ (as will be evident, the measure of this set is always less than mes $K$). Let these be the points $d_{l_m}^k$ $(k = 1, \cdots, i_m)$. Taking $l_m$ sufficiently large it may be assumed that for any $x \in K\setminus\sum_{t=1}^{m-1}\sum_{k=1}^{i_t}\sigma_{l_t}^k$ there exists a point of $\{d_{l_m}^k\}$ for which $|x - d_{l_m}^k| < \tau_{l_m}$. We construct cubes $\sigma_{l_t}^k$ $(k = 1, \cdots, i_m)$ with centers $d_{l_m}^k$ in such a way that they do not intersect each other or the cubes $\sigma_{l_t}^k$ $(t = 1, \cdots, m-1)$ constructed earlier. We assume that $\operatorname{diam}\sigma_{l_m}^k < \mu_{l_m}$ and $\sum_{k=1}^{i_m}\operatorname{mes}\sigma_{l_m}^k < \operatorname{mes} K/4 \cdot 2^m$.

It is clear that for any $m$ the inequality [1])

$$i_m \leqslant \frac{C_n \operatorname{mes} K}{\tau_{l_m}^n} \tag{26}$$

1) The constant $C_n$ can be chosen so that it will depend only on the dimension $n$ of the space and not on $l_m$.

is valid. Introducing the set $D = \sum_{m=1}^{\infty}\sum_{k=1}^{i_m}\sigma_{l_m}^k$, we have the estimate

$$\operatorname{mes} D < \sum_{m=1}^{\infty} \frac{\operatorname{mes} K}{4\cdot 2^m} = \frac{\operatorname{mes} K}{4}. \tag{27}$$

Note that for every $x \in K\setminus D$ there exist arbitrarily large values $l_m$ for which it is possible to find a point $d_{l_m}^k$ satisfying the inequality $|x - d_{l_m}^k| < \tau_{l_m}$ ($\tau_{l_m}$ is the diameter of the network $F_{l_m}$).

We construct a continuous function $\phi(x)$ on the cube $K$: $\phi(d_{l_m}^k) = r_{l_m}$ ($m = 1, 2, \cdots; k = 1, \cdots, i_m$), $\phi(x) = 0$ for $x \in CD$; on every cube $\sigma_{l_m}^k$ it is defined as follows: starting with the values $\phi(d_{l_m}^k) = r_{l_m}$ at the center of the cube $\sigma_{l_m}^k$ and $\phi(x) = 0$ on the boundary of this cube, the function is filled in linearly on each segment joining the center of the cube with a point of the boundary of the cube $\sigma_{l_m}^k$; it is clear that the function $\phi(x)$ is continuous on $\sigma_{l_m}^k$ and that $r_{l_m}$ is the greatest value of the function $\phi(x)$ on $\sigma_{l_m}^k$. Taking into account the fact that for each $l_m$ there are only a finite number of points $d_{l_m}^k$ and that $\lim_{l_m\to\infty}(d_{l_m}^k) = \lim_{l_m\to\infty} r_{l_m} = 0$, we may conclude that $\phi(x)$ is continuous on $K$.

We introduce the sets $M_{l_m}^k = \{x \in K,\ x + h = d_{l_m}^k,\ h \in T_{l_m}\}$. The set $M_{l_m}^k$ is congruent to the set $T_{l_m}$ reflected about the origin of the coordinates (or to the part of it lying in $K$), so, by inequality (25), we obtain

$$\operatorname{mes} M_{l_m}^k \leqslant \operatorname{mes} T_{l_m} \leqslant \frac{9}{8}\alpha_{l_m} a_{l_m}^n. \tag{28}$$

We define the set $R = \sum_{m=1}^{\infty}\sum_{k=1}^{i_m} M_{l_m}^k$; using (28), (26), (24) and (23), we obtain

$$\operatorname{mes} R \leqslant \sum_{m=1}^{\infty}\sum_{k=1}^{i_m} \operatorname{mes} M_{l_m}^k < \sum_{m=1}^{\infty}\sum_{k=1}^{i_m} \frac{9}{8}\alpha_{l_m} a_{l_m}^n$$

$$= \sum_{m=1}^{\infty} \frac{9}{8} i_m \alpha_{l_m} a_{l_m}^n \leqslant \frac{9}{8} C_n \sum_{m=1}^{\infty} \frac{\operatorname{mes} K}{\tau_{l_m}^n} \alpha_{l_m} a_{l_m}^n$$

$$= \frac{9}{8} C_n \operatorname{mes} K \cdot \sum_{m=1}^{\infty} \alpha_{l_m} \frac{a_{l_m}^n r_{l_m}^n}{r_{l_m}^n \tau_{l_m}^n} = \frac{9}{8} C_n \operatorname{mes} K \sum_{m=1}^{\infty} \alpha_{l_m} \alpha_{l_m}^{-\frac{2}{3}}$$

$$= \frac{9}{8} C_n \operatorname{mes} K \cdot \sum_{m=1}^{\infty} \alpha_{l_m}^{\frac{1}{3}} < \frac{9}{16} \operatorname{mes} K. \tag{29}$$

We consider the set $E = K\setminus D\setminus R$. According to (27) and (29) $\operatorname{mes} E > \operatorname{mes} K - (1/4)\operatorname{mes} K - (9/16)\operatorname{mes} K > \operatorname{mes} K/8 > 0$.

Let $x$ be a arbitrary point of the set $E$. We shall show that $\lim_{h\to 0, h\in Q}(\phi(x+h)-\phi(x))/|h| = 0$. If the expression $(\phi(x+h_0)-\phi(x))/|h_0|$ does not equal 0 for some $h_0 \in Q$, then $x + h_0 \in \sigma_{l_m}^k$ for certain $l_m$ and $k$. Let $h_1$ be such that $x + h_1 = d_{l_m}^k$; then $h_1 \notin T_{l_m}$, insofar as $x \notin M_{l_m}^k$. Since the diameter of $\sigma_{l_m}^k$ is less than $\mu_{l_m}$, while $x + h_0 \in \sigma_{l_m}^k$ and $x + h_1 \in \sigma_{l_m}^k$, we have $|h_0 - h_1| < \mu_{l_m}$. As $h_1 \notin T_{l_m}$, we have $h_0 \notin QS_{a_{l_m}}$, for $QS_{a_{i_m}} \subset T_{l_m}$, and the distance from $QS_{a_{l_m}}$ to the boundary of the open set $T_{l_m}$ equals $\mu_{l_m}$. Since $h_0 \in QS_{a_{l_m}}$ we have $|h_0| > a_{l_m}$.

Thus $(\phi(x+h_0)-\phi(x))/|h_0| = \phi(x+h_0)/|h_0| < r_{l_m}/a_{l_m} = \alpha_{l_m}^{1/3n}$, for $\phi(x+h_0) \le r_{l_m}$ for $x + h_0 \in \sigma_{l_m}^k$. If $h_0 \to 0$ $(h_0 \in Q)$, then the index $l_m$ of the cube $\sigma_{l_m}^k$ in which the point $x + h_0$ can fall tends to infinity. Therefore (since $\alpha_{l_m} \to 0$) $(\phi(x+h_0)-\phi(x))/|h_0| \to 0$ as $h_0 \to 0$ $(h_0 \in Q)$.

Again let $x \in E$. As already noted, it is possible to find arbitrarily large $l_m$ for which there exist points $d_{l_m}^k$ which satisfy the inequality $|x - d_{l_m}^k| < \tau_{l_m}$. Then

$$\frac{\varphi(d_{l_m}^k)-\varphi(x)}{\left|d_{l_m}^k - x\right|} = \frac{\varphi(d_{l_m}^k)}{\left|d_{l_m}^k - x\right|} > \frac{r_{l_m}}{\tau_{l_m}} = \frac{1}{\alpha_{l_m}^{\frac{1}{3n}}} \to \infty \quad (\text{as } l_m \to \infty),$$

i.e. $\phi(x)$ does not have a total differential at the point $x$.

Thus a function $\phi(x)$ has been constructed continuous in the cube $K$ and having the following properties at each point of a set $E$ of positive measure: 1) $\lim_{h\to 0, h\in Q}(\phi(x+h)-\phi(x))/|h| = 0$; 2) $d\phi(x)$ does not exist. It may be assumed that the continuous function $\phi(x)$ is defined in the whole region $G$; for this one must continuously extend from $K$ onto $G$ in an arbitrary manner.

Hence we may conclude that there exists a continuous function $\Phi(x)$ on the closed region $\overline{G}$ which has the following properties almost everywhere on $G$: 1) $\lim_{h\to 0, h\in Q}(\Phi(x+h)-\Phi(x))/|h| = 0$, 2) $d\Phi(x)$ does not exist. This assertion follows from a theorem of the author published earlier [10]:

*Assume that in some region $G$ of $n$-dimensional Euclidean space it is possible to construct a continuous function $\phi(x)$, $x = (x_1, \cdots, x_n)$, having some specified differential property of type $(\alpha)$, not depending on $x$, at every point of some set $E \subset G$ of positive measure. Then a continuous function $\Phi(x)$ can be constructed on the closed region $\overline{G}$ which has the same differential property almost everywhere on $G$. Here $\Phi(x)$ can be taken to have any continuous values given in* *advance on the boundary of G.*

Here by a differential property of type $(\alpha)$ is understood any description of the behavior of the difference quotient $(f(x+h)-f(x))/|h|$ as $h \to 0$ which satisfies the condition: any two functions $f_1(x)$ and $f_2(x)$ connected by the relation

$$\frac{f_1(x+h)-f_1(x)}{|h|}-\frac{f_2(x+h)-f_2(x)}{|h|}\to 0 \quad \text{as} \quad h\to 0,$$

must simultaneously either have or not have this property.

The collection of properties 1)–2) is a differential property of type $(\alpha)$ not depending on $x$. A property is considered independent of $x$ if it is not violated by parallel translation of the domain of the independent variables (for more details see [10]).

Theorem 2 is proved.

Corollary of Theorem 2. *There exists a continuous function* $f(x_1, \cdots, x_n)$ (*in any region* $G$) *which almost everywhere does not have a total differential but has partial differentials with respect to any set of* $k$ $(k < n)$ *variables* $x_{i_1}, \cdots, x_{i_k}$ *of the* $n$ *variables* $x_1, \cdots, x_n$.

In fact, let $Q$ be the set consisting of all coordinates of hyperplanes of dimension $1, 2, \cdots, n-1$ which intersect at the origin of the coordinates. The set $Q = Q$ has $n$-dimensional measure equal to zero, i.e. $Q$ does not belong to $(C^{(n)})$. Therefore the stated assertion follows from the second part of Theorem 2.

## §4

Definition 3. The function $f(x)$ is called $Q$-continuous at the point $x = (x_1, \cdots, x_n)$ if $\lim_{h\to 0, h\in Q} f(x+h) = f(x)$. We naturally assume that the origin is a limit point of the set $Q$; the set $Q$ does not depend on $x$.

Definition 4. Measurable sets $Q$ having the property $\operatorname{mes} Q > 0$ in any neighborhood of the origin of the coordinates compose class $(A)$.

In [9] we proved that the condition $Q \in (A)$ is necessary and sufficient for ordinary continuity of the measurable function $f(x)$ almost everywhere on some set $E$ to follow from $Q$-continuity of the function $f(x)$ on $E$.

It turns out that in this theorem the requirement of measurability of the function $f(x)$ can be eliminated: the theorem is valid for an arbitrary function.[1] We shall state this theorem and develop a brief proof of it which differs from the proof adduced in [9] only in details.

---

1) Precisely in that form the theorem was used in the proof of Lemma 2.

**Theorem.** *If a function $f(x)$, $x = (x_1, \cdots, x_n)$ defined in a region $G \subset R^{(n)}$ is $Q$-continuous at every point of the set $E \subset G$, where the set $Q$ belongs to class $(A)$, then $f(x)$ is continuous in the usual sense at almost every point of $E$. For every set $Q$ not belonging to $(A)$ it is possible to construct an everywhere discontinuous (measurable) function which will be $Q$-continuous almost everywhere.*

**Proof.** We shall naturally assume that $\overline{\text{mes}}\, E > 0$. Suppose that the assertion of the first part of the theorem is false. Then there exists a set $E_1 \subset E$ ($\overline{\text{mes}}\, E_1 > 0$) at each point of which $f(x)$ is discontinuous. We designate by $F_m$ the set of all points of $E_1$ where the fluctuation of the function $f(x)$ is not less than $1/m$. Then $E_1 = \Sigma_{m=1}^{\infty} F_m$. There exists a natural number $m_0$ such that $\overline{\text{mes}}\, F_{m_0} > 0$. For any natural numbers $l$ and $k$ we define the set $R_{l,k} \equiv \{x \in F_{m_0}, |f(x+h) - f(x)| \leq 1/l, h \in QS(O, 1/k)\}$, where $S(O, 1/k)$ is the sphere with center at $O$ and radius $1/k$. Since $f(x)$ is $Q$-continuous on $F_{m_0}$ ($F_{m_0} \subset E$), every point of $F_{m_0}$ belongs to some set $R_{l,k}$, where $l$ is any number and $k$ depends on $l$; $F_{m_0} = \Sigma_{k=1}^{\infty} R_{l,k}$ for any $l$.

We set $l_0 = 10\, m_0$ and find a $k = k(l_0)$ such that $\text{mes}\, \Sigma_{k=1}^{k(l_0)} R_{l_0,k} > 0$. We introduce the notation $D = \Sigma_{i=1}^{k(l_0)} R_{l_0,i}$. If $x \in D$, then, considering the definition of the set $R_{l_0,k}$, we obtain

$$|f(x+h) - f(x)| \leqslant \frac{1}{l_0} = \frac{1}{10m_0} \quad \text{for} \quad h \in QS\left(O, \frac{1}{k(l_0)}\right). \tag{30}$$

For any integer $t$ we define the set $M_t = \{x \in D, t/10\, m_0 < f(x) < (t+1)/10m_0\}$. It is clear that $D = \Sigma_{-\infty}^{+\infty} M_t$. Since $\overline{\text{mes}}\, D > 0$, for some $t = t_0$ we have $\overline{\text{mes}}\, M_{t_0} > 0$. If $x$ and $x' \in M_{t_0}$, then

$$|f(x') - f(x)| < \frac{1}{10m_0}. \tag{31}$$

There exist a point $x_0 \in M_{t_0}$ of outer density of the set $M_{t_0}$ and a point $\tilde{h} \in Q$ which is a density point of the set $Q$, such that $x_0 + \tilde{h} \in M_{t_0}$, where it may be assumed that $0 < |\tilde{h}| < 1/k(l_0)$. This simple property of sets of positive outer measure, concerning its intersections with a shifted set, is proved in Lemma 1 of [9].

Since $x_0 + \tilde{h} \in F_{m_0}$ (by the definition of $F_{m_0}$), there exists a sequence of points $c_i \to x_0 + \tilde{h}$ $(i \to \infty)$ for which

$$|f(c_i) - f(x_0 + \tilde{h})| > \frac{1}{4m_0}. \tag{32}$$

We define sets $A_i \equiv \{x = c_i - h, h \in Q, |h| < 1/k(l_0)\}$ $(i = 1, 2, \cdots)$, which are congruent for all $i$. Since $\tilde{h}$ is a density point of $Q$, $c_i - \tilde{h}$ is a density point of $A_i$. Since $c_i \to x_0 + \tilde{h}$, $c_i - \tilde{h} \to x_0$ as $i \to \infty$.

As $x_0$ is a point of outer density of the set $M_{t_0}$, while $c_i - \tilde{h}$ is a density point of $A_i$ and $c_i - \tilde{h} \to x_0$, for a sufficiently large $i = i_0$ the sets $A_{i_0}$ and $M_{t_0}$ intersect, and, moreover, they do so at any prescribed nearness to the point $x_0$. Namely, for some $h' \in Q$, $|h'| < 1/k(l_0)$, we have $c_{i_0} - h' \in M_{t_0}$; here we shall assume that

$$|(c_{i_0} - h') - x_0| < \frac{1}{k(l_0)}$$

Then

$$|(c_{i_0} - h') - (x_0 + \tilde{h})| < |(c_{i_0} - h') - x_0| + |x_0 - (x_0 + \tilde{h})|$$
$$< \frac{1}{k(l_0)} + |\tilde{h}| < \frac{2}{k(l_0)}.$$

We compare the following values of the function: $f(c_{i_0} - h')$, $f(c_{i_0})$, $f(x_0 + \tilde{h})$. We have $|f(c_{i_0}) - f(c_{i_0} - h')| \leq 1/10m_0$, according to (30), since $c_{i_0} - h' \in D$ and $h' \in QS(O, 1/k(l_0))$; also $|f(c_{i_0} - h') - f(x_0 + \tilde{h})| < 1/10m_0$ according to (31), since $c_{i_0} - h' \in M_{t_0}$, $x_0 + \tilde{h} \in M_{t_0}$. From these two inequalities it follows that

$$|f(c_{i_0}) - f(x_0 + \tilde{h})| \leqslant |f(c_{i_0}) - f(c_{i_0} - h')| + |f(c_{i_0} - h') - f(x_0 + \tilde{h})|$$
$$< \frac{1}{10m_0} + \frac{1}{10m_0} < \frac{1}{4m_0}. \tag{33}$$

On the other hand, it was established above (see (32)) that for all $i$, and consequently for $i = i_0$, the inequality

$$|f(c_{i_0}) - f(x_0 + \tilde{h})| > \frac{1}{4m_0}$$

is valid. This contradicts (33). Thus the first part of the theorem is proved.

The second part of the theorem is proved in [9].

## BIBLIOGRAPHY


[1] G. H. Sindalovskiĭ, *Continuity and differentiability with respect to congruent sets*, Dokl. Akad. Nauk SSSR 134 (1960), 1305–1306 = Soviet Math. Dokl. 1 (1960), 1217–1218. MR 23 #A1757.

[2] ———, *Continuity and differentiability with respect to congruent sets*, Izv. Akad. Nauk SSSR Ser. Mat. 26 (1962), 125–142. (Russian) MR 25 #1248.

[3] ———, *On the equivalence between ordinary derivatives and the derivatives with respect to congruent sets of a certain class*, Izv. Akad. Nauk SSSR Ser. Mat. 29 (1965), 987–996. (Russian) MR 33 #231.

[4] ———, *Congruent and asymptotic differentiability*, Dokl. Akad. Nauk SSSR 150 (1963), 995–997 = Soviet Math. Dokl. 4 (1963), 807–809. MR 27 #2588.

[5] ———, *Differentiability with respect to congruent sets*, Izv. Akad. Nauk SSSR Ser. Mat. 29 (1965), 11–40. (Russian) MR30 #3185.

[6] V. V. Stepanov, *Über totale Differenzierbarkeit*, Math. Ann. 90 (1923), 318–320.

[7] H. Rademacher, *Über partielle und totale Differenzierbarkeit*. I, Math. Ann. 79 (1919), 340–359.

[8] S. Saks, *Théorie de l'intégrale*, Monogr. Mat., vol. II, PWN, Warsaw, 1933; English transl., Dover, New York, 1964; Russian transl., IL, Moscow, 1949. MR 29 #4850.

[9] G. H. Sindalovskiĭ, *On the continuity of functions of several variables with respect to congruent sets*, Uspehi Mat. Nauk 19 (1964), no. 4 (118), 201–207. (Russian) MR 29 #4843.

[10] ———, *On the construction of continuous functions with given differential properties on a set of complete measure and certain applications to Fourier series*, Mat. Sb. 69 (111) (1966), 640–657. (Russian) MR 34 #1463.

Translated by:
N. Koblitz

# DIFFERENTIATION AND INTEGRATION IN ABSTRACT SPACES

UDC 517.29+517.397

G. P. TOLSTOV

We discuss an operation of differentiation for set functions in an abstract space generating, in a natural way, a certain operation of integration which, as we shall see, coincides with integration in the sense of Lebesgue. The present article follows closely on my preceding articles [1], [2], [3].

Let $\mu(X)$ be a measure of Lebesgue type, defined on the ring $K$ of sets in the space $X_0$. This means that: a) $\mu(X)$ is finite, nonnegative and countably-additive; b) if $X_n \in K$ $(n - 1, 2, \cdots)$ and $\Sigma_n \mu(X_n) < +\infty$, then $\bigcup_n X_n \in K$; c) if $X_n \in K$ $(n = 1, 2, \cdots)$, then also $\bigcap_n X_n \in K$ (note that c) follows from a) and b)).

It is natural to say that the sets of the ring $K$ are sets of *finite* measure $\mu$. The sets which are not in $K$ but are representable in the form $X = \bigcup_n X_n$, $X_n \in K$ $(n = 1, 2, \cdots)$, will be called sets of *infinite* measure $\mu$. The totality of all sets of finite and infinite measure $\mu$ form, as is easily seen, a Borel ring $L \supset K$.* The sets belonging to $L$ will be said to be $\mu$-measurable.

By a $\mu$-function we shall mean an arbitrary function, defined for $X \in L$, of the form

$$F(X) = \sum_n t_n \mu(X \cap X_n), \tag{1}$$

where $X_n$ $(n = 1, 2, \cdots)$ are arbitrary sets of finite measure $\mu$, and $t_n$ $(n = 1, 2, \cdots)$ are arbitrary real numbers with the single condition

$$\sum_n |t_n| \mu(X_n) < +\infty \quad ^{**} \tag{2}$$

(From (2) it follows that the series (1) converges for arbitrary $X \in L$.) By the $\mu$-derivative of the function (1) we shall mean a finite point function defined almost everywhere on $X_0$ (with respect to the measure $\mu$) by the equality

$$f(x) = \sum_n t_n \varphi(x, X_n), \tag{3}$$

---

*If $K$ is an algebra of sets, then of course no set of *infinite* measure will exist; in this case $L = K$. A Borel ring is invariant under countable sums and intersections of its sets.

**In general, the sets $X_n$ will intersect. For $t_n = 0$ it will be a fact that, from a certain index on, the sum (1) (for arbitrary $X_n$) will be *finite*.

where $\phi(x, X_n)$ is the characteristic function of the set $X_n$. The fact that the series (3) converges almost everywhere in $X_0$ is an easy consequence of (2) and the theorem on the integration (in the sense of Lebesgue) of positive series. There are infinitely many of these derivatives of $\mu$-functions (1), if only because deletion from an arbitrary $X_n$ of an arbitrary set of $\mu$-measure zero does not change the function (1) but *will* in general change the function (3). Moreover, it follows from the theorem proved immediately below that the class of *all* $\mu$-derivatives of the function (1) precisely coincides with the class of all functions defined almost everywhere on $X_0$ and coinciding there with $f(x)$.

The function (1) will be called the *$\mu$-primitive* of $f(x)$.

**Theorem.** *Let the function $g(x)$ be defined on a $\mu$-measurable set $E$. In order for this function to be integrable in the sense of Lebesgue with respect to the measure $\mu$ it is necessary and sufficent that it coincide on $E$ with some $\mu$-derivative. If $f(x)$ is this $\mu$-derivative and $F(X)$ is its $\mu$-primitive, then for the Lebesgue integral we have*

$$\int_E g(x)\,d\mu = \int_E f(x)\,d\mu = F(E).$$

Let us prove the necessity. Let $g(x)$ be integrable on $E$ in the sense of Lebesgue. It is sufficient to consider the case of a function $g(x)$ satisfying the condition $g(x) \geq 0$ everywhere on $E$. For each natural number $n$ we consider the sets

$$E_{nk} = \left\{x : x \in E,\ \frac{k-1}{2^n} \leqslant g(x) < \frac{k}{2^n}\right\} \quad (k = 1, 2, 3, \ldots, n2^n),$$
$$E_n = \{x : x \in E,\ g(x) \geqslant n\}.$$

These sets are pairwise disjoint. Let us write

$$g_n(x) = \sum_{k=1}^{k=n2^n} \frac{k-1}{2^n}\,\varphi(x, E_{nk}) + n\varphi(x, E_n).$$

The function $g_n(x)$ is a nonnegative function which assumes a finite number of values. It is easy to verify that everywhere on $E$

$$g_1(x) \leqslant g_2(x) \leqslant \ldots \leqslant g_n(x) \leqslant \ldots, \quad \lim_n g_n(x) = g(x),$$

and therefore $g(x) = g_1(x) + \Sigma_n [g_{n+1}(x) - g_n(x)]$.

From the properties of the functions $g_n(x)$ it follows that $g_1(x)$ and each of the functions $g_{n+1}(x) - g_n(x)$ has the form $\Sigma t_k\, \phi(x, X_k)$, where the sum is finite, $t_k \geq 0$, and $X_k \subset E$. Thus $g(x)$ can be represented as a series

$$g(x) = \sum_n t_n \varphi(x, X_n), \tag{4}$$

where $x \in E$, $X_n \in E$, $t_n \geq 0$ (the sets $X_n$ may intersect). By the properties of positive series we have

$$\int_E g(x)\, d\mu = \sum_n t_n \int_E \varphi(x, X_n)\, d\mu = \sum_n t_n \mu(X_n). \tag{5}$$

Thus for the numbers $t_n$ $(n = 1, 2, \cdots)$ the condition (2) is satisfied. Consequently we may speak of the corresponding $\mu$-function (1) and its $\mu$-derivative (3). From (4) it follows that $g(x) = f(x)$ on $E$, where (let us recall that $X_n \subset E$)

$$F(E) = \sum_n t_n \mu(E \cap X_n) = \sum_n t_n \mu(X_n) = \int_E g(x)\, d\mu$$

(see (5)), which completes the proof of the necessity of the conditions of the theorem. The sufficiency follows easily from (2) and from the theorem on integration (in the sense of Lebesgue) of positive series.

Corollaries. a) The class of $\mu$-functions and the class of their derivatives exhaust the class of indefinite Lebesgue integrals and the class of functions integrable (with respect to $\mu$) in the sense of Lebesgue, respectively. From the equalities (1) and (3) we obtain a rather clear picture of the structure of these integrals (to supplement their well-known property of absolute continuity) and of these functions. b) The algorithm for passage from a $\mu$-function to its derivative is very simple: the functions $\mu(X \cap X_n)$ on the right side of (1) are replaced by the functions $\phi(x, X_n)$.

Let us make the following remarks. Our theory of the integral, as it turns out, can be developed quite simply on the basis of the concepts of $\mu$-function and $\mu$-derivative *without appeal to the Lebesgue integral*. For in fact, if $f(x)$ is the $\mu$-derivative of the function $F(X)$, then we set

$$F(X) = \int_X f(x)\, d\mu, \tag{6}$$

and for $X$ we let the number (6) be called the $\mu$-integral of $f(x)$ with respect to the set $X$; a function $g(x)$ is said to be $\mu$-integrable on the set $X$ if on the set it coincides with some $\mu$-derivatives. If $f(x)$ is this $\mu$-derivative, then we set $\int_X g(x)\, d\mu = \int_X f(x)\, d\mu$. By the methods of [3] it is possible to prove that this definition is correct and to obtain, step by step, the basic properties of the $\mu$-integral.

While this article was in press the author learned that a construction similar to (1) was employed by J. Mikusiński in [4] (with respect to integration in $n$-dimensional Euclidean space, without reference to differentiation).

## BIBLIOGRAPHY

[1] G. P. Tolstov, *The derivative and the integral from a general point of view*, Dokl. Akad. Nauk SSSR 142 (1962), 1040–1042 = Soviet Math. Dokl. 3 (1962), 253–255. MR 25 #154.

[2] ———, *Three types of abstract integrals*, Dokl. Akad. Nauk SSSR 158 (1964), 536–539 = Soviet Math. Dokl. 5 (1964), 1275–1278. MR 30 #2118.

[3] ———, *Derivative and integral (the axiomatic approach)*, Mat. Sb. 66 (108) (1965), 608–630. (Russian) MR 30 #4895.

[4] J. Mikusiński, *Sur une définition de l'intégrale de Lebesgue*, Bull. Acad. Polon. Sci. Sér. Sci. Math. Astronom. Phys. 12 (1964), 203–204. MR 29 #4857.

Translated by:
S. H. Gould

# INTEGRATION OF ALMOST PERIODIC FUNCTIONS WITH VALUES IN A BANACH SPACE

UDC 517.5

B. M. LEVITAN

An example of an almost periodic function $f(t)$ with values in the Banach space $l^\infty$ of bounded sequences having a bounded non-almost-periodic integral $F(t) = \int_0^t f(\eta)d\eta$ for which the mean $\lim_{T\to\infty} T^{-1}\int_0^T F(t)dt$ exists is constructed.

The following theorem is proved: If the mean $\lim_{T\to\infty} T^{-1}\int_x^{x+T} F(t)dt$ of the integral $F(t)$ of an almost periodic function $f(t)$ with values in a Banach space exists uniformly in $x$ $(-\infty < x < \infty)$ and $\sup_x L^{-1}\int_x^{x+L}\|F(t)\|dt < \infty$ for each $L > 0$, then $F(t)$ is almost periodic.

## Introduction

1. Let $f(t)$ be a numerical-valued almost periodic function (henceforth abbreviated as a.p. function) with Fourier series

$$f(t) \sim \sum_{n=1}^{\infty} A_n e^{i\mu_n t}. \tag{0.1}$$

One of the deepest theorems in the theory of a.p. functions is the Bohl-Bohr theorem on the indefinite integral of an a.p. function. It states that if the indefinite integral of $f(t)$,

$$F(t) = C + \int_0^t f(\eta)\,d\eta$$

is bounded, then $F(t)$ is an a.p. function.

The theorem does not carry over to a.p. functions with values in a Banach space (hereafter such functions will be referred to as abstract a.p. functions).

Indeed, [1)] let $l^\infty$ be the space of bounded sequences $x = (\xi_1, \cdots, \xi_n, \cdots)$ with norm

$$\|x\| = \sup_n |\xi_n|.$$

Consider a function $f(t)$ with values in $l^\infty$,

$$f(t) = \{\lambda_n \cos \lambda_n t\}, \tag{0.2}$$

1) The example cited below has been adopted from Amerio's article [1].

where $\lambda_n > 0$ and $\lambda_n \downarrow 0$. It is not hard to see that $f(t)$ is an a.p. function. For, take an arbitrary positive $\epsilon$ and let

$$\lambda_n < \frac{1}{2}\varepsilon.$$

for $n > n_\epsilon$. Then for arbitrary real $\tau$ and $n > n_\epsilon$

$$\sup_t |\lambda_n \cos \lambda_n (t+\tau) - \lambda_n \cos \lambda_n t| < \varepsilon. \tag{0.3}$$

Since a finite number of a.p. functions are equi-almost periodic, there exists a relatively dense set of numbers $E_\epsilon$ such that if $\tau \in E_\epsilon$, then

$$\sup_t |\lambda_n \cos \lambda_n (t+\tau) - \lambda_n \cos \lambda_n t| < \varepsilon \quad (n = 1, 2, \ldots, n_\varepsilon). \tag{0.4}$$

This and (0.3) imply that

$$\sup_t \| f(t+\tau) - f(t) \| < \varepsilon,$$

i.e. $f(t)$ is an a.p. function.

Consider now the integral of $f(t)$:

$$F(t) = \int_0^t f(\eta)\, d\eta = \{\sin \lambda_n t\}.$$

$F(t)$ is a bounded function, namely,

$$\|F(t)\| = \sup_n |\sin \lambda_n t| \leqslant 1.$$

We shall show, however, that $F(t)$ is not an a.p. periodic function (with values in $l^\infty$). To this end, consider the linear functionals

$$a_n(x) = \xi_n,$$

with norms equal to 1. We have

$$|a_n(F(t+\tau)) - a_n(F(t))| \leqslant \|F(t+\tau) - F(t)\|,$$

which implies that if $F(t)$ were an a.p. function the set of numerical-valued a.p. functions $\{\sin \lambda_n t\}$ would be equicontinuous and equi-almost periodic, and hence compact in $C(-\infty, \infty)$. However, this is impossible since the limit of the sequence $\{\sin \lambda_n t\}$ must equal zero and zero cannot be the uniform limit of this sequence since for arbitrary $n$ there exists a $t$ such that $\lambda_n t = \pi/2$ and hence $\sin \lambda_n t = 1$.

We next observe that not only is $F(t)$ not an a.p. function but it does not even have a mean. Consider

$$\frac{1}{T}\int_0^T F(t)\,dt = \left\{\frac{2\sin^2\frac{\lambda_n T}{2}}{\lambda_n T}\right\}.$$

If

$$\lim_{T\to\infty}\frac{1}{T}\int_0^T F(t)\,dt \tag{0.5}$$

existed (in the $l^\infty$ norm), it would equal zero (because a weak limit equal to zero exists). But given any $T_0 > 0$, one can choose a $T > T_0$ and $\lambda_n$ such that

$$\lambda_n T = \pi$$

and therefore

$$\frac{2\sin^2\frac{\lambda_n T}{2}}{\lambda_n T} = \frac{2}{\pi}.$$

Thus the limit (0.5) cannot be equal to zero and hence does not exist.

In connection with this there arises the following question: Does there exist an a.p. function (with values in $l^\infty$) having a bounded non-almost periodic integral whose mean (0.5) exists? The question has an affirmative answer and an appropriate example will be constructed in §1.

It is natural to pose the following question: Could the indefinite integral $F(t)$ of an a.p. function $f(t)$ be bounded and not a.p. and at the same time the mean of $F(t)$ exist uniformly, i.e. could

$$\lim_{T\to\infty}\frac{1}{T}\int_x^{x+T} F(t)\,dt \tag{0.6}$$

exist uniformly in $x\,(-\infty < x < \infty)$? This question has a negative answer, as is evident from the following theorem (proved in §2).

Let $f(t)$ be an abstract a.p. function. Let

$$F(t) = \int_0^t f(\eta)\,d\eta, \quad \sup_x \frac{1}{L}\int_x^{x+L} \|F(t)\|\,dt < \infty \quad (L > 0),$$

and let the mean (0.6) exist uniformly. Then $F(t)$ is an a.p. function.

This theorem appears to have been unknown even for numerical-valued a.p. functions. However, as the reader will see, its proof does not depend on the nature of the function $f(t)$.

We point out that for numerical-valued functions this theorem implies in particular that the integral of an a.p. function cannot be an a.p. Weyl function without being an a.p. Bohr function.

2. In the above example the Fourier exponents for $f(t)$ were the numbers $\lambda_n$, which by assumption tend to zero. This is not haphazard, because if the Fourier exponents $\mu_n$ of an abstract a.p. function satisfy the condition

$$\inf_n |\mu_n| \geqslant \alpha > 0,$$

its integral $F(t)$ is an a.p. function. This theorem was proved for numerical-valued a.p. functions by Favard and its proof goes over verbatim to abstract functions (see [2]). Favard's theorem will be used in the sequel.

Since boundedness of the integral of an abstract a.p. function does not imply its almost periodicity, it would be desirable to clarify under what conditions (besides boundedness of the integral) the almost periodicity of the integral would follow. The first such condition was indicated by Bochner [3] and it states that if the integral $F(t)$ of an abstract a.p. function $f(t)$ has a compact trajectory, then it is an a.p. function.

Amerio [1] recently obtained a deeper result by proving the following theorem: If $f(t)$ is an a.p. function with values in a uniformly convex Banach space and if its integral $F(t)$ is bounded, then $F(t)$ has a compact trajectory and so by Bochner's theorem it is an a.p. function.

The following theorem, also due to Amerio, plays an important part in the proof of his theorem: If $f(t)$ is weakly a.p. and its trajectory is compact, then $f(t)$ is strongly a.p. A simple proof of this theorem is given in the appendix to the present paper.

## §1

In this section we shall construct an example of an a.p. function with values in $l^\infty$ having a bounded non-almost-periodic integral the mean of which exists.

Consider a sequence $\phi_n(t)$ $(n = 1, 2, \cdots)$ of periodic functions each having the same period 1 and satisfying the following obviously consistent conditions:

1) each $\phi_n(t)$ is continuous and has a continuous first derivative;
2) $\phi_n(t) = 0$ for $0 \leq t \leq \delta_0 < 1$;
3) $|\phi_n(t)| \leq 1$;
4) $\int_0^1 \phi_n(t)\, dt = 0$;

5) $\lim_{n\to\infty} \int_0^1 |\phi_n(t)|\, dt = 0$;

6) for each $n$ there exist (infinitely many) points $t_n$ such that $\phi_n(t_n) = \pm 1$.

Let

$$a_n = \max_{0 \le t \le 1} |\varphi_n(t)|,$$

and let $\lambda_n > 0$, $\lambda_n \downarrow 0$, and

$$\lambda_n a_n = o(1). \tag{1.1}$$

Set

$$f(t) = \{\lambda_n \varphi_n'(\lambda_n t)\}.$$

From condition (1.1) it easily follows that $f(t)$ is an a.p. function with values in $l^\infty$ (see the example in the Introduction).

Consider the integral of $f(t)$:

$$F(t) = \int_0^t f(\eta)\, d\eta = \{\varphi_n(\lambda_n t)\}.$$

By condition 3) $F(t)$ is a bounded function. $F(t)$ cannot be an a.p. function (with values in $l^\infty$) because in that event the sequence $\{\phi_n(\lambda_n t)\}$ would converge uniformly to zero over the whole line (see the Introduction), and this is impossible by condition 6).

It remains to show that the limit (convergence being in $l^\infty$)

$$\lim_{T\to\infty} \frac{1}{T} \int_0^T F(t)\, dt \tag{1.2}$$

exists.

Let us show that the limit (1.2) is equal to zero. We have

$$\frac{1}{T} \int_0^T F(t)\, dt = \left\{ \frac{1}{T} \int_0^T \varphi_n(\lambda_n t)\, dt \right\} = \left\{ \frac{1}{\lambda_n T} \int_0^{\lambda_n T} \varphi_n(u)\, du \right\}.$$

If $\lambda_n T \le \delta_0$, then by 2)

$$\int_0^{\lambda_n T} \varphi_n(u)\, du = 0. \tag{1.3}$$

Let $\lambda_n T > \delta_0$ and set $\lambda_n T = N + r$, where $N$ is a positive integer and $0 \le r < 1$. Then the periodicity and condition 4) imply

$$\int_0^{\lambda_n T} \varphi_n(u)\, du = \int_0^r \varphi_n(u)\, du. \tag{1.4}$$

Let $\epsilon$ be an arbitrary positive number and choose $n = n_\epsilon$ so that for $n > n_\epsilon$

$$\int_0^1 |\varphi_n(u)|\, du < \varepsilon\delta_0. \tag{1.5}$$

This is possible by virtue of condition 5).

From (1.4) and (1.5) we find that for $n > n_\epsilon$

$$\left| \frac{1}{T} \int_0^T \varphi_n(\lambda_n t)\, dt \right| < \frac{\varepsilon\delta_0}{\lambda_n T} < \varepsilon. \tag{1.6}$$

For each $n = 1, \cdots, n_\epsilon$ one can choose a $T_0 = T_0(\epsilon)$ such that for $T > T_0$

$$\left| \frac{1}{T} \int_0^T \varphi_n(\lambda_n t)\, dt \right| < \varepsilon. \tag{1.7}$$

This results from the fact that for each fixed $n$

$$\lim_{T\to\infty} \frac{1}{T} \int_0^T \varphi_n(\lambda_n t)\, dt = \lim_{T\to\infty} \frac{1}{\lambda_n T} \int_0^{\lambda_n T} \varphi_n(u)\, du = \int_0^1 \varphi_n(u)\, du = 0.$$

From (1.3), (1.6) and (1.7) it follows that for $T > T_0$ and $n = 1, 2, \cdots$

$$\left| \frac{1}{T} \int_0^T \varphi_n(\lambda_n t)\, dt \right| < \varepsilon. \tag{1.8}$$

Since $\epsilon$ is arbitrary, (1.8) yields

$$\lim_{T\to\infty} \frac{1}{T} \int_0^T F(t)\, dt = 0.$$

It is not hard to verify that the limit

$$\lim_{T\to\infty} \frac{1}{T} \int_x^{x+T} F(t)\, dt \tag{1.9}$$

at least exists, though not uniformly in $x$. However this also follows from the theorem of the next section.

Remark. On the basis of the above example one can construct an example of an a.p. function with values in $C(0, \lambda_1)$ having the same properties as $f(t)$. In fact it is sufficient to define the function $x(\lambda;\, t)$ $(0 \le \lambda \le \lambda_1;\ -\infty < t < \infty)$ as follows (see [1]):

$$x(0;\, t) = 0,$$

$$x(\lambda_n;\, t) = \lambda_n \varphi_n(\lambda_n t),$$

$$x(\lambda;\, t) = x(\lambda_n;\, t) \frac{\lambda - \lambda_{n+1}}{\lambda_n - \lambda_{n+1}} + x(\lambda_{n+1};\, t) \frac{\lambda_n - \lambda}{\lambda_n - \lambda_{n+1}}, \qquad \lambda_{n+1} \leqslant \lambda \leqslant \lambda_n.$$

## §2

In this section we shall prove the following assertion.

**Theorem.** *Let $f(t)$ be an abstract a.p. function and let*

$$F(t)=\int_0^t f(\eta)\,d\eta.$$

*If the mean (1.9) exists uniformly in $x$ $(-\infty<x<\infty)$ and if for each $L>0$*

$$\sup_x \frac{1}{L}\int_x^{x+L} \|F(t)\|\,dt<\infty, \tag{2.1}$$

*then $F(t)$ is an a.p. function.*

**Proof.** [1] Let $\epsilon$ be an arbitrary positive number and set

$$\varphi_\varepsilon(\lambda)=\begin{cases} 1, & 0\leqslant\lambda\leqslant\varepsilon,\\ \frac{1}{\varepsilon}(2\varepsilon-\lambda), & \varepsilon\leqslant\lambda\leqslant 2\varepsilon,\\ 0, & \lambda\geqslant 2\varepsilon,\end{cases}$$

$$\varphi_\varepsilon(-\lambda)=\varphi_\varepsilon(\lambda).$$

Let $\psi_\epsilon(u)$ denote the Fourier transform of $\phi_\epsilon(\lambda)$. We have

$$\psi_\varepsilon(u)=\frac{1}{2\pi}\int_{-\infty}^{\infty}\varphi_\varepsilon(\lambda)e^{i\lambda u}d\lambda=\frac{1}{\pi}\int_0^{2\varepsilon}\varphi_\varepsilon(\lambda)\cos\lambda u\,d\lambda$$

$$=\frac{1}{\pi\varepsilon u^2}\sin\frac{3}{2}\varepsilon u\sin\frac{1}{2}\varepsilon u.$$

The Fourier inversion formula yields

$$\varphi_\varepsilon(\lambda)=\int_{-\infty}^{\infty}\psi_\varepsilon(u)e^{-i\lambda u}du.$$

In particular, if $\lambda=0$,

$$1=\int_{-\infty}^{\infty}\psi_\varepsilon(u)\,du. \tag{2.2}$$

Let $f(t)$ be an abstract a.p. function with Fourier series

$$f(t)\sim\sum_n A_n e^{i\lambda_n t}.$$

Set

1) See [4], Chapter 2, where a similar idea is used in a proof.

$$f_\varepsilon^{(1)}(t) = \int_{-\infty}^{\infty} f(t+u)\psi_\varepsilon(u)\,du.$$

It is not hard to show that $f_\epsilon^{(1)}(t)$ is an a.p. function with Fourier series

$$f_\varepsilon^{(1)}(t) \sim \sum_n \varphi_\varepsilon(\lambda_n) A_n e^{i\lambda_n t}.$$

Further let

$$f_\varepsilon^{(2)}(t) = f(t) - f_\varepsilon^{(1)}(t) \sim \sum_n [1 - \varphi_\varepsilon(\lambda_n)] A_n e^{i\lambda_n t}.$$

Then

$$f(t) = f_\varepsilon^{(1)}(t) + f_\varepsilon^{(2)}(t)$$

and hence

$$F(t) = \int_0^t f(\eta)\,d\eta = \int_0^t f_\varepsilon^{(1)}(\eta)\,d\eta + \int_0^t f_\varepsilon^{(2)}(\eta)\,d\eta = F_\varepsilon^{(1)}(t) + F_\varepsilon^{(2)}(t).$$

Since the Fourier exponents for $f_\epsilon^{(2)}(t)$ are not less than $\epsilon$ in magnitude, by Favard's theorem $F_\epsilon^{(2)}(t)$ is an a.p. function for each fixed $\epsilon$. Therefore our theorem is proved if we can show that $F_\epsilon^{(1)}(t)$ tends to zero as $\epsilon \to 0$ uniformly in $t$.

We have

$$F_\varepsilon^{(1)}(t) = \int_0^t f_\varepsilon^{(1)}(\eta)\,d\eta = \int_0^t \left[\int_{-\infty}^{\infty} f(\eta+u)\psi_\varepsilon(u)\,du\right] d\eta \tag{2.3}$$

$$= \int_{-\infty}^{\infty} \psi_\varepsilon(u)\left[\int_0^t f(\eta+u)\,d\eta\right] du = \int_{-\infty}^{\infty} F(t+u)\psi_\varepsilon(u)\,du - \int_{-\infty}^{\infty} F(u)\psi_\varepsilon(u)\,du.$$

Let

$$\lim_{T\to\infty} \frac{1}{T}\int_t^{t+T} F(u)\,du = \lim_{T\to\infty} \int_0^T F(t+u)\,du = B,$$

the limit existing uniformly in $t$ (by the hypothesis of the theorem). Observe first of all that the existence of this uniform limit implies that

$$\lim_{T\to-\infty} \int_t^{t+T} F(u)\,du = \lim_{T\to-\infty} \frac{1}{T}\int_0^T F(u+t)\,du$$

$$= \lim_{(-T)\to+\infty} \frac{1}{-T}\int_{t+T}^{(t+T)-T} F(u)\,du = B,$$

the latter limit also existing uniformly in $t$.

Let $\eta$ denote an arbitrary positive number and choose $U_0 = U_0(\eta)$ so large that for $|u| \geq U_0$ and for all $t$ the inequality

$$\left\| \frac{1}{u} \int_0^u F(t+u)\,du - B \right\| < \eta \tag{2.4}$$

holds. Now, using (2.2), we set

$$\int_{-\infty}^{\infty} F(t+u)\psi_\varepsilon(u)\,du - B = \int_{-\infty}^{\infty} [F(t+u) - B]\psi_\varepsilon(u)\,du$$

$$= \int_{|u|\leq U_0} [F(t+u) - B]\psi_\varepsilon(u)\,du + \int_{-\infty}^{-U_0} [F(t+u) - B]\psi_\varepsilon(u)\,du$$

$$+ \int_{U_0}^{\infty} [F(t+u) - B]\psi_\varepsilon(u)\,du = I_\varepsilon^{(1)}(t) + I_\varepsilon^{(2)}(t) + I_\varepsilon^{(3)}(t).$$

$I_\epsilon^{(2)}(t)$ and $I_\epsilon^{(3)}(t)$ are estimated in a similar way. Consider, for example, $I_\epsilon^{(3)}(t)$. Integrating by parts, we obtain

$$I_\varepsilon^{(3)}(t) = -\left\{ \frac{1}{U_0} \int_0^{U_0} F(t+u)\,du - B \right\} U_0 \psi_\varepsilon(U_0)$$

$$- \int_{U_0}^{\infty} \left\{ \frac{1}{u} \int_0^u F(t+u)\,du - B \right\} u \frac{d}{du} \psi_\varepsilon(u)\,du.$$

Therefore (2.4) implies that

$$\|I_\varepsilon^{(3)}(t)\| \leqslant \frac{\eta}{2\pi} + \eta \int_{-\infty}^{\infty} \left| u \frac{d}{du} \psi_\varepsilon(u) \right| du$$

$$= \frac{\eta}{2\pi} + \eta \int_{-\infty}^{\infty} \left| u \frac{d}{du} \frac{\sin \frac{3u}{2} \sin \frac{1}{2} u}{\pi u^2} \right| du = C\eta, \tag{2.5}$$

where $C$ is a constant. A similar estimate (with the same constant $C$) holds for $I_\epsilon^{(2)}(t)$. Let us now estimate $I_\epsilon^{(1)}(t)$. We have

$$\|I_\varepsilon^{(1)}(t)\| \leqslant \frac{3}{4\pi} \varepsilon \int_{-U_0}^{U_0} \|F(t+u)\|\,du + \frac{3}{2\pi} \varepsilon U_0 \|B\|.$$

From this estimate and (2.1) it follows that for fixed $U_0$ one can choose $\epsilon_0$ so small that for $\epsilon < \epsilon_0$ and for all $t$ the inequality

$$\|I_\varepsilon^{(1)}(t)\| < \eta \tag{2.6}$$

holds. From (2.5) and (2.6) we find that

$$\left\| \int_{-\infty}^{\infty} F(t+u)\psi_\varepsilon(u)\,du - B \right\| < (2C+1)\eta$$

for $\epsilon < \epsilon_0$ and all $t$. Therefore

$$\lim_{\varepsilon\to 0} \int_{-\infty}^{\infty} F(t+u)\psi_\varepsilon(u)\,du = B \tag{2.7}$$

uniformly in $t\ (-\infty < t < \infty)$. In particular, setting $t = 0$ in (2.7), we obtain

$$\lim_{\varepsilon\to 0} \int_{-\infty}^{\infty} F(u)\psi_\varepsilon(u)\,du = B. \tag{2.8}$$

From (2.3), (2.7) and (2.8) it follows that

$$\lim_{\varepsilon\to 0} F_\varepsilon^{(1)}(t) = 0$$

and moreover uniformly in $t$. The theorem is proved.

Remark. Condition (2.1) may be weakened slightly. Integrating the expression for $I_\epsilon^{(1)}(t)$ by parts, we obtain

$$\int_{-U_0}^{U_0} \{F(t+u) - B\}\,\psi_\varepsilon(u)\,du = \left\{\int_{-U_0}^{U_0} F(t+u)\,du - 2BU_0\right\}\psi_\varepsilon(U_0)$$

$$-\int_{-U_0}^{U_0}\left\{\int_{-U_0}^{u} [F(t+u) - B]\,du\right\}\frac{d}{du}\psi_\varepsilon(u)\,du. \tag{2.9}$$

Therefore if instead of (2.1) we impose the condition that

$$\lambda(X) = \sup_x \max_{0\leqslant u\leqslant X} \left\| \int_x^{x+u} F(u)\,du \right\| < \infty$$

for arbitrary $X > 0$, then from (2.4) and (2.9) we find that

$$\left\| \int_{-U_0}^{U_0} F(t+u)\,\psi_\varepsilon(u)\,du \right\| \leqslant \frac{1}{\pi}\eta + [\lambda(2U_0) + \|B\|\,2U_0] \int_{-U_0}^{U_0} \left|\frac{d}{du}\psi_\varepsilon(u)\right| du$$

$$= \frac{1}{\pi}\eta + C\varepsilon\,[\lambda(2U_0) + 2\|B\|\,U_0].$$

This estimate implies that $I_\epsilon^{(1)}(t)$ tends to zero as $\epsilon \to 0$ uniformly in $t$.

## Appendix

In Amerio's work on integration of abstract a.p. functions, an important part is played by the following theorem.

*If $f(t)$ is weakly a.p. (with values in a Banach space $E$) and if the trajectory of $f(t)$ is compact, then $f(t)$ is strongly a.p.*

We shall give the simple proof of this theorem in its entirety.

Let $\Phi$ denote the unit sphere of the dual space $\overline{E}$. Let $\phi$ be an arbitrary element of $\Phi$ and consider the function of two variables

$$F(\varphi, t) = \varphi[f(t)].$$

Let us show that the family of functions $F(\phi, t)$ ($t$ the parameter) is compact under uniform convergence on $\Phi$. For if $f(t_1), \cdots, f(t_n)$ is a finite $\epsilon$-net for $f(t)$, then $F(\phi, t_1), \cdots, F(\phi, t_n)$ is a finite $\epsilon$-net for the family $F(\phi, t)$ ($t$ the parameter). Hence (see [2], Russian p. 256) the family $F(\phi, t)$ ($\phi$ the parameter) is compact under uniform convergence on the whole line. Thus for each $\epsilon > 0$ there exist $\phi_1, \cdots, \phi_m \in \Phi$ such that for any $\phi \in \Phi$ one can find an element $\phi_k$ $(k = 1, \cdots, m)$ such that

$$\sup_t |F(\varphi, t) - F(\varphi_k, t)| < \varepsilon. \tag{1}$$

Since $F(\phi_k, t)$ is an a.p. function for any fixed $k$, (1) implies that the family $F(\phi, t)$, $\phi \in \Phi$, is equicontinuous and equi-almost periodic. Let $\tau = \tau_\epsilon$ be the common $\epsilon$-almost period for the family $F(\phi, t)$, i.e. let

$$\sup_t |F(\varphi, t+\tau) - F(\varphi, t)| < \varepsilon \quad \text{for all} \quad \varphi \in \Phi. \tag{2}$$

Then by a corollary to the Hahn-Banach Theorem, for fixed $t$ and $\tau$ we have

$$\|f(t+\tau) - f(t)\| = \sup_{\varphi\in\Phi} |F(\varphi, t+\tau) - F(\varphi, t)| < \varepsilon.$$

The strong continuity of $f(t)$ is proved in a similar way. Thus $f(t)$ is strongly a.p.

**Remark 1.** It is not hard to see that the same proof goes through if one takes as a basis Bochner's definition of almost periodicity. Therefore the theorem remains true for abstract a.p. functions defined on an arbitrary group.

**Remark 2.** It is also not hard to see that the theorem holds true for abstract $N$-a.p. functions (if $f(t)$ is an abstract weakly $N$-a.p. function and the trajectory of $f(t)$ is compact, then $f(t)$ is a strongly $N$-a.p. function; for the definition of numerical-valued $N$-a.p. functions (see [2], Chapter 3).

## BIBLIOGRAPHY

[1] L. Amerio, *Sull'integrazione delle funzioni quasi-periodiche astratte*, Ann. Mat. Pura Appl. (4) 53 (1961), 371–382. MR 24 #A807.

[2] B. M. Levitan, *Almost periodic functions*, GITTL, Moscow, 1953. (Russian) MR 15, 700.

[3] S. Bochner, *Abstrakte fastperiodische Funktionen*, Acta Math. 61 (1933), 149–184.

[4] ———, *Vorlesungen über Fouriersche Integrale*, Akademische Verlagsgesellschaft, Leipzig, 1932; English transl., Ann. of Math. Studies, no. 42, Princeton Univ. Press, Princeton, N. J., 1959; Russian transl., Fizmatgiz, Moscow, 1962. MR 21 #5851.

Translated by:
B. Seckler

# ON COMPLETENESS OF A SET OF ANALYTIC FUNCTIONS*

UDC 517.531

Ju. F. KOROBEĬNIK

Recent studies on the question of the completeness of a set of analytic functions on a curve and in a domain have begun to make use of the techniques of differential equations of infinite order. Such an approach was apparently first employed in the papers of A. F. Leont'ev [1], [2] which dealt with the question of the completeness of the set $\{z^{\lambda_n}\}$ on a curve. A little later Ju. A. Kaz'min [3], [4] used differential equations of infinite order to study the completeness of the set of successive derivatives $\{f^{(n)}(z)\}$ of an analytic function in a simply connected domain under certain assumptions on $f(z)$. V. P. Gromov [6], [7] obtained analogous results for a set of generalized derivatives (in the sense of Gel'fond-Leont'ev [5]).

Our note considers the question of the completeness of the set $\{f^{(\lambda_n)}(z)\}$, where $\lambda_n$ is a sufficiently "dense" sequence of positive integers. In so doing, we shall use on the one hand the method of reducing a set of analytic functions to a differential equation of infinite order developed in the papers of Kaz'min and Gromov, and on the other hand certain results concerning the properties of analytic solutions of differential equations of infinite order derived in [8].

Let $y(z)$ be an arbitrary analytic function. Let $A(y)$ be a domain in the plane in which $y(z)$ is single-valued and analytic and such that $y(z)$ cannot be single-valued and analytic in any domain containing $A(y)$ and at least one boundary point of $A(y)$. Such a domain $A(y)$ will be referred to as the *maximal domain of analyticity* of $y(z)$.

If $y(z)$ is single-valued everywhere the domain $A(y)$ is the same as $E(y)$—the complete Weierstrass domain of existence of the function $y(z)$. But in general $A(y) \subseteq E(y)$. Suppose $y(z)$ has, for instance, two singularities (critical points) at $a$ and $\infty$. Draw a cut joining the points (for example, a ray emanating from the point $a$) and denote by $G$ the domain obtained by removal of the cut from the extended plane. If we select a particular branch $y_0(z)$ of $y(z)$ which is single-valued in $G$, then $A(y_0) = G$. The definition of $A(y)$ leads to the following property: if $y(z)$ is single-valued and analytic in some region $G_1$, then a maximal domain of analyticity $A(y)$ can always be constructed so that $G_1 \subseteq A(y)$.

A domain $G$ (bounded or not) will be said to be *simply connected* if

**Editor's note.* The present translation incorporates suggestions made by the author.

the interior of any closed bounded Jordan curve in $G$ is also in $G$. For bounded domains this definition is equivalent to the customary definition of simply connected domain as a domain with connected boundary. For unbounded domains these two definitions need not coincide. For example, the exterior of the unit disc is not simply connected according to the definition made here, although its boundary is a connected set.

Let $G$ be an arbitrary domain and let $G^0$ denote the smallest simply connected domain containing $G$. It may be obtained for example by adjoining to $G$ the interior of all closed Jordan curves lying in $G$.

Consider the differential equation of infinite order given by

$$y(x) + \sum_{k=1}^{\infty} P_k(x) y^{(k)}(x) = F(x), \tag{1}$$

where $P_k(x)$ is a polynomial of at most degree $k-1$ ($k = 1, 2, \cdots$). The following theorem is proved in [8].

**Theorem A.** *Suppose $y(x)$ is an analytic function in $G$ and that the series*

$$\sum_{k=1}^{\infty} P_k(x) y^{(k)}(x) \tag{2}$$

*converges uniformly within $G$ to a function $F(x)$ analytic in $G^0$. Then $y(x)$ is analytic in $G^0$ and the series (2) converges uniformly to $F(x)$ within $G^0$ (i.e. $y(x)$ is a solution of equation (1) in $G^0$).*

**Remark.** The phrase "an analytic function in $G$" in the statement of the theorem is understood to mean (as is accepted and customary) a single-valued analytic function in this domain.

## § 1. Main theorem

Let $A(f)$ be the maximal domain of analyticity of a function $f(z)$ and $E_d(f)$ the open set of points $z_1$ in $A(f)$ such that $y(z)$ is analytic in the disc $|z - z_1| \le d$; $E_d(f) = \sum_{n=1}^{\infty} E_d^n(f)$, where the $E_d^n(f)$ are the components of $E_d(f)$.

**Theorem B.** *Suppose $\{\lambda_n\}$ is an increasing sequence of positive integers such that $\underline{\lim}_{n\to\infty} \lambda_n/n = 1$. Suppose for some $d > 0$ and $m \ge 1$ the component $E_d^m(f)$ is multiply connected. Then the set $\{f^{(\lambda_n)}(z)\}$ is complete in any disc of radius less than $d$ with center in $E_d^m(f)$.*

**Proof.** Suppose that the set $\{f^{(\lambda_n)}(z)\}$ is incomplete in a disc $G$ of radius $d-2b$ $(b>0)$ with center at a point $z_0 \in E_d^m(f)$ (hereafter we shall denote $E_d(f)$ by $E_d$ and $E_d^m(f)$ by $E_d^m$).

Without loss of generality we may take $z_0=0$. To pass to this case, we can set $z=t+z_0$ and $f(z)=f(z_0+t)=\phi(t)$. Then $f^{(n)}(z)=\phi^{(n)}(t)$ $(n=0, 1, \cdots)$ and the domains $E_d^k(\phi)$ are obtained through parallel translation of $E_d^k(f)$ by $z_0$, $E_d^k(\phi)$ being either simply connected or not together with $E_d^k(f)$.

Thus we shall assume that the set $\{f^{(\lambda_n)}(z)\}$, $\underline{\lim}_{n\to\infty}\lambda_n/n=1$, is not complete in the disc $|z|<d-2b$ and that $0\in E_d^m(f)$. Then for all values of $\alpha$ sufficiently small in absolute value, $|\alpha|\le\rho$, the relation

$$\Phi(\alpha)\equiv\sum_{k=0}^{\infty} b_k\alpha^k=\sum_{n=0}^{\infty}\frac{f^{(n)}(\alpha)}{n!}c_n \qquad (3)$$

is known to hold (see [4], [6]), where

$$b_i=\int_C f^{(i)}(z)g(z)\,dz,\quad c_n=\int_C z^n g(z)\,dz,$$

$C$ is the circle $|z|=d-2b-h$ $(h>0)$, and $g(z)$ is an analytic function in the domain $|z|\ge d-2b-h$ equaling zero at infinity.

Since the distance from any point $z$ in $E_d^m$ to the closed set of all singular points of $f(z)$ is greater than $d$, the series $\sum_{n=0}^{\infty}f^{(n)}(z)c_n/n!$ is uniformly convergent in a neighborhood of each point of $E_d^m$ by virtue of the inequality $\overline{\lim}_{n\to\infty}\sqrt[n]{|c_n|}\le d-2b-h$, and hence is uniformly convergent within $E_d^m$.

Equation (3) is satisfied for all sufficiently small $\alpha$. Since the right-hand side of the equation is an analytic function in a domain $E_d^m$ containing the origin, the function $\Phi(\alpha)$ can be continued analytically by means of the series $\sum_{n=0}^{\infty}c_n f^{(n)}(\alpha)/n!$ into the whole domain $E_d^m$.

Pólya's Theorem [12]. *The series $\sum_{n=0}^{\infty}A_n x^n$, where $\overline{\lim}_{n\to\infty}\sqrt[n]{|A_n|}<\infty$, represents a single-valued analytic function with a simply connected domain of existence $E(y)$ if $\underline{\lim}_{n\to\infty}t_n/n=0$, where $t_n$ is the number of nonvanishing coefficients among the first $n$ coefficients $A_i$: $A_0, \cdots, A_{n-1}$.*

Let us estimate the quantity $t_n$ for the function $\Phi(\alpha)=\sum_{k=0}^{\infty}b_k\alpha^k$. Observe that $b_m=0$ for $m=\lambda_k$ $(k=1, 2, \cdots)$. Let $\{\mu_n\}$ denote the sequence obtained from $0, 1, 2, \cdots$ by removal of the sequence $\{\lambda_n\}$. Let us prove that if $\overline{\lim}_{n\to\infty}n/\lambda_n=1$, then $\underline{\lim}_{n\to\infty}t(n)/n=0$.

Indeed, $t(\lambda_n)=\lambda_n-n+1$ and $\underline{\lim}_{n\to\infty}t(\lambda_n)/\lambda_n=1-\overline{\lim}_{n\to\infty}n/\lambda_n=0$. But since $\underline{\lim}_{n\to\infty}t(n)/n$ is always nonnegative, it follows from this that $\underline{\lim}_{n\to\infty}t(n)/n=0$.

Thus if $\overline{\lim}_{n\to\infty} n/\lambda_n = 1$, Pólya's theorem is applicable to the function $\Phi(t) = \Sigma_{k=0}^{\infty} b_k t^k$.

On the other hand, since $\Phi(t)$ is analytic in $E_d^m$, Pólya's theorem implies that it will be analytic in $(E_d^m)^0 = H_0$ (i.e. in the smallest extension of $E_d^m$ to a simply connected domain).

By Theorem A, $f(z)$ is an analytic function in $H_0$ and the series on the left-hand side of equation (3) is uniformly convergent in $H_0$.

Let $C_1$ be an arbitrary closed Jordan curve lying in $E_d^m$. The interior of the curve $C_1$ clearly belongs to $H_0$. Let $z_1$ be any point in the interior of $C_1$ and let $\rho_1 = \rho(z_1, C_1)$ and $\delta_1 = \min_{z\in C_1} \rho(z)$, where $\rho(z)$ is the distance from $z$ to the closest singular point of $f(z)$. Since $C_1 \in E_d^m$ and $C_1$ is a closed set, $\delta_1 > d$. The function $f(x)$ is clearly analytic within the disc $|x - z_1| < \rho_1 + \delta_1$ and by the same token in the closed disc $|x - z_1| \leq d$. Moreover, $z_1 \in H_0$, a subset of $A(f)$ by Theorem A. Hence $z_1 \in E_d^m$ and so $E_d^m$ is simply connected, contradicting the hypothesis.

## § 2. Corollaries

1. Corollary 1. *Suppose that $f(z)$ is a single-valued analytic function, $E(f)$ is its complete Weierstrass domain of existence, $E_d(f)$ is the set of points $z$ in $E(f)$ such that $f(x)$ is analytic in the disc $|x - z| \leq d$, and $E_d(f) = \Sigma_{n=1}^{\infty} E_d^n(f)$, where the $E_d^n(f)$ are the components of $E_d(f)$. Suppose that for some $d > 0$ and $m \geq 1$ the component $E_d^m$ is multiply connected. Then the set $\{f^{(\lambda_n)}(z)\}$, where $\underline{\lim}_{n\to\infty} \lambda_n/n = 1$ and $\lambda_n$ is a subsequence of the positive integers, is complete in any disc of radius less than $d$ with center lying in $E_d^m(f)$.*

Example. Let $f(z) = \cot z$. The domain $E(f)$ is the entire plane with the exception of the points $z = k\pi$ ($k = 0, \pm 1, \pm 2, \cdots$). For $d \geq \pi/2$ the domain $E_d(f)$ decomposes into two simply connected domains while for $d < \pi/2$ it is an infinitely connected domain (and thus consists of a single component). Therefore (by what was proved above) the set $(\cot z)^{(\lambda_n)}$, where $\underline{\lim}_{n\to\infty} \lambda_n/n = 1$, is complete in any disc of radius less than $d$ ($d < \pi/2$) with center lying in $E_d(f)$. Further, if $f(z) = \cot z$ is analytic in the disc $|z - z_1| \leq d$, then the distance from $z_1$ to the singular points of $f(z)$ is greater than $d$ and $z_1 \in E_d(f)$. On account of this the following statement is valid.

The set $(\cot z)^{(\lambda_n)}$, $\underline{\lim}_{n\to\infty} \lambda_n/n = 1$, is complete in any disc of radius less than $\pi/2$ that does not include points of the form $z = k\pi$ ($k = 0, \pm 1, \pm 2, \cdots$) in it or on its boundary.

For the case $\lambda_n = n$ this result was obtained by Kaz'min [4] using another method.

2. Corollary 2. *Let $f(z)$ be a single-valued analytic function having at most a countable number of isolated singularities $z_1, z_2, \cdots$ for which*

$$\alpha = \inf_{\substack{k\neq l\\ k,l=1,2,\ldots}} |z_k - z_l| > 0.$$

*Let $\{\lambda_n\}$ be a subsequence of the positive integers such that $\underline{\lim}_{n\to\infty} \lambda_n/n = 1$. Then the set $\{f^{(\lambda_n)}(z)\}$ is complete in any disc of radius less than $\alpha/2$ not containing any of the points $z_1, z_2, \cdots$ within it or on its boundary.*

For if $B$ is the center of such a disc and $r$ is its radius, then $B \subset E_r(f)$. Moreover for $d < \alpha/2$ the set $E_d(f)$ is a multiply connected domain obtained by removal from the whole plane of discs of radius $d$ centered at the points $z_l$, $l = 1, 2, \cdots$ (the connectedness of $E_d(f)$ is easy to establish if one observes that for $\alpha > 0$ the set of singularities $\{z_k\}$ has no limit point in the finite plane).

The corollary is valid particularly for a single-valued analytic function with a finite number of singularities $z_1, \cdots, z_m$. In this case

$$\alpha = \min_{\substack{l\neq k\\ 1\leqslant l,\, k\leqslant m}} |z_l - z_k|. \tag{4}$$

3. Let us consider the case where $f(z)$ has a finite number of isolated singularities $z_1, \cdots, z_m$ in the entire extended plane, without necessarily being of single-valued nature. We shall however assume that at least one of these singularities is an isolated singularity of single-valued nature.

Let $\alpha$ be defined by (4) and let $K$ be a disc of radius $d < \alpha/2$ centered at point $B$ containing no singularities of $f(z)$ within it or on its boundary. Choose an analytic element $e_0$ of $f(z)$ (for instance, the Taylor expansion in a neighborhood of an interior point of $K$). The function obtained by analytic continuation of the element $e_0$ along any paths lying in $K$ is by the monodromy theorem (see [9]) a single-valued analytic function in $K$. We denote the branch of $f(z)$ so singled out by $f_0(z)$.

Draw cuts (rays) to infinity from the singularities $z_j$ which are not singularities of a single-valued nature. Denote the collection of such cuts and all singularities by $F_0$. It is not hard to see that the cuts can be drawn so that $\rho(B, F_0) > d$. The function $f_0(z)$ is clearly analytically continuable to the whole region $G$ obtained by removal of the set $F_0$ from the whole plane and is single-valued and analytic in $G$. However, if $G$ is broadened by the adjunction of at least one of its boundary points (i.e. a point of $F_0$), then $f_0(z)$ can no longer be a single-valued analytic function in the resulting domain. Therefore $G = A(f_0)$.

Finally the cuts may always be drawn so that the distance between the various individual components of the set $F_0$ (points and rays) is not less than $\alpha$. Then for any $h < \alpha/2$ (in particular for $h = d$), the domain $E_h(f_0)$ is multiply connected. For since $F_0$ contains at least one isolated point $\gamma$ of single-valued nature, $E_h(f_0)$ contains a neighborhood $|z-\gamma| = d_1$ for any $d_1$ in the interval $(h, \alpha/2)$ but does not contain the interior point $\gamma$ of such a neighborhood.

Finally the point $B$ belongs to $E_d(f_0)$ and (by Theorem B) the set $\{f_0^{(\lambda_n)}(z)\}$, where $\underline{\lim}_{n\to\infty} \lambda_n/n = 1$, is complete in $K$. This gives us

Corollary 3. *Suppose $f(z)$ is analytic everywhere with the exception of a finite number of singularities at $z_1, \cdots, z_m$ of which at least one is a singularity of single-valued nature. Suppose further that $\alpha$ is defined as in (4), $K$ is a disc of radius $d < \alpha/2$ containing no singularities within it or on its boundary, $e_0$ is a prescribed arbitrary analytic element of $f(z)$ in a neighborhood of some point of $K$ and $f_0(z)$ is the function obtained by analytic continuation of $e_0$ to the whole of $K$. Then the set $\{f_0^{(\lambda_n)}(z)\}$, where $\lambda_n$ is an increasing sequence of positive integers for which $\underline{\lim}_{n\to\infty} \lambda_n/n = 1$, is complete in $K$.*

4. Let $f(z)$ be analytic (and single-valued) in the annulus $D_0: R_1 < |z-a| < R_2$, and let at least one singularity of $f(z)$ lie in the disc $|z-a| \le R_1$. Consider a disc $K_0: |z-z_0| < r$ lying entirely in the annulus $D_0$ so that $r < \frac{1}{2}(R_2 - R_1)$. Choose positive numbers $\delta$ and $h$ so small that the disc $K_\delta: |z-z_0| \le r+\delta$ lies inside the annulus $D_h: R_1 + h < |z-a| < R_2 - h$ (clearly $\delta + r + h < \frac{1}{2}(R_2 - R_1)$). Then the point $z_0$ will belong to the set $E_h/2 + r(f)$ together with the annulus $D_h/2 + r: R_1 + h/2 + r < |z-a| < R_2 - h/2 - r$ (since if $K_\delta$ belongs to $D_h$ the distance from $z_0$ to the boundary of $D_0$ is no less than $r + \delta + h > h/2 + r$).

The annulus $D_h/2 + r$ belongs to some component of the open set $E_h/2 + r(f)$. This component is multiply connected since it contains the circle $|z-a| = \frac{1}{2}(R_1 + R_2)$ but does not contain all of the interior of the circle, in which there is at least one singularity of $f(z)$ by assumption. On the basis of Theorem B we can make the following statement.

Corollary 4. *If $f(z)$ is analytic in the annulus $R_1 < |z-a| < R_2$ and* *has at least one singularity in the disc* $|z-a| \le R_1$, *then the set* $\{f^{(\lambda_n)}(z)\}$, $\overline{\lim}_{n\to\infty} n/\lambda_n = 1$, *is complete in any disc lying entirely inside the annulus.*

For the case $\lambda_n = n$ this corollary was obtained by Kaz'min [4] using another method.

The circular annulus $R_1 < |z-a| < R_2$ may be replaced by an "elliptic" annulus, a domain included between two confocal ellipses.

To consider a more general case, we take a closed convex smooth Jordan curve $C_1$ and we carry out the following construction. Lay off along the normal to $C_1$ at each point $t$ of the curve a segment of length $h$ (the segments are to be laid off on the same side of the curve, for instance along its exterior normals). The endpoints of these segments determine a curve $C_2$. Then, exactly as for the case of a circular annulus, we can show that if $f(z)$ is analytic in the curvilinear annulus determined by the curves $C_1$ and $C_2$ and has at least one singularity inside $C_1$, then the set $\{f^{(\lambda_n)}(z)\}$, where $\overline{\lim}_{n\to\infty} n/\lambda_n = 1$, is complete in any disc lying entirely inside this annulus.

## § 3. Remarks on the exactitude of the results

It was assumed above that $\{\lambda_n\}$ is an increasing subsequence of positive integers such that $\overline{\lim}_{n\to\infty} n/\lambda_n = 1$ (note that for any increasing subsequence $\lambda_n$ of positive integers we always have $0 \le \overline{\lim}_{n\to\infty} n/\lambda_n \le 1$). This requirement that the numbers $\lambda_n$ be of sufficiently great "denseness" is essential for Theorem B and its corollaries to be valid; namely, it will be shown below that the condition $\overline{\lim}_{n\to\infty} n/\lambda_n = 1$ cannot be replaced by the condition $\overline{\lim}_{n\to\infty} n/\lambda_n = 1-\sigma$ $(\sigma > 0)$ no matter how small $\sigma$ is $(\sigma < 1)$.

We first prove the following lemma.

**Lemma.** *Let* $\phi_n(z)$ $(n = 0, 1, \cdots)$ *be analytic in the disc* $|z| < r$ *and let none of the Taylor expansions* $\phi_n(z) = \sum_{m=0}^{\infty} a_{n,m} z^m$ *contain* $z^k$ *for some fixed* $k \ge 0$. *Then the set* $\phi_n(z)$ *is incomplete in the disc* $|z| < r$.

**Proof.** If $\rho < r$ is an arbitrary number, then

$$\int_{|z|=\rho} \varphi_n(z)\, z^{-k-1}\, dz = 0 \quad (n = 0, 1, \ldots)$$

and so by the known completeness criterion [10], [11], the set $\phi_n(z)$ is incomplete in the disc $|z| < r$.

Consider the function $\phi_l(z) = 1/(1-z^l)$, $l \ge 2$. In this case the maximal domain of analyticity $A(\phi_l)$ coincides with the Weierstrass domain of existence $E(\phi_l)$ of the single-valued function $\phi_l(z)$. The domain $E(\phi_l)$ is obtained by removal from the whole plane of $l$ points, the poles $\alpha_j$ of $\phi_l(z)$ located at a distance $d_l = 2\sin(\pi/l)$ from each other. By Theorem B and Corollaries 1 and 2, the set $\{\phi_l^{(\lambda_n)}(z)\}$, where $\overline{\lim}_{n\to\infty} n/\lambda_n = 1$, is complete in any disc of radius $r < \sin(\pi/l)$ not containing (within it or on its boundary) the points $\alpha_j$. And in particular it is complete in the disc $|z| < d$, where $d$ is an arbitrary number less than $\sin(\pi/l)$.

At the same time, from the relations

$$\left(\frac{1}{1-z^l}\right)^{(s)} = \sum_{m=\left\{\frac{s}{l}\right\}}^{\infty} a_{s,m} z^{lm-s} \quad (s = 0, 1, \ldots)$$

we find that if $s \neq nl$ $(n = 0, 1, \cdots)$, then $lm - s \not\equiv 0 \pmod{l}$ and the Taylor expansion of $(1/(1-z^l))^{(s)}$ $(s = 1, 2, \cdots;\ s \neq nl;\ n = 0, 1, \cdots)$ does not contain the powers of $z$ with exponents a multiple of $l$. Therefore the set of functions $(1/(1-z^l))^{(\lambda_n)}$ $(n = 1, 2, \cdots)$, where the sequence $\{\lambda_n\}$ is obtained from $0, 1, 2, \cdots$ by removal of all numbers $0, l, 2l, \cdots$, is incomplete in any disc $|z| < r$.

Let us detemine $\overline{\lim}_{n\to\infty} n/\lambda_n$. Suppose that among the numbers $0, \cdots, \lambda_n$ there is a number $\mu_{s_n}$ of the form $ml$. Then $\mu_{s_n}$ is clearly equal to the integral part of $(\lambda_n + 1)/l$: $\mu_{s_n} = E((\lambda_n + 1)/l)$. Hence $(\lambda_n + 1)/l - 1 < \mu_{s_n} \leq (\lambda_n + 1)/l$. At the same time there are $n$ numbers among $0, \cdots, \lambda_n$ that are not multiples of $l$ (the numbers $\lambda_1, \cdots, \lambda_n$) and the $\mu_{s_n}$ that are. Therefore

$$\mu_{s_n} + n = \lambda_n + 1, \quad n = \lambda_n + 1 - E\left(\frac{\lambda_n + 1}{l}\right),$$

$$\lambda_n + 1 - \frac{\lambda_n + 1}{l} \leqslant n < \lambda_n + 2 - \frac{\lambda_n + 1}{l},$$

from which we conclude that $\lim_{n\to\infty} n/\lambda_n = 1 - 1/l$.

Thus the set $(1/(1-z^l))^{(\lambda_n)}$, where $\lim_{n\to\infty} n/\lambda_n = 1 - 1/l$ is incomplete in any disc $|z| < r$ for arbitrary $r$ (in paritcular, $r$ can be less than $\sin(\pi/l)$). Moreover, $l$ may be chosen to be arbitrarily large.

## Appendix

### Remarks added by the author for the English translation

1. In the particular case when $\lambda_n = n$ Theorem B may be considerably strengthened by using the method and results recently published by A. F. Leont'ev [13].

Let $W(f)$ be the Weierstrass domain of existence of the function $f(z)$, let $W_d(f)$ be the open set of points $z_1$ of $W(f)$ such that $f(z)$ is analytic in the closed circle $|z - z_1| \leq d$; and let $W_d(f) = \Sigma_{n=1}^{\infty} W_d^n(f)$; here the $W_d^n(f)$ are the components of $W_d(f)$. Then we have

**Theorem C.** *Let the component $W_d^m(f)$ be multiply connected with some given $d > 0$ and $m > 0$. Then the system $\{f^{(n)}(z)\}$ is complete in any circle of radius $r < d$ with center in the domain $W_d^m(f)$.*

2. Let $E(G)$ be the space of functions analytic in the domain $G$ with the topology of uniform convergence in each closed subset of $G$; let $E_0(G)$ be the subspace of $E(G)$ (with the same topology) consisting of all the functions of $E(G)$ that vanish at infinity, if the point at infinity belongs to $G$ (for finite domains $E(G) = E_0(G)$).

Shortly after this paper was prepared for publication the author reported a number of results on completeness of the system $\{f^{(n)}(z)\}$ in infinite and multiply connected domains [14]. In particular let us mention the following results.

**Theorem 1.** *Let $f(z) \in E(G)$. The system $\{f^{(n)}(z)\}$ is incomplete in any infinite domain $G$ which does not coincide with the extended plane.*

**Theorem 2.** *Let $f(z) \in E(G)$ where the domain $G$ contains infinity and does not coincide with the extended plane. Then for the system $\{f^{(n)}(z)\}$ to be complete it is necessary that $f(z) \in E_0(G)$ and $\lim_{z\to\infty} zf(z) \neq 0$. If $f(z) \in E_0(G)$, then the system $\{f^{(n)}(z)\}$ will be complete if and only if*

1. *the complement of $G$ in the extended plane is connected, and*

2. *$f(z) = a/(z - b)$, $b \in F$.*

**Theorem 3.** *If $G$ is a multiply connected infinite domain then the system $\{f^{(n)}(z)\}$ is incomplete in $G$.*

**Theorem 4.** *The system $\{f^{(n)}(z)\}$ is incomplete in any domain of connectivity not less than three.*

The case of a doubly connected finite domain was also investigated, but not completely.

## BIBLIOGRAPHY

[1] A. F. Leont′ev, "On application of differential equations of infinite order to questions of the completeness of a set of analytic functions," in *Studies on contemporary problems in the theory of functions of a complex variable,* Fizmatgiz, Moscow, 1961, pp. 41–49.

[2] ———, *On the completeness of the system* $\{z^{\lambda_k}\}$ *on curves in the complex plane,* Dokl. Akad. Nauk SSSR 121 (1958), 797–800. (Russian) MR 21 #1396.

[3] Ju. A. Kaz′min, *On the completeness of a system of analytic functions.* I, Vestnik Moskov. Univ. Ser. I Mat. Meh. 1960, no. 5, 3–13. (Russian) MR 23 #A3268.

[4] ———, *On the completeness of a set of analytic functions.* II, Vestnik Moskov. Univ. Ser. I Mat. Meh. 1960, no. 6, 11–19. (Russian) MR 23 #A3268.

[5] A. O. Gel′fond and A. F. Leont′ev, *On a generalization of Fourier series,* Mat. Sb. 29 (71) (1951), 477–500. (Russian) MR 13, 638.

[6] V. P. Gromov, *The completeness of systems of derivatives of an analytic function,* Izv. Akad. Nauk SSSR Ser. Mat. 25 (1961), 543–556. (Russian) MR 29 #6030.

[7] ———, *Completeness of systems of analytic functions in a domain,* Mat. Sb. 62 (104) (1963), 320–334. (Russian) MR 28 #1310.

[8] Ju. F. Korobeĭnik, *On the domain of definition of an analytic solution of a differential equation of infinite order,* Mat. Sb. 64 (106) (1964), 153–170. (Russian) MR 31 #1433.

[9] A. I. Markuševič, *Theory of analytic functions.* Vol. I: *Origin of theories,* GITTL, Moscow, 1950; 2nd ed., "Nauka", Moscow, 1967; English transl., Prentice-Hall, Englewood Cliffs, N. J., 1965. MR 12, 87; MR 30 #2125; MR 31 #5965.

[10] B. Ja. Levin, *Distribution of zeros of entire functions,* GITTL, Moscow, 1956; English transl., Transl. Math. Monographs, vol. 5, Amer. Math. Soc., Providence, R. I., 1964. MR 19, 402; MR 28 #217.

[11] M. G. Haplanov, *Linear functionals in the space of single-valued analytic functions,* Trans. Sem. on Functional Analysis, Voronež. Univ., 1960, no. 3–4, pp. 115–121.

[12] G. Pólya, *Eine Verallgemeinerung des Fabryschen Lückensatzes,* Nachr. Ges. Wiss. Göttingen 2 (1927), 187–195.

[13] A. F. Leont′ev, *On a property of uniqueness*, Mat. Sb. 72 (114) (1967), 237–249 = Math. USSR Sbornik 1 (1967), 209–220. MR 34 #7780.

[14] Ju. F. Korobeĭnik, *Completeness in multiply connected domains*, Seventh All-Union Conference on the Theory of Functions of a Complex Variable, Rostov-on-Don, 1964, Abstracts of reports, pp. 85–87.

Translated by:
B. Seckler

# ON BOUNDS OF CONVEXITY FOR STARLIKE FUNCTIONS OF ORDER $\alpha$ IN THE CIRCLE $|z| < 1$ AND IN THE CIRCULAR REGION $0 < |z| < 1$

UDC 517.53

V. A. ZMOROVIČ

§ 1

Let $\alpha \in [0; 1)$. By $S_\alpha^*$ we denote the class of univalent functions $f(z)$ ($f(0) = 0$, $f'(0) = 1$) starlike in the circle $|z| < 1$ and satisfying in this circle the condition

$$\operatorname{Re}\left(\frac{zf'(z)}{f(z)}\right) > \alpha,$$

and by $\Sigma_\alpha^*$, the class of univalent functions $F(z) = z^{-1} + a_0 + a_1 z + \cdots$ starlike in the circular region $0 < |z| < 1$ and satisfying in this region the condition

$$\operatorname{Re}\left(-\frac{zF'(z)}{F(z)}\right) > \alpha.$$

The aim of the present note is to establish explicit expressions (as functions of $\alpha$) for the bounds of convexity for the classes $S_\alpha^*$ and $\Sigma_\alpha^*$. Up to now this has been successfully done only when $\alpha = 0$ and $\alpha = 1/2$ (see [1-4]).

By $P$ we denote the class of functions $p(z)$ ($p(0) = 1$), regular in the circle $|z| < 1$, satisfying in this circle the condition $\operatorname{Re} p(z) > 0$. If $f \in S_\alpha^*$ there exists a function $p \in P$ such that

$$\frac{zf'}{f} = \alpha + (1-\alpha)p, \tag{1}$$

whence

$$1 + \frac{zf''}{f'} = (1-\alpha)(p+h) + \frac{zp'}{p+h}, \tag{2}$$

where $h = \alpha/(1-\alpha)$. Let

$$Q_\alpha(r) = \min_{p \in P} \min_{|z|=r<1} \operatorname{Re}\left[(1-\alpha)(p+h) + \frac{zp'}{p+h}\right]. \tag{3}$$

Then the bound $r_\alpha$ of convexity for the class $S_\alpha^*$ is defined as the smallest positive root of the equation $Q_\alpha(r) = 0$.

Analogously, when $F \in \Sigma_\alpha^*$ we can find a function $p \in P$ such that

$$-\frac{zF'}{F} = \alpha + (1-\alpha)p, \tag{4}$$

$$-\left(1+\frac{zF''}{F'}\right) = (1-\alpha)(p+h) - \frac{zp'}{p+h}. \tag{5}$$

If

$$\widetilde{Q}_\alpha(r) = \min_{p\in P}\ \min_{|z|=r<1} \operatorname{Re}\left[(1-\alpha)(p+h) - \frac{zp'}{p+h}\right], \tag{6}$$

then for the bound $\widetilde{r}_\alpha$ of convexity for the class $\Sigma^*_\alpha$ we obtain the equation $\widetilde{Q}_\alpha(r) = 0$.

Thus both problems are reduced to seeking the quantity

$$Q(r) = \min_{p\in P}\ \min_{|z|=r<1} \operatorname{Re}\Psi(p; zp'), \tag{7}$$

where $\Psi(\omega; w)$ is an analytic function of the variables $\omega$ and $w$ in the $w$-plane and in the halfplane $\operatorname{Re}\omega > 0$. By $P_2$ we denote the subclass of $P$ consisting of functions of the form

$$p(z) = \lambda_1 \frac{1+ze^{-i\theta_1}}{1-ze^{-i\theta_1}} + \lambda_2 \frac{1+ze^{-i\theta_2}}{1-ze^{-i\theta_2}}, \tag{8}$$

where $\theta_1$ and $\theta_2$ are arbitrary real constants which obviously can be taken to lie in $[0; 2\pi]$, while $\lambda_1$ and $\lambda_2$ are any nonegative constants whose sum is unity. Then, as is well known [5, 6], the minimum in (7) is reached in $P_2$, so that everything is reduced to substituting into formula (7) the function $p(z)$ from (8) and to finding the minimum of the resulting differentiable function of three real variables (we can set $\lambda_1 = (1+\lambda)/2$, $\lambda_2 = (1-\lambda)/2$, where $-1 \le \lambda \le 1$). However, this is precisely where the basic difficulties arise. The following theorem turns out to be useful for surmounting them.

Theorem 1. *Let* $\Psi(\omega; w) = M(\omega) + N(\omega)w$, *where* $M(\omega)$ *and* $N(\omega)$ *are defined and are finite in the halfplane* $\operatorname{Re}\omega > 0$. *We set*

$$\omega = \lambda_1\frac{1+z_1^m}{1-z_1^m} + \lambda_2\frac{1+z_2^m}{1-z_2^m}, \quad w = \lambda_1\frac{2mz_1^m}{(1-z_1^m)^2} + \lambda_2\frac{2mz_2^m}{(1-z_2^m)^2}, \tag{9}$$

*where* $z_1$ *and* $z_2$ *are any points on the circumference* $|z| = r < 1$, $m$ *is an integer* $(m \ge 1)$, $\lambda_1 \ge 0$, $\lambda_2 \ge 0$, $\lambda_1 + \lambda_2 = 1$. *Then the function* $\Psi(\omega; w)$ *can be put in the form*

$$\Psi(\omega; w) = M(\omega) + \frac{m}{2}(\omega^2 - 1)N(\omega) + \frac{m}{2}(\rho^2 - \rho_0^2)N(\omega)e^{2i\psi}, \tag{10}$$

*where*

$$\frac{1+z_k^m}{1-z_k^m} = a + \rho e^{i\psi_k} \ (k = 1, 2),$$

$$\omega = a + \rho_0 e^{i\psi_0} \quad (0 \leqslant \rho_0 \leqslant \rho), \ |z_1| = |z_2| = r,$$

$$a = \frac{1+r^{2m}}{1-r^{2m}}, \ \rho = \frac{2r^{2m}}{1-r^{2m}}, \ e^{i\psi} = ie^{\frac{i}{2}(\psi_1+\psi_2)}.$$

**Proof.** We note that

$$\frac{2mz^m}{(1-z^m)^2} = z\left(\frac{1+z^m}{1-z^m}\right)' = \frac{m}{2}\left[\left(\frac{1+z^m}{1-z^m}\right)^2 - 1\right].$$

Therefore (9) implies that

$$w - \frac{m}{2}(\omega^2 - 1) = -2m\lambda_1\lambda_2\rho^2 \sin^2\frac{\psi_2-\psi_1}{2} e^{i(\psi_1+\psi_2)}.$$

On the other hand

$$\rho_0 e^{i\psi_0} = \rho\,(\lambda_1 e^{i\psi_1} + \lambda_2 e^{i\psi_2}). \tag{11}$$

Hence

$$\rho_0^2 = \rho^2(\lambda_1^2 + \lambda_2^2 + 2\lambda_1\lambda_1 \cos(\psi_2 - \psi_1)) = \rho^2\left(1 - 4\lambda_1\lambda_2 \sin^2\frac{\psi_2-\psi_1}{2}\right).$$

Therefore

$$w = \frac{m}{2}(\omega^2 - 1) - \frac{m}{2}(\rho^2 - \rho_0^2)\, e^{i(\psi_1+\psi_2)}.$$

It is obvious that this equality proves Theorem 1.

**Remark.** For a fixed value of $\omega$ ($|\omega - a| < \rho$), the angle $2\psi$ in formula (10) can take any value from the interval $[0; 2\pi]$. To show this it suffices to consider that for suitably defined angles $\psi_1$ and $\psi_2$ the angle $\psi = (\pi + \psi_1 + \psi'_2)/2$ defines the slope (with the positive real axis) of the secant passing through the fixed point $a + \rho_0 e^{i\psi_0}$ ($\rho_0 < \rho$) and intersecting the circumference $|\omega - a| = \rho$ at the points $a + \rho e^{i\psi_k}$ ($k = 1, 2$). By placing positive masses $\lambda_1$ and $\lambda_2$ ($\lambda_1 + \lambda_2 = 1$) at these points such that $\lambda_1/\lambda_2 = |\rho e^{i\psi_2} - \rho_0 e^{i\psi_0}| \times |\rho e^{i\psi_1} - \rho_0 e^{i\psi_0}|^{-1}$, we satisfy condition (11). Since the slope angle of the secant can take any value in the interval $[0; \pi]$, it is clear that the angle $2\psi = \pi + \psi_1 + \psi_2$ can take any value in the interval $[0; 2\pi]$. Using the arbitrariness of the angle $2\psi$, for a fixed value of $\omega$ ($|\omega - a| < \rho$) we obtain from (10)

$$\min \operatorname{Re} \Psi(\omega; w) \equiv \Psi_\rho(\omega)$$
$$= \operatorname{Re}\left[M(\omega) + \frac{m}{2}(\omega^2 - 1) N(\omega)\right] - \frac{m}{2}|N(\omega)|(\rho^2 - \rho_0^2). \tag{12}$$

This minimum is reached when

$$\exp[i(2\psi + \arg N(\omega))] = -1. \tag{13}$$

An analogous equality can be obtained also for $\max \operatorname{Re} \Psi(\omega; w)$.

Thus, thanks to formula (12), the determination of the value of (7) is reduced to finding the minimum of the function $\Psi_\rho(\omega)$ in the circle $|\omega - a| \leq \rho$. By the same token Theorem 1 reduces the number of variables in the minimum-problem to two, which in a number of cases is a significant simplification. Theorem 1 can be generalized considerably, but we do not need this generalization in the present paper. In addition to the applications of Theorem 1 pointed out in the present note, many others are also possible. Some of them have been presented in [7].

## § 2

In this section we solve the problem of the bound $r_\alpha$ of convexity for the class $S_\alpha^*$. By applying formula (12) to the function

$$\Psi(\omega; w) = (1-\alpha)(\omega + h) + \frac{w}{\omega + h}, \quad h = \frac{\alpha}{1-\alpha}, \quad \alpha \in [0, 1),$$

for $m = 1$ we get:

$$\min \operatorname{Re} \Psi(\omega; w) \equiv \Psi_\rho(\omega)$$
$$= \operatorname{Re}\left[(1-\alpha)(\omega + h) + \frac{1}{2}\frac{\omega^2 - 1}{\omega + h}\right] - \frac{1}{2} \cdot \frac{\rho^2 - \rho_0^2}{|\omega + h|}. \tag{14}$$

This minimum is reached when the point $\omega$ ($|\omega - a| < \rho$) is fixed, and the chord passing through it and through the points $a + \rho e^{i\psi_k}$ ($k = 1, 2$) is perpendicular to the vector $e^{i\vartheta/2}$, where $\omega + h = |\omega + h| e^{i\vartheta}$ (this follows from formula (13)). By setting $\omega = a + \xi + i\eta$ and $\rho_0^2 = \xi^2 + \eta^2 \leq \rho^2$, from (14) we get

$$\Psi_\rho(\omega) \equiv \Psi_\rho(\xi; \eta) = \left(\frac{3}{2} - \alpha\right)(\xi + a + h)$$
$$+ \frac{1}{2}(h^2 - 1)(\xi + a + h) R^{-2} - \frac{1}{2}(\rho^2 - \xi^2 - \eta^2) R^{-1} - h. \tag{15}$$

Here we have set $R^2 = (\xi + a + h)^2 + \eta^2$.

Let us prove that the minimum of the function $\Psi_\rho(\xi;\eta)$ in the circle $\xi^2 + \eta^2 \le \rho^2$ is achieved on the diameter $\eta = 0$. We have

$$\frac{\partial}{\partial\eta}\Psi_\rho(\xi;\eta) = \frac{1}{2}\eta R^{-4} S(\xi;\eta), \tag{16}$$

where

$$S(\xi;\eta) = [\xi^2 + 4(a+h)\xi + \rho^2 + \eta^2 + 2(a+h)^2]R - 2(h^2-1)(\xi+a+h)$$

$$\geqslant [\xi^2 + 4(a+h)\xi + \rho^2 + 2(a+h)^2 - 2(h^2-1)](\xi + a + h).$$

It is easily verified that the minimum on the interval $-\rho \le \xi \le \rho$ of the expression in the last square brackets is reached at the point $\xi = -\rho$ and equals $2(a-\rho)^2 + 4h(a-\rho) + 2 > 0$. Therefore, as follows from (16), the minimum of $\Psi_\rho(\xi;\eta)$ on every chord $\xi = \text{const}$ is reached when $\eta = 0$. But then it is obvious that the minimum of $\Psi_\rho(\xi;\eta)$ in the circle $\xi^2 + \eta^2 \le \rho^2$ also is reached on the diameter $\eta = 0$, which is what we had to prove.

By setting $\eta = 0$ in (15) we arrive finally at the following problem: to find the minimum of the function

$$l(R) \equiv \Psi_\rho(\xi;\eta) = (2-\alpha)R + (h^2 + ah)R^{-1} - a - 2h, \tag{17}$$

where $R = a + h + \xi$, on the interval $a + h - \rho \le R \le a + h + \rho$. It is easily verified that the absolute minimum of $l(R)$ on the interval $(0, \infty)$ is reached at the point

$$R_0 = \sqrt{\frac{h^2 + ah}{2-\alpha}} \tag{18}$$

and equals

$$l(R_0) = 2\sqrt{(2-\alpha)(h^2 + ah)} - a - 2h. \tag{19}$$

We can show that always $R_0 < a + h + \rho$, but that it is not always true that $R_0 \ge a + h - \rho$. Therefore we should consider the case when $R_0 \notin [a + h - \rho;\ a + h + \rho]$. In this case the minimum of $l(R)$ on the interval $[a + h - \rho;\ a + h + \rho]$ is reached at the point

$$R_1 = a + h - \rho. \tag{20}$$

It follows from what has been said that the bound $r_\alpha$ of convexity for the class $S^*_\alpha$ is determined either from the equation

$$l(R_0) = 0, \tag{21}$$

or from the equation

$$l(R_1) = 0; \tag{22}$$

these two equations coincide when $\alpha \in (0;\ 1)$, which is determined from the condition (see (18) and (20))

$$R_0 = R_1. \tag{23}$$

Equations (21) and (22) may be reduced to the following equations:

$$a^2 - 4\alpha a - 4\alpha h = 0, \tag{24}$$

$$(1-2\alpha)^2 r^2 - 2(2-3\alpha) r + 1 = 0. \tag{25}$$

From (24) and (25) we get

$$r_\alpha = \sqrt{\frac{2\sqrt{\Delta} + 2\alpha - 1}{2\sqrt{\Delta} + 2\alpha + 1}}, \quad \Delta = a^2 + \alpha h, \tag{26}$$

$$r_\alpha = [2 - 3\alpha + \sqrt{(1-\alpha)(3-5\alpha)}]^{-1}. \tag{27}$$

We can show that it is impossible to use formula (26) when $0 \le \alpha \le 1/5$, and formulas (27) when $\alpha \ge 1/2$. Therefore it is clear that the value $\alpha = \alpha_0$ which determines the transition from formula (27) to formula (26) lies in the interval $(1/5,\ 1/2)$. To obtain the equation which determines $\alpha_0$ we must eliminate $a$ from equations (23) and (24). We get

$$20\alpha^4 - 52\alpha^3 + 15\alpha^2 + 12\alpha - 4 = 0. \tag{28}$$

The roots of this equation lie in the intervals $(-1;\ 0)$, $(1/3;\ 2/5)$, $(1/2;\ 1)$, $(1;\ 3)$, and therefore $1/3 < \alpha_0 < 2/5$. Thus we must use formula (27) when $0 \le \alpha \le \alpha_0$ and formulas (26) when $\alpha_0 \le \alpha < 1$.

Let us determine the form of the extremal functions $f_0(z)$ of class $S_\alpha^*$ which realize the mappings with the radii of convexity (26) and (27). Taking into account (13) and the fact that the minimum in case (26) is realized at a point on the diameter $\eta = 0$, we conclude that $p(z)$ (see (1)) should in this case be taken in the form

$$p(z) = \frac{1}{2}\cdot\frac{1 + ze^{-i\theta}}{1 - ze^{-i\theta}} + \frac{1}{2}\cdot\frac{1 + ze^{i\theta}}{1 - ze^{i\theta}},$$

where $\cos\theta$ is found from the equation

$$h + (1 - r_\alpha^2)(1 - 2r_\alpha \cos\theta + r_\alpha^2)^{-1} = R_0 \tag{29}$$

in which the quantities $r_\alpha$ and $R_0$ are determined by formulas (26) and (18). On the basis of what we have said the extremal function $f_0(z) \in S^*_\alpha$ can be represented by the formula

$$f_0(z) = z(1-2z\cos\theta + z^2)^{-1+\alpha}. \tag{30}$$

In the case of (27) we must take into account that the minimum of (17) is reached at one of the endpoints of the interval $[a+h-\rho;\ a+h+\rho]$, i.e. on the circumference $|\omega - a| = \rho$. In this case the extremal function $f_0(z)$ can be obtained from formula (1) if we set

$$p(z) = \frac{1+z}{1-z}.$$

Then

$$f_0(z) = z(1-z)^{-2(1-\alpha)}. \tag{31}$$

A more general form for the extremal functions is $e^{-i\gamma} f_0(e^{i\gamma} z)$, where $\gamma$ is an arbitrary real constant.

We have thus proved the next theorem.

Theorem 2. *If* $\alpha_0$ *lies in the interval* $(1/3, 2/5)$ *of the root of equation* (28), *then the bound* $r_\alpha$ *of convexity for the class* $S^*_\alpha$ *is determined from formula* (26) *when* $\alpha_0 \le \alpha < 1$ *and from formula* (27) *when* $0 \le \alpha \le \alpha_0$. *In the first case the extremal function can be taken in the form* (30); *in the second, in the form* (31).

Remark. When $\alpha = 0$, from (27) we obtain the well-known classical result: $r_0 = 2 - \sqrt{3}$, while when $\alpha = 1/2$, from (26) there follows the result of MacGregor [2]: $r_{1/2} = \sqrt{2\sqrt{3} - 3}$, obtained more simply by Robertson [4].

By noting that the extremal functions (30) and (31) belong to the subclass of typically-real starlike functions of order $\alpha$ in $S^*_\alpha$, we see that they also solve the corresponding minimum problem in this subclass.

## § 3

Let us find the solution to the problem of the bound of convexity for the class $\Sigma^*_\alpha$. Using formulas (5) and (12) and retaining the previous notation, we are led to the problem of finding the minimum of the function

$$\Psi_\rho(\omega) = \operatorname{Re}\left[(1-\alpha)(\omega + h) - \frac{1}{2}\cdot\frac{\omega^2 - 1}{\omega + h}\right] - \frac{1}{2}\cdot\frac{\rho^2 - \rho_0^2}{|\omega + h|} \tag{32}$$

in the circle $|\omega - a| \le \rho$. We are interested in the value of $r \in (0;\ 1)$

($a = (1+r^2)/(1-r^2)$; $\rho = 2r/(1-r^2)$) for which the minimum of the function $\Psi_\rho(\omega)$ equals zero (we make use of this remark later on). If we set $\omega + h = \mathrm{Re}^{i\vartheta}$ in (32) we obtain

$$\Psi_\rho(\omega) \equiv L(R;\ \vartheta) = \left(\frac{1}{2} - \alpha\right) R\cos\vartheta + \frac{1}{2}(1-h^2)R^{-1}\cos\vartheta$$
$$+\frac{1}{2}R + \frac{1}{2}R^{-1}(1+h^2+2ah) - (a+h)\cos\vartheta + h. \tag{33}$$

Hence

$$\frac{\partial}{\partial\vartheta} L(R;\ \vartheta) = T(R;\ \vartheta)\sin\vartheta, \tag{34}$$

where

$$T(R;\ \vartheta) = a + h + \left(\alpha - \frac{1}{2}\right) R + \frac{1}{2}(h^2-1)R^{-1}.$$

When $\alpha \geq 1/2$ (and consequently $h \geq 1$) we obviously have $T(R;\vartheta) > 0$, and therefore, as is seen from (34), the minimum of $\Psi_\rho(\omega)$ on every arc $R = \text{const}$ inside the circle $|\omega - a| \leq \rho$ for any $r \in (0;\ 1)$ is reached on the diameter $\vartheta = 0$; hence the minimum of $\Psi_\rho(\omega)$ in the whole of this circle also is reached on this same diameter. When $0 \leq \alpha < 1/2$ (and consequently $0 \leq h < 1$) we can assert only that the minimum of $\Psi_\rho(\omega)$ on any arc $R = \text{const}$ inside the circle $|\omega - a| \leq \rho$ is reached either when $\vartheta = 0$ or at the endpoints of this arc, situated on the circumference $|\omega - a| = \rho$. By setting $\rho_0 = \rho$ in (32) we get

$$L(R;\ \vartheta) = \left(\frac{1}{2} - \alpha\right) R\cos\vartheta + \frac{1}{2}(1-h^2)R^{-1}\cos\vartheta + h > 0,$$

where $r \in (0;\ 1)$, $0 \leq \alpha < 1/2$, $0 \leq h < 1$. This signifies that the minimum of $L(R;\ \vartheta)$, equal to zero, can be reached only on the diameter $\vartheta = 0$. Thus the matter reduces to the investigation of the function

$$l(R) \equiv L(R;\ 0) = (1-\alpha)R + \frac{1+ah}{R} - a$$

on the interval $a + h - \rho \leq R \leq a + h + \rho$; moreover, we are required to find such a value of $r \in (0;\ 1)$ for which the minimum of this function on this interval equals zero. Since the plan of the investigation is basically the same as in § 2, and all the calculations are elementary, we shall merely cite the ready results.

The absolute minimum of $l(R)$ on $(0\ ;\ \infty)$ is reached at the point

$$R_0 = \sqrt{\frac{1+ah}{1-\alpha}}. \tag{35}$$

We can show that always $R_0 > a + h - \rho$, but that it is not always true that $R_0 \leq a + h + \rho$. Therefore it is possible that the minimum of $l(R)$ on $[a + h - \rho;\ a + h + \rho]$ will be reached at the point

$$R_1 = a + h + \rho. \tag{36}$$

Both the minima coincide for the value of $\alpha \in (0;\ 1)$ for which

$$R_0 = R_1. \tag{37}$$

By solving the equations $l(R_0) = 0$ and $l(R_1) = 0$ with respect to $r$, we get the following two formulas for the bound $\tilde{r}_\alpha$ of convexity for the class $\Sigma_\alpha^*$:

$$\tilde{r}_\alpha = \sqrt{\frac{\sqrt{\Delta_1} + 2\alpha - 1}{\sqrt{\Delta_1} + 2\alpha + 1}}, \quad \Delta_1 = (1 - 2\alpha)^2 + 3, \tag{38}$$

$$\tilde{r}_\alpha = [\alpha + \sqrt{(1 - \alpha)(3\alpha - 1)}]^{-1} \tag{39}$$

Formula (39) cannot be used when $0 \leq \alpha \leq 1/2$, since when $0 \leq \alpha < 1/3$ it yields imaginary values, while when $1/3 \leq \alpha \leq 1/2$ it implies that $\tilde{r}_\alpha^* \geq 1$, which is impossible. Therefore formula (38) is always valid when $0 \leq \alpha \leq 1/2$. In order to obtain an equation to determine the "transition" value of $\alpha \in (0;\ 1)$ for going from formula (38) to formula (39), we must make use of relation (37) and of the equations $l(R_0) = 0$, $l(R_1) = 0$. We finally obtain the following results. If we introduce the notation $\tilde{\alpha}_0 = (x_0^2 - 4)(4x_0 - 4)^{-1}$, where $x_0$ is the single positive root of the equation

$$x^4 - 4x^3 + 2x^2 - 8 = 0, \tag{40}$$

then formula (38) is true when $0 \leq \alpha \leq \tilde{\alpha}$ while formula (39) is true when $\tilde{\alpha}_0 \leq \alpha < 1$. Computations show that $2 + \sqrt{2} < x_0 < 4$ and $(3 - \sqrt{2})/2 = 0.79289\cdots < \tilde{\alpha}_0 < 1$.

Let us determine the form of the extremal function $F_0(z)$ of class $\Sigma_\alpha^*$ in each of the cases (38) and (39). Since formula (38) is obtained when $l(R)$ reaches its own minimum at an interior point of the interval $[a + h - \rho;\ a + h + \rho]$, while formula (39) is obtained when this minimum is reached at one of the endpoints of this interval, in the first case the function $p(z)$ in formula (4) belongs to the subclass $P_2$, while in the second it belongs to the subclass $P_1$ for which one of the numbers $\lambda_1$, $\lambda_2$ in (8) equals zero. If we also take condition (13) into account, we finally get that in (8) we must set $\theta_1 = \pi$ and $\theta_2 = 2\pi$. Therefore the extremal function $F_0(z) \in \Sigma_\alpha^*$ in the case of (38) is determined by the formula

$$F_0(z) = \frac{1}{z}[(1-z)^{1+\lambda}(1+z)^{1-\lambda}]^{1-\alpha}, \tag{41}$$

where $R_0 = a_\alpha + \lambda\rho_\alpha + h$; moreover, $R_0$ is determined by equality (35), while $a_\alpha$ and $\rho_\alpha$ are the values of $a$ and $\rho$ when $r = \tilde{r}_\alpha$ from (38).

In the case of (39) the extremal function can be represented in the form

$$F_0(z) = \frac{1}{z}(1-z)^{2(1-\alpha)}. \tag{42}$$

A more general form of the extremal functions is $e^{i\gamma}F_0(ze^{i\gamma})$, where $\gamma$ is an arbitrary real constant.

We have thus proved the next theorem.

Theorem 3. *If* $\tilde{\alpha}_0 = (x_0^2 - 4)(4x_0 - 4)^{-1}$, *where* $x_0$ *is the single positive root of equation* (40), *then the bound* $\tilde{r}_\alpha$ *of convexity for the class* $\Sigma_\alpha^*$ *is determined by formula* (38) *when* $0 \le \alpha \le \tilde{\alpha}_0$, *and by formula* (39) *when* $\tilde{\alpha}_0 \le \alpha < 1$. *The extremal functions may be taken in the forms* (41) *and* (42), *respectively.*

Remark. When $\alpha = 0$ and $\alpha = 1/2$ the results of Robertson [3, 4] are obtained from formula (38). Since the extremal functions (41) and (42) belong to the subclass of typically-real starlike functions of order $\alpha$ in the class $\Sigma_\alpha^*$, they solve the corresponding minimum problem in this subclass also. It should be noted that by applying an analogous method we have succeeded in establishing also the bounds of convexity for subclasses of $m$-fold symmetric functions ($m \ge 2$ is an integer) in the classes $S_\alpha^*$ and $\Sigma_\alpha^*$.

## BIBLIOGRAPHY

[1] A Colloquium on Classical Function Theory (Cornell Univ., 17–21 August 1961), Notices Amer. Math. Soc. 8 (1961), 483–485. (Problem 12.)

[2] T. H. MacGregor, *The radius of convexity for starlike functions of order* ½, Proc. Amer. Math. Soc. 14 (1963), 71–76. MR 27 #283.

[3] M. S. Robertson, *Extremal problems for analytic functions with positive real part and applications*, Trans. Amer. Math. Soc. 106 (1963), 236–253. MR 26 #325.

[4] ———, *Some radius of convexity problems*, Michigan Math. J. 10 (1963), 231–236. MR 27 #1580.

[5] ———, *Variational methods for functions with positive real part*, Trans. Amer. Math. Soc. 102 (1962), 82–93. MR 24 #A3288.

[6] V. A. Zmorovič, *On certain theorems of the theory of extremal estimates in special classes of analytic functions*, Dopovīdī Akad. Nauk Ukraïn. RSR 1965, 980–984. (Ukrainian) MR 33 #5874.

[7] T. G. Èzrohi, *Certain estimates in special classes of univalent functions regular in the circle* $|z| < 1$, Dopovīdī Akad. Nauk Ukraïn. RSR 1965, 984–988. (Ukrainian) MR 33 #7518.

Translated by:
N. H. Choksy

# ON A CLASS OF EXTREMAL PROBLEMS ASSOCIATED WITH REGULAR FUNCTIONS WITH POSITIVE REAL PART IN THE CIRCLE $|z|<1$

V. A. ZMOROVIČ

## §1

Let $P$ denote the class of functions $p(z)$, $p(0)=1$, which are regular in the circle $|z|<1$ and for which $\operatorname{Re} p(z)>0$. By $P_n$ ($n$ is a positive integer) we denote the subclass of $P$ formed by functions of the type

$$\sum_{k=1}^{n} \lambda_k \frac{1+ze^{-i\theta_k}}{1-ze^{-i\theta_k}},$$

where $\theta_k$ are arbitrary real numbers and $\lambda_k$ are arbitrary nonnegative numbers connected by the relation $\Sigma_{k=1}^{n}\lambda_k=1$. With the aid of a variational method specially worked out by him, M. S. Robertson [1] proved the following theorem.

Theorem 1. *If $F(\zeta; w)$ is an analytic function of the complex variables $\zeta$ and $w$ in the region defined by the inequalities* $\operatorname{Re}\zeta>0$, $|w|<\infty$, *then*

$$\min_{p\in P}\ \min_{|z|=r<1} \operatorname{Re} F(p(z);\ zp'(z))$$

*is achieved in the subclass $P_2$.*

In the present note we prove very simply a more general theorem which in a number of cases makes it possible to construct a process for obtaining exact estimates for the functionals encountered in the theory of special classes of analytic functions associated with the functions of class $P$. Certain applications of this theorem are cited in the next section; other applications have been given in [2,3].

We introduce the class $M$ of functions $\mu(\theta)$, nondecreasing on $(-\infty, +\infty)$, satisfying the condition $\mu(\theta+2\pi)-\mu(\theta)=1$, $-\infty<\theta<+\infty$, and by $M_n$ ($n$ is a positive integer) we denote the subclass of $M$ consisting of all functions each of which possesses no more than $n$ points of growth on the interval $[0; 2\pi]$. If in addition we require, for example, that the functions in $M$ be continuous from the right, then we can establish a one-to-one correspondence between the classes $P$ and $M$ by means of the well-known Riesz-Herglotz formula

$$p(z)=\int_0^{2\pi}\frac{1+ze^{-i\theta}}{1-ze^{-i\theta}}\,d\mu(\theta). \tag{1}$$

The subclass $M_n$ here corresponds to the subclass $P_n$. We introduce the notation

$$\gamma_k = \int_0^{2\pi} e^{-ik\theta} d\mu(\theta) \quad (k = 1, 2, 3, \ldots). \tag{2}$$

As is well known, the quantities $\gamma_k$ (the moments of the distribution function $\mu(\theta)$) satisfy the inequalities

$$C_n = \begin{vmatrix} 1, & \gamma_1, & \gamma_2, \ldots, & \gamma_n \\ \bar{\gamma}_1, & 1, & \gamma_1, \ldots, & \gamma_{n-1} \\ \bar{\gamma}_2, & \bar{\gamma}_1, & 1, \ldots, & \gamma_{n-2} \\ \cdots & \cdots & \cdots & \cdots \\ \bar{\gamma}_n, & \bar{\gamma}_{n-1}, & \bar{\gamma}_{n-2}, \ldots, & 1 \end{vmatrix} \geqslant 0 \quad (n = 1, 2, 3, \ldots), \tag{3}$$

which are the necessary and sufficient conditions for the sequence $\{\gamma_k\}$ to be the sequence of moments (2) for some function $\mu(\theta) \in M$. If we take the numbers $\gamma_1, \cdots, \gamma_{n-1}$ as being fixed and such that $C_k > 0$ $(k = 1, \cdots, n-1)$, then it is not difficult to verify that inequality (3) with $n$ expresses the fact that under these conditions the domain of $\gamma_n$ is a certain circle $K_n$: $|\gamma_n - a_n| < R_n$ whose center affix $a_n$ and radius $R_n$ are continuous functions of the quantities $\gamma_1, \cdots, \gamma_{n-1}$ [9]. In particular, $a_2 = \gamma_1^2$ and $R_2 = 1 - |\gamma_1|^2$, where $|\gamma_1| < 1$.

It is important to note that the points on the circumference of the circle $K_n$ are the affixes of the values of the quantity $\gamma_n$ for the functions $\mu(\theta) \in M$ possessing not less than $n$ points of growth; moreover, it suffices that there be precisely $n$ of them.

Let us show this for $n = 2$, i.e. for the case which will be of particular interest for us in what follows. In this case we have

$$\gamma_1 = \lambda_1 e^{-i\theta_1} + \lambda_2 e^{-i\theta_2}, \quad \gamma_2 = \lambda_1 e^{-2i\theta_1} + \lambda_2 e^{-2i\theta_2},$$

where

$$\theta_1, \theta_2 \in (0; 2\pi], \lambda_1 > 0, \lambda_2 > 0, \lambda_1 + \lambda_2 = 1.$$

We set

$$e^{-i\theta_k} = \gamma_1 + \sigma_k, \text{ where } \sigma_k \neq 0, \ k = 1, 2.$$

Then

$$\lambda_1\sigma_1 + \lambda_2\sigma_2 = 0 \text{ and } \gamma_2 = \gamma_1^2 + \lambda_1\sigma_1^2 + \lambda_2\sigma_2^2.$$

Hence

$$\lambda_1\sigma_1^2 + \lambda_2\sigma_2^2 = -\sigma_1\sigma_2 = |\sigma_1|\cdot|\sigma_2|\, e^{i\vartheta},$$

where $\vartheta = 2 \arg \sigma_1$, while $|\sigma_1| \cdot |\sigma_2| = 1 - |\gamma_1|^2$ on the basis of a well-known theorem on the geometry of circles. Thus,

$$\gamma_2 = \gamma_1^2 + (1 - |\gamma_1|^2)\, e^{i\vartheta},$$

where we can take $\vartheta \in [0; 2\pi]$.

Consequently the circumference $|\gamma_2 - \gamma_1^2| = 1 - |\gamma_1|^2$ is covered by the values of $\gamma_2$ for functions in class $M_2$ with two points of growth, whereas the values of $\gamma_2$ for the entire class $M$ cover the whole circle $|\gamma_2 - \gamma_1^2| \leq 1 - |\gamma_1|^2$, where $|\gamma_1| < 1$, and only this circle.

The next theorem is easily proved on the basis of the foregoing.

Theorem 2. *Let the real function* $\Phi(\zeta; w)$ *of the complex variables* $\zeta$ *and* $w$ *be given uniquely and be finite in the region defined by the inequalities* $\operatorname{Re}\zeta > 0$, $|w| < \infty$. *Let us assume that this function has the following property: in any circle* $|w - w_0| \leq R$, *for any fixed value of* $\zeta$ *from* $\operatorname{Re}\zeta > 0$, *it reaches a certain specific limit (the same one for all circles) from among its two exact limits in this circle, on the circumference of this circle. Then, denoting this exact limit by "extr", we can assert that*

$$I = \operatorname*{extr}_{p \in P}\ \operatorname*{extr}_{|z| = r < 1} \Phi(p(z);\ zp'(z)) \tag{4}$$

*is reached in the subclass* $P_2$ *and, moreover, that*

$$I = \operatorname*{extr}_{|\tau| \leq \varrho}\ \operatorname*{extr}_{\vartheta \in [0;\, 2\pi]} \Phi(a + \tau;\ b(\tau) + R(\tau)\, e^{i\vartheta}), \tag{5}$$

*where*

$$a = \frac{1 + r^2}{1 - r^2}, \quad \varrho = \frac{2r}{1 - r^2}, \quad b(\tau) = \frac{1}{2}\,[(a + \tau)^2 - 1],$$

$$R(\tau) = \frac{1}{2}(\varrho^2 - |\tau|^2).$$

Proof. It is easy to verify that when $z = re^{i\phi}$ we have

$$\frac{1 + ze^{-i\theta}}{1 - ze^{-i\theta}} = a + \varrho e^{-i\psi}, \tag{6}$$

where $a$ and $\rho$ are as indicated in the theorem, while

$$\psi = \theta - \varphi + 2\operatorname{arctg}\frac{r \sin(\theta - \varphi)}{1 - r\cos(\theta - \varphi)},$$

$$\frac{\partial \psi}{\partial \theta} = \frac{1 - r^2}{1 - 2r\cos(\theta - \varphi) + r^2}.$$

Hence for fixed values of $r$ and $\phi$ the function $\psi = \psi(\theta)$ is strictly increasing on $[0; 2\pi]$, and therefore the inverse function $\theta = \theta(\psi)$ also is strictly increasing on $[0; 2\pi]$, and, moreover, $\theta(\psi + 2\pi) = \theta(\psi) + 2\pi$. We need this remark in what follows. From (1) we get

$$zp'(z) = \frac{1}{2}\int_0^{2\pi}\left[\left(\frac{1+ze^{-i\theta}}{1-ze^{-i\theta}}\right)^2 - 1\right]d\mu(\theta). \tag{7}$$

On the basis of (6), from (1) and (7) we obtain

$$p(z) = a + \tau, \; zp'(z) = \frac{1}{2}(a^2 - 1) + a\tau + t,$$

where

$$\tau = \varrho\int_0^{2\pi} e^{-i\psi(\theta)}d\mu(\theta), \; t = \frac{1}{2}\varrho^2\int_0^{2\pi} e^{-2i\psi(\theta)}d\mu(\theta). \tag{8}$$

Taking the remark made above into consideration, we can rewrite formula (8) in the form

$$\tau = \varrho\gamma_1, \quad t = \frac{1}{2}\varrho^2\gamma_2,$$

where

$$\gamma_k = \int_0^{2\pi} e^{-ik\psi}d\tilde{\mu}(\psi), \; k = 1, 2, \; \tilde{\mu}(\psi) = \mu(\theta(\psi)).$$

Since, as we noted earlier,

$$|\gamma_2 - \gamma_1^2| \leqslant 1 - |\gamma_1|^2,$$

we have

$$\left|t - \frac{1}{2}\tau^2\right| \leqslant \frac{1}{2}(\varrho^2 - |\tau|^2),$$

and moreover, for functions in $P_2$ we have

$$t = \frac{1}{2}\tau^2 + \frac{1}{2}(\varrho^2 - |\tau|^2)e^{i\vartheta},$$

where $\vartheta \in [0; 2\pi]$.

If we recall the property of the function $\Phi(\zeta; w)$ stated precisely in the hypothesis of the theorem, we conclude that (4) is indeed achieved in subclass $P_2$, and moreover that equality (5) is valid.

Theorem 2 is proved.

We make several remarks.

Remark 1. If $F(\zeta; w)$ is an analytic function of the complex variables $\zeta$ and $w$ in the region $\operatorname{Re}\zeta > 0$, $|w| < \infty$, then as the function $\Phi(\zeta; w)$ in Theorem 2 we can take any of the quantities $\operatorname{Re} F(\zeta; w)$, $\operatorname{Im} F(\zeta; w)$, $|F(\zeta; w)|$, $\arg F(\zeta; w)$.

We then obtain a straightforward generalization of Theorem 1.

As the function $\Phi(\zeta; w)$ we can select also any subharmonic or superharmonic function of the variable $w$, regular in the region $|w| < \infty$, depending also on a complex parameter $\zeta$, Re $\zeta > 0$, or we can select simply a differentiable function of the variables $\xi$, $\eta$, $u$, $v$, where $\zeta = \xi + i\eta$, $w = u + iv$, under the condition that grad $\Phi \neq 0$ in the appropriate region.

Other cases are also possible.

Remark 2. If the quantity

$$\operatorname*{extr}_{\vartheta \in [0;\ 2\pi]} \Phi(a + \tau;\ b(\tau) + R(\tau) e^{i\vartheta})$$

can be easily determined, as, for example, in those cases when $\Phi(\zeta; w) =$ Re $F(\zeta; w)$ or $\Phi(\zeta; w) =$ Im $F(\zeta; w)$, where $F(\zeta; w) = S(\zeta) + T(\zeta)w$ and, moreover, $S(\zeta)$ and $T(\zeta)$ are unique and finite in Re $\zeta > 0$ and $T(\zeta) \neq 0$ there, then Theorem 2 allows us to get rid of one of the three variables in the problem and to reduce the latter to the determination of the extremum of a function of two variables in a circle. In a number of cases this is an essential simplification.

Remark 3. Theorem 2 is easily generalized to the case when we have been given a real function $\Phi(\zeta; w_1; \cdots; w_n)$ of the complex variables $\zeta$, $w_1, \cdots, w_n$, defined in the region Re $\zeta > 0$, $|w_k| < \infty$, $k = 1, \cdots, n$, and we have to obtain its extremum on the circumference $|z| = r < 1$ when $\zeta = p(z)$, $w_k = z^k p^{(k)}(z)$ $(k = 1, \cdots, n)$, and $p(z)$ ranges over the class $P$.

If for fixed values of the variables $\zeta$, $w_1, \cdots, w_{n-1}$ this function has with respect to $w_n$ the very same property which was mentioned in the hypothesis of Theorem 2, then it is easy to prove that

$$I = \operatorname*{extr}_{p \in P} \operatorname*{extr}_{|z| = r < 1} \Phi(p(z);\ zp'(z); \ldots; z^n p^{(n)}(z))$$

is achieved in the subclass $P_{n+1}$, and moreover

$$I = \operatorname*{extr}_{\zeta} \operatorname*{extr}_{w_1} \ldots \operatorname*{extr}_{w_n} \Phi(\zeta;\ w_1; \ldots; w_n),$$

where the domain of $w_n$ is a certain circumference whose center affix and radius are continuously dependent on the variables $\zeta$, $w_1, \cdots, w_{n-1}$, while the domain of each variable $w_k$, $k = 1, \cdots, n-1$, is a certain circle whose center affix and radius are continuously dependent on the variables $\zeta, w_1 \cdots \cdots, w_{k-1}$. Finally, $\zeta$ varies in the circle $|\zeta - a| \leq \rho$.

All these circles can be determined with the aid of the circles defined by inequalities (3) if we make use of formula (1) and of the formulas obtained from it for $z^k p^{(k)}(z)$, $k = 1, \cdots, n$.

It is also not difficult to generalize Theorem 2 to the case when the function $\Phi$ depends on several vector-valued arguments $(w_{0,j}; \cdots; w_{n,j})$, $j = 1, \cdots, m$, where $w_{k,j} = z^k p_j^{(k)}(z)$, $k = 1, \cdots, n$, while the $p_j(z)$ are mutually independent functions in $P$.

Remark 4. If by $P(m)$ $(m \geq 2$ is an integer) we denote the subclass of

$P$ consisting of those functions $p(z)$ which for every $z$ from $|z|<1$ satisfy the condition $p(\epsilon_m z) = p(z)$, where $\epsilon_m = \exp(2\pi i/m)$, and if by $P_n(m)$ we denote the corresponding subclass in $P_n$, then Theorem 2 and Remarks 1–3 are easily extended to the case when $p(z)$ ranges over $P(m)$.

The matter is reduced to replacing $P_n$ by $P_n(m)$ and to inconsequential changes in the writing of several formulas. In particular, formula (5) retains its own form, except only that now

$$a = \frac{1+r^{2m}}{1-r^{2m}}, \quad \varrho = \frac{2r^m}{1-r^{2m}}, \quad b(\tau) = \frac{m}{2}[(\alpha+\tau)^2 - 1], \tag{9}$$

$$R(\tau) = \frac{m}{2}(\varrho^2 - |\tau|^2).$$

Remark 5. The method proposed in the present note differs both from the variational method developed in [4, 5] and from the method in [6].

## §2

In this section we present some applications of Theorem 2. First we shall prove one theorem, from which, as a corollary, we can obtain the greatest lower bound for the curvature of level curves in the class $S^*(m)$ of regular univalent $m$-fold symmetric functions, starlike in the circle $|z|<1$, which was obtained previously by other methods.

1) If $f(z) \in S^*(m)$, then, as we know,

$$\frac{zf'(z)}{f(z)} = p(z) \in P(m) \tag{10}$$

(see Remark 4 to Theorem 2 for the definition of $P(m)$).

The curvature of the image of the circumference $|z| = r < 1$ under the mapping $w = f(z)$ of the circle $|z|<1$ is determined by the formula

$$K_r(z) = \frac{1}{r|f'(z)|} \operatorname{Re}\left(1 + \frac{zf''(z)}{f'(z)}\right).$$

On the basis of (10) this formula can be rewritten as

$$K_r(z) = \frac{(1-r^{2m})^{\frac{1}{m}}}{r} \cdot \frac{\operatorname{Re}\widetilde{F}(p(z);\ zp'(z))}{[G(z)]^{1/m}}, \tag{11}$$

where

$$F(\zeta;\ w) = \frac{1}{|\zeta|}\left(\zeta + \frac{w}{\zeta}\right), \quad p(z) = \int_0^{2\pi} \frac{1+z^m e^{-i\theta}}{1-z^m e^{-i\theta}}\, d\mu(\theta),$$

$$\mu(\theta) \in M, \quad G(z) = \exp\left\{\int_0^{2\pi} \ln \frac{1-r^{2m}}{|1-z^m e^{-i\theta}|^2}\, d\mu(\theta)\right\}.$$

Taking into account the well-known inequality between the geometric and arithmetic means,

$$G(z) \leqslant A(z) = \int_0^{2\pi} \frac{1-r^{2m}}{|1-z^m e^{-i\theta}|^2}\, d\mu(\theta) = \operatorname{Re} p(z) \leqslant |p(z)|,$$

we obtain an estimate from below for $K_r(z)$:

$$K_r(z) \geqslant \frac{(1-r^{2m})^{\frac{1}{m}}}{r} \operatorname{Re} F_1(p(z);\ zp'(z)), \tag{12}$$

where

$$F_1(\zeta; w) = \frac{1}{|\zeta|^\lambda}\left(\zeta + \frac{w}{\zeta}\right), \quad \lambda = 1 + \frac{1}{m}.$$

We denote

$$K_r = \min_{p \in P(m)} \min_{|z|=r<1} K_r(z).$$

Then (12) implies that

$$K_r \geqslant \frac{(1-r^{2m})^{1/m}}{r} \min_{p\in P(m)} \min_{|z|=r<1} \operatorname{Re} F_1(p(z);\ zp'(z)). \tag{13}$$

It is easy to see that the function $F_1(\zeta; w)$ satisfies the hypothesis of Theorem 2 with due regard to Remark 4.

Applying equality (9), we prove the next theorem with the help of elementary calculations.

Theorem 3. *If*

$$F_1(\zeta; w) = \frac{1}{|\zeta|^\lambda}\left(\zeta + \frac{w}{\zeta}\right), \quad 0 \leqslant \lambda \leqslant 2,$$

*then the following estimate is valid when* $m \geq 1$:

$$\min_{p\in P(m)} \min_{|z|=r<1} \operatorname{Re} F_1(p;\ zp') = \left(\frac{1+r^m}{1-r^m}\right)^\lambda \frac{1-2(m+1)r^m + r^{2m}}{1-r^{2m}}.$$

We can take the extremal function in the form

$$p_0(z) = \frac{1+z^m}{1-z^m}.$$

As a corollary of this theorem, by using (13) we get

$$K_r \geqslant \frac{1-2(m+1)r^m + r^{2m}}{r} \frac{(1+r^m)^{2/m}}{(1-r^m)^2}, \tag{14}$$

since $1 + (1/m) \leq 2$ when $m \geq 1$.

Since the quantity $K_r(z)$ for the function

$$f_0(z) = z(1+z^m)^{-\frac{2}{m}} \in S^*(m)$$

equals, as we can easily show, the expression occurring on the right-hand side of (14), the estimate given by this inequality for $K_r$ is exact, as is obvious; in other words, we thus obtain an explicit expression for $K_r$ as a function of $r$, $r \in (0; 1)$ and we have simultaneously indicated a function in class $S^*(m)$ which realizes this estimate, although the question of the uniqueness of such a function remains open.

This estimate was first obtained for $m = 1$ by I. E. Bazilevič and for $m \geq 2$ by G. V. Korickiĭ [7] by the method of "imbedding" the class $S^*(m)$ into a wider class of functions analytic in the circle $|z| < 1$. Korickiĭ has even succeeded in proving that this same estimate is valid also in the whole class $S(m)$ of regular univalent functions $m$-fold symmetric in the circle $|z| < 1$ [8].

Recently I. A. Aleksandrov and V. V. Černikov have stated without proof [5] that this same estimate (and others, in a number of cases) can be obtained also by their method.

The method for obtaining estimate (14) indicated above turns out to be useful in many other cases also.

2) The least upper bound for $K_r(z)$ in the class $S^*(m)$ also has been presented without proof in [5]. For a long time there had been no success in obtaining such an estimate, although it has been known that to the extremal function $f_0(z) \in S^*(m)$ there corresponds the function $\mu_0(\theta) \in M_3$ (indeed we even have $\mu_0(\theta) \in M_2$).

We now prove a more general theorem on the least upper bound of the curvature of level curves in the class $S^*_\alpha(m)$, where $\alpha \in [0; 1)$. When $\alpha = 0$ the theorem announced in [5] is obtained from it.

We again make use of Theorem 2 in the case when $p(z)$ ranges over the class $P(m)$ (see (9)), and, using the very same elementary calculations as in the case of Theorem 3 (which we omit because they are easily reproduced), we begin by proving a preliminary theorem.

Theorem 4. *If*

$$F(\zeta;\ w) = \frac{1}{|\zeta + h| \cdot \mathrm{Re}^{\lambda}(\zeta + h)} \left( \frac{\zeta + h}{1 + h} + \frac{w}{\zeta + h} \right),$$

*where* $h = \alpha/(1 - \alpha)$, $\alpha \in [0; 1)$, $0 < \lambda < 1$, *then*

$$\max_{p \in P(m)} \max_{|z| = r < 1} \mathrm{Re}\, F(p(z);\ zp'(z)) = \frac{1}{R_0^{2+\lambda}} [(1 - \alpha) R_0^2 + m a R_0 - m(1 + ah)],$$

*where*

$$R_0 = 2m(2+\lambda)t_\alpha[ma(1+\lambda)+\sqrt{m^2a^2(1+\lambda)^2+4m\lambda(2+\lambda)t_\alpha}]^{-1}, \quad (15)$$

*when*
$$t_\alpha = (1-\alpha)(1+ah),$$

$$\lambda \geqslant \frac{m[ah+(a-2\varrho)(a+\varrho)]}{(1-\alpha)(a+h+\varrho)^2+m\varrho(a+\varrho)}, \quad a = \frac{1+r^{2m}}{1-r^{2m}},$$

$$\varrho = \frac{2r^m}{1-r^{2m}}, \quad (16)$$

*and* $R_0 = a + h + \rho$ *if the inequality in* (16) *is not fulfilled.*

In the first case the extremal function $p_0(z) \in P(m)$ can be taken in the form
$$p_0(z) = \frac{1}{1+h}\left(h + \mu_1 \frac{1-z^m}{1+z^m} + \mu_2 \frac{1+z^m}{1-z^m}\right),$$
where
$$\mu_1 = \frac{a+h+\varrho-R_0}{2\varrho}, \quad \mu_2 = \frac{R_0-a-h+\varrho}{2\varrho},$$

while in the second, in the form
$$p_0(z) = \frac{1}{1+h}\left(h + \frac{1+z^m}{1-z^m}\right).$$

Next, it is simple to prove the following lemma by an approximate method usual in such cases (see for example [7]), using functions of the class $M_n$.

Lemma. *Let* $\Psi(z;\mu) = \mathrm{Re}\,(p+h)/[G(z)]^{1/\beta}$, *where* $\beta > 0$ *and* $h$ *is a real constant; the meaning of* $p(z)$ *and* $G(z)$ *have been indicated in* (11). *We set*
$$s_0 = \frac{\beta}{\ln(a+\varrho)} - \frac{a+h}{\varrho}, \quad (17)$$
*where* $a$ *and* $\rho$ *are the same as in Theorem* 4. *Then the quantity*
$$J = \min_{\mu\in M}\min_{|z|=r<1}\Psi(z;\mu)$$
*is achieved in* $M_2$, *and moreover the points of growth of the extremal function* $\mu_0(\theta)$ *are* $\theta = \pi$ *and* $\theta = 2\pi$.

If $-1 < s_0 < 1$, then the jumps in $\mu_0(\theta)$ at these points are equal to $(1-s_0)/2$ and $(1+s_0)/2$, respectively; if $s_0 \le -1$, then 1 and 0; finally, if $s_0 \ge 1$, then 0 and 1.

In all these three cases we do not write out the explicit expressions for $J$ even though they are not complicated. The following theorem can already be proved on the basis of Theorem 4 and of the lemma.

Theorem 5. *Let* $r_0$ *denote the root, unique on the half-open interval* $(0; 1]$, *of the equation*

$$\frac{a+h+\varrho}{a+\varrho}\,\frac{\ln(a+\varrho)}{\varrho}=\frac{m^2(1+h)[(1+ah)(a-\varrho)-\varrho]}{(1-\alpha)(a+\varrho+h)^2+m\varrho(a+\varrho)}, \tag{18}$$

*where* $h=\alpha/(1-\alpha)$, $\alpha\in[0;1)$, $m\geq 1$, *while* $a$ *and* $\rho$ *are the same as in Theorem* 4. *Then when* $m=1$ *we have* $r_0=0$, *while when* $m\geq 2$ *we always have* $0<r_0<1$, *and on the interval* $r_0<r<1$ *the least upper bound* $\overline{K}_r$ *of the curvature of the level curves in the class* $S^*_\alpha(m)$ *is determined by the formula*

$$\overline{K}_r=\frac{(1+h)(1-r^{2m})^{1/\gamma}}{r}[(1-\alpha)+maR_0^{-1}-m(1+ah)R_0^{-2}]$$

$$\times\exp\left[\frac{a+h-R_0}{\gamma\varrho}\ln(a+\varrho)\right],\quad \gamma=m(1+h), \tag{19}$$

*where* $R_0$ *is the root, unique on the interval* $(a+h-\rho,\ a+h+\rho)$, *of the equation*

$$m(2+AR)(1+ah)-ma(1+AR)R-(1-\alpha)AR^3=0,$$
$$A=\frac{\ln(a+\varrho)}{\gamma\varrho}. \tag{20}$$

*When* $0<r\leq r_0$ *(only when* $m>1$), *in formula* (19) *we must set*

$$R_0=a+h+\varrho.$$

*The extremal function in class* $S^*_\alpha(m)$ *which realizes estimate* (19), *can in the first case take the form*

$$f_0(z)=z[(1+z^m)^{\mu_1}(1-z^m)^{\mu_2}]^{-\frac{2}{m}(1-\alpha)}$$

*where*

$$\mu_1=\frac{a+h+\varrho-R_0}{2\varrho},\quad \mu_2=\frac{R_0-a-h+\varrho}{2\varrho},$$

*while in the second case, the form*

$$f_0(z)=z(1-z^m)^{-\frac{2}{m}(1-\alpha)}.$$

Proof of Theorem 5. If $f(z)\in S^*_\alpha(m)$, then

$$\frac{zf'}{f}=\frac{p+h}{1+h},\text{ where } p\in P(m),\quad h=\frac{\alpha}{1-\alpha},\quad \alpha\in[0;1).$$

Therefore the formula for $K_r(z)$ can be written in the form

$$K_r(z)=\frac{1+h}{r}(1-r^{2m})^{1/\gamma}U(\mu;\lambda)V(\mu;\lambda),$$

where $\gamma=m(1+h)$, $0<\lambda<1$, $|z|=r<1$, $\mu\in M$,

$$U(\mu;\lambda)=\frac{\operatorname{Re}^{\lambda}(p+h)}{[G(z)]^{1/\gamma}},\quad V(\mu;\lambda)=\frac{1}{|p+h|\cdot\operatorname{Re}^{\lambda}(p+h)}\operatorname{Re}\left(\frac{p+h}{1+h}+\frac{zp'}{p+h}\right),$$

and $p$ and $G(z)$ are the same as in (11).

Since by Theorem 4 and the lemma, for any $\lambda\in(0;1)$, the absolute maxima as $\mu(\theta)$ ranges over $M$ of both the functionals $U$ and $V$, on the circumference $|z| = r < 1$, are realized by functions in the subclass $M_2$ with points of growth at $\theta = \pi$ and $\theta = 2\pi$, it only remains to choose a value of $\lambda$ such that both extremal functions coincide, i.e. such that we would have the same jumps at the points of growth.

To do this, as follows from Theorem 4 and the lemma, the condition

$$\frac{\gamma\varrho\lambda}{\ln(a+\varrho)} = R_0, \tag{21}$$

where $R_0$ is determined by formula (15), should be satisfied.

We can easily see that equation (21) always has a unique root on the interval (0; 1) when $0 \le \alpha < 1$, $0 < r < 1$, $m \ge 1$. We arrive at equation (20) by eliminating $\lambda$ from (15) and (21).

As regards equation (18), it is obtained from the conditions $R_0 \le a + h + \rho$ and $s_0 \le 1$, where $s_0$ is determined by formula (17).

These conditions delineate the "effective areas" of the two-jump and one-jump extremal function $\mu_0(\theta) \in M_2$ when $m > 1$. The theorem is proved.

Remark. The theorem in [5] is obtained when $\alpha = 0$ and $R_0 = 1/2(a+\rho)x_0$. We should keep in mind that the quantity $a$ which occurs in the formulation of the corresponding theorem in [5] equals $(a+\rho)^2$ in our notation.

It is considerably more difficult to obtain the greatest lower bound for $K_r(z)$ in the class $S^*_\alpha(m)$ when $\alpha > 0$. In the corresponding class in the circular region $0 < |z| < 1$ the lower bound when $\alpha > 0$ is more easily obtainable.

When $\alpha = 0$ the converse phenomenon is observed in both cases.

Aside from these cases, the method indicated turns out to be useful also for solving a number of other problems.

## BIBLIOGRAPHY

[1] M. S. Robertson, *Variational methods for functions with positive real part*, Trans. Amer. Math. Soc. **102** (1962), 82–93. MR **24** #A3288.

[2] T. G. Èzrohi, *Certain estimates in special classes of univalent functions regular in the circle* $|z| < 1$, Dopovīdī Akad. Nauk Ukraïn. RSR 1965, 984–988. (Ukrainian) MR **33** #7518.

[3] V. A. Zmorovič, *On certain theorems of the theory of extremal estimates in special classes of analytic functions*, Dopovīdī Akad. Nauk Ukraïn. RSR 1965, 980–984. (Ukrainian) MR **33** #5874.

[4] I. A. Aleksandrov and V. V. Černikov, *Extremal properties of starlike mappings*, Sibirsk. Mat. Ž. 4 (1963), 241–267. (Russian) MR **26** #6387.

[5] ———, *Extremal properties of univalent starlike mappings*, Sibirsk. Mat. Ž. 4(1963), 1201–1207. (Russian) MR 28 #1284.

[6] V. G. Lozovik, *Functionals defined on certain classes of analytic functions*, Ukrain. Mat. Ž. 15(1963), 95–100. (Russian) MR 27 #1575.

[7] G. V. Korickiĭ, *On curvature of level curves and of their orthogonal trajectories under conformal mappings*, Mat. Sb. 37(79) (1955), 103–116. (Russian) MR 17, 26.

[8] ———, *On the curvature of level curves under univalent conformal mappings*, Uspehi Mat. Nauk 15(1960), no. 5(95), 179–182. (Russian) MR 23 #A1793.

[9] Ja. L. Geronimus, *On polynomials orthogonal on a circle, on trigonometric moment-problem and on allied Carathéodory and Schur functions*, Mat. Sb. 15(57) (1944), 99–130. (Russian) MR 7, 63.

Translated by:
N. H. Choksy

# ON THE BOUNDS OF STARLIKENESS AND OF UNIVALENCE IN CERTAIN CLASSES OF FUNCTIONS REGULAR IN THE CIRCLE $|z| < 1$

V. A. ZMOROVIČ

In the present note we consider classes of functions $f(z)$, $f(0) = 0$, $f'(0) = 1$, regular in the circle $|z| < 1$ (denoted by $E$ in the following) and representable in the form

$$f(z) = F'(z) \cdot q(z), \tag{1}$$

where $F(z)$ ranges over a certain class $R$ of functions regular in $E$ and normalized by the conditions $F(0) = 0$, $F'(0) = 1$, while $q(z)$ ranges over a second class $R_1$ of functions regular in $E$ and normalized by the condition $q(0) = 1$. The definitions of these classes are given in §1. The class of the functions $f(z)$ in (1) itself will be denoted by $Q(R, R_1)$. The aim of the present note is to establish the bounds of starlikeness and of univalence in certain classes of type $Q(R, R_1)$. From the theorems obtained, as very special corollaries there ensue the theorems established by MacGregor [1, 2] and their special refinements by Krzyż and Reade [3]. Certain lemmas, which themselves are of independent interest, are used in proving the theorems.

## §1.

We introduce the classes $R$ and $R_1$.

Definition 1. We agree to designate as class $R$ any class of functions $F(z)$, regular in the circle $E$ and normalized as stated above, which satisfy the following conditions: 1) If $F(z) \in R$ and $\epsilon$ is a constant, $|\epsilon| = 1$, then $\epsilon^{-1} F(\epsilon z) \in R$; 2) there exists a number $r_0 \in (0; 1]$ such that on any circumference $|z| = r \leq r_0$, for any function $F(z) \in R$, the estimate, exact in $R$,

$$\operatorname{Re} \frac{zF'(z)}{F(z)} \geqslant \lambda(r) \tag{2}$$

is satisfied, where $\lambda(r)$ is some function defined for the given class $R$, strictly decreasing and continuous on $[0; r_0]$, $\lambda(0) = 1$, $\lambda(r_0) \geq 0$.

Remark. We agree to denote by $F(z; r)$ the function in $R$ which realizes estimate (2) at the point $z = r$ (the existence of at least one such function follows from the conditions defining $R$). We shall sometimes denote the class $R$ itself by

$$R[\lambda(r);\ r_0],$$

although it should be noted that $R$ is not completely specified by giving $r_0$ and $\lambda(r)$. However, in what follows it is precisely these characteristics of $R$ which play an essential role, so that two different classes $R$ for which these characteristics are the same will in a number of cases be equivalent (when establishing the bounds of starlikeness and of univalence). In the case when apart from the characteristics mentioned the corresponding extremal functions of both classes also will coincide, these classes are completely equivalent from the point of view of those theorems we shall consider here.

**Definition 2.** We agree to designate as class $R_1$ any class of functions $q(z)$, regular in $E$, which apart from the normalization $q(0) = 1$ further satisfy the following conditions: 1) If $q(z) \in R_1$ and $\epsilon$ is a constant, $|\epsilon| = 1$, then $q(\epsilon z) \in R_1$; 2) for every function $q(z) \in R_1$ one has on the circumference $|z| = r < 1$ the estimate, exact in $R_1$,

$$\operatorname{Re}\frac{zq'(z)}{q(z)} \geqslant \sigma(r), \tag{3}$$

where $\sigma(r)$ is a certain function defined for the given class $R_1$, strictly decreasing on $[0; 1]$, continuous at least on $[0;1)$, and if $\sigma(r)$ is bounded on $[0; 1)$, then it is continuous also on $[0; 1]$.

From the condition of exactness of estimate (3) it follows that $\sigma(0) = 0$ and therefore $\sigma(r) < 0$ on $(0; 1)$.

**Remark.** We agree to denote by $q(z; r)$ those functions in $R_1$ which realize estimate (3) at the point $z = r$. The existence of at least one such function follows from the definition of $R_1$.

Two classes $R_1$ for which estimates (3) coincide, as also do the corresponding extremal functions, are completely equivalent from the point of view of the theorems to be established in this note.

**Definition 3.** We agree to call the two classes $R$ and $R_1$ concordant if

$$\lambda(r_0) + \sigma(r_0) \leqslant 0.$$

The equation $\lambda(r) + \sigma(r) = 0$ has a root (obviously unique) on $(0; r_0)$ in this and only this case.

**Definition 4.** We shall say that the root $\tilde{r}$ of the equation $\lambda(r) + \sigma(r) = 0$ of two concordant classes $R$ and $R_1$ possesses the $l$-property if

$$\operatorname{Im}\left[\frac{zF'(z;\tilde{r})}{F(z;\tilde{r})}+\frac{zq'(z;\tilde{r})}{q(z;\tilde{r})}\right]_{z=\tilde{r}}=0.$$

This certainly holds if the extremal functions $F(z;\tilde{r})$ and $q(z;\tilde{r})$ take conjugate values at the points $z$ and $\bar{z}$ of circle $E$, but it occurs in other cases also.

The next proposition follows from what we said in the Introduction and from Definitions 1–4.

**Theorem 1.** *If the classes $R[\lambda(r); r_0]$ and $R_1[\sigma(r)]$ are concordant, then the root $\tilde{r}$ of the equation $\lambda(r)+\sigma(r)=0$ is the bound of starlikeness for the class $Q(R, R_1)$. If, however, the number $\tilde{r}$ possesses the l-property, then it also equals the radius of univalence for the class $Q(R, R_1)$.*

The proof of this theorem is not particularly difficult and we shall omit it. Many corollaries are possible from this theorem. We consider some of them, but first we introduce in the next section special classes of functions regular in $E$, from among which we shall select the classes $R$ and $R_1$.

## §2.

Let $P$ denote the class of functions $p(z)$, $p(0)=1$, regular in $E$, for which $\operatorname{Re} p(z)>0$ in $E$.

**Definition 5.** Let $\operatorname{Re} h \geq 0$. We agree to denote by $P_{[h]}$ and $P_{(h)}$ the classes of functions $q(z)$ regular in $E$ defined by the formulas

$$q(z)=\frac{p(z)+h}{1+h}, \tag{4}$$

$$q(z)=\frac{1+h}{p(z)+h}, \tag{5}$$

where $p(z)$ ranges over $P$.

**Remark.** Obviously $P_{[0]}=P_{(0)}=P$.

It is easy to see that if $hh_1 \neq 0$, $\operatorname{Re} h \geq 0$ and $\operatorname{Re} h_1 \geq 0$, then $P_{[h]} \neq P_{(h_1)}$.

Next, if we denote

$$\operatorname{tg}\gamma=-\frac{\operatorname{Im} h}{1+\operatorname{Re} h}, \qquad -\frac{\pi}{2}<\gamma<\frac{\pi}{2},$$

then we can prove that $q(z)\in P_{[h]}$ if and only if $\operatorname{Re}(e^{i\gamma}q(z))>\alpha\cos\gamma$ in $E$, where $\operatorname{Re} h=\alpha/(1-\alpha)$, $\alpha\in[0;1)$. The following assertion is valid for the class $P_{(h)}$.

Lemma 1. *If* Re $h > 0$, *then* $q(z)$ *belongs to the class* $P_{(h)}$ *if and only if the inequality* $|q(z) - \tau| < |\tau|$, *where* $\tau = (1 + h)/(h + \bar{h})$, *is satisfied in the circle* $E$.

The proof of this lemma also is omitted because it is easily reproduced.

From this lemma, among other things, it follows that the class of functions introduced by MacGregor by means of the inequality $|q(z) - 1| < 1$ coincides with the class $P_{(1)}$ in our notation. We shall make use of this observation later on.

Definition 6. Let $f(z)$ be regular in $E$. We denote

$$D_1(f) = f'(z), \quad D_2(f) = \frac{zf'(z)}{f(z)}, \quad D_3(f) = 1 + \frac{zf''(z)}{f'(z)}.$$

If Re $h \geq 0$, then by $P_{\{h\}}$ we shall mean either of the classes $P_{[h]}$ and $P_{(h)}$ (see Definition 5). We agree to denote by $S^{(k)}_{\{h\}}$, $k = 1, 2, 3$, the class of functions $f(z)$, $f(0)$, $f'(0) = 1$, regular in $E$, defined by the formula $D_k(f) = q(z)$, $k = 1, 2, 3$, where $q(z)$ ranges over the class $P_{\{h\}}$.

Remark. The functions in the classes $S^{(k)}_{\{h\}}$, $k = 1, 2$, are univalent in $E$. Among the functions in the classes $S^{(3)}_{\{h\}}$ there are some which are not univalent in $E$ if Im $h \neq 0$. When Im $h = 0$, setting $h = \alpha/(1 - \alpha)$, $\alpha \in [0; 1)$, instead of the notation

$$S^{(1)}_{[h]}, \ S^{(2)}_{[h]}, \ S^{(2)}_{(h)}, \ S^{(3)}_{[h]}, \ S^{(3)}_{(h)}$$

we shall adopt the notation

$$S^{\mathrm{v}}_{\alpha}, \ S^{\mathrm{v}}_{(\alpha)}, \ S^{*}_{\alpha}, \ S^{*}_{(\alpha)}, \ S^{0}_{\alpha}, \ S^{0}_{(\alpha)}. \tag{6}$$

All functions in the classes (6) are univalent in $E$, so that these classes are subclasses of the class $E$ of functions which are regular, normed and univalent in $E$. The classes $S^{*}_{\alpha}$ and $S^{0}_{\alpha}$ are usually called, respectively, the class of starlike functions of order $\alpha$ and the class of convex functions of order $\alpha$, all functions being univalent in $E$. By analogy with these we shall call $S^{*}_{(\alpha)}$ and $S^{0}_{(\alpha)}$, respectively, the class of starlike functions of type $\alpha$ and the class of convex functions of type $\alpha$, all functions being univalent in $E$.

By $K_{\alpha}$ we denote any of classes (6). Then, when $0 \leq \alpha' < \alpha'' < 1$ we have $K_{\alpha'} \supset K_{\alpha''}$. If $L_{\beta}$ is again one of the classes (6) different from the one designated by $K_{\alpha}$, then $K_{\alpha} \neq L_{\beta}$ if $\alpha \neq 0$ and $\beta \neq 0$. When $\alpha = 0$ the classes

(6) coincide pairwise. The classes obtained under such coincidence will be denoted $S^v$, $S^*$, $S^0$. We consider certain lemmas.

Lemma 2. *The equalities*

$$S^*_\alpha = R\left[\frac{1+(2\alpha-1)r}{1+r};1\right],\quad S^*_{(\alpha)} = R\left[\frac{1-r}{1+(1-2\alpha)r};1\right],$$

$$S^0 = R\left[\frac{1}{1+r};1\right],$$

*are valid.*

This follows directly from the definitions of classes $R$, $S^*_\alpha$, $S^*_{(\alpha)}$, $S^0$ and from known estimates.

Lemma 3. *Let* $q(z) \in P_{[h]}$, *where* $\operatorname{Re} h = h_1 \geq 0$. *Then the estiamte*

$$\operatorname{Re}\frac{zq'(z)}{q(z)} \leqslant \frac{2r}{(1-r)[1+r+h_1(1-r)]} \tag{7}$$

*is valid on every circumference* $|z| = r < 1$. *This estimate is reached only when* $\operatorname{Im} h = 0$, *and the extremal function realizing it at the point* $z = r$ *on the indicated circumference (in what follows we shall call such a function a normalized extremal function) has the form*

$$q(z;r) = \frac{1+(1-2\alpha)z}{1-z}. \tag{8}$$

This lemma is well known in the literature on extremal estimates [6].

Lemma 4. *Let* $h \geq 0$. *By* $r(h)$ *we denote the root, unique on* $(2-\sqrt{3};1]$, *of the equation*

$$h(1+r)(4r-1-r^2) = (1-r)^3. \tag{9}$$

*Then on every circumference* $|z| = r < 1$, *for every function* $q(z) \in P_{[h]}$, *the estimate*

$$\operatorname{Re}\frac{zq'(z)}{q(z)} \geqslant \sigma(r) \tag{10}$$

*is valid, where*

$$\sigma(r) = \frac{-2r}{(1+r)[1-r+h(1+r)]} \tag{11}$$

*when* $0 \leq r \leq r(h)$ *and*

$$\sigma(r) = -(\sqrt{a+h} - \sqrt{h})^2, \quad a = \frac{1+r^2}{1-r^2}, \tag{12}$$

*when* $r(h) \leq r < 1$. *Estimate* (10) *is exact. The normalized extremal function for* (11) *is determined by the formula*

$$q(z; r) = \frac{1 + (2\alpha - 1) z}{1+z}, \tag{13}$$

*and for* (12), *by the formula*

$$q(z; r) = \frac{1 - 2\alpha\lambda z + (2\alpha - 1) z^2}{1 - 2\lambda z + z^2}, \tag{14}$$

*where*

$$\alpha = \frac{h}{1+h}, \quad \lambda = \frac{1}{a \cdot \varrho \sqrt{h}} (\varrho^2 \sqrt{h} - \sqrt{a+h}), \quad \varrho = \frac{2r}{1-r^2}, \quad h > 0.$$

**Proof.** We apply the same method we used in [4, 5]. We denote

$$\Phi(\omega; w) = \frac{w}{\omega + h}, \quad \omega + h = Re^{i\vartheta}, \quad R > 0, \quad |\omega - a| = \varrho_0 \leqslant \varrho,$$

where $|z| = r \in (0; 1)$, and we consider the function

$$l(R; \vartheta) = \min_{\psi \in [0; 2\pi]} \operatorname{Re} \Phi(\omega; w_\psi), \quad w_\psi = \frac{1}{2}(\omega^2 - 1) + \frac{1}{2}(\varrho^2 - \varrho_0^2) e^{i\psi}.$$

By means of elementary calculations we obtain

$$l(R; \vartheta) = A(R) \cos\vartheta + B(R),$$

where

$$A(R) = \frac{1}{2} R + \frac{1}{2} \frac{h^2 - 1}{R} - a - h,$$

$$B(R) = \frac{1}{2R}(R^2 + 1 + h^2 + 2ah) - h.$$

Then, as was shown in [4, 5],

$$\sigma(r) = \min_{\Gamma_r} l(R; \vartheta), \tag{15}$$

if $\Gamma_r$ denotes the circle

$$R^2 - 2(a+h) R \cos\vartheta + 1 + h^2 + 2ah \leqslant 0.$$

It is not difficult to prove that (15) is achieved on the diameter $\vartheta = 0$ of the circle $\Gamma_r$, so that the problem is reduced to seeking the minimum of the function

$$l_0 \equiv l(R;\,0) = R + \frac{h^2 + ha}{R} - a - 2h$$

on the segment

$$a - \varrho + h \leqslant R \leqslant a + \varrho + h.$$

Since

$$\frac{dl_0}{dR} = 1 - \frac{h^2 + ha}{R^2}, \quad \frac{d^2 l_0}{dR^2} = \frac{2(h^2 + ha)}{R^3},$$

on setting $R_0 = \sqrt{h^2 + ha}$ and keeping in mind that

$$R_0 < a + \varrho + h$$

for any $h \geq 0$ and $r \in (0;1)$, we conclude that

$$\sigma(r) = \frac{h^2 + ha}{a - \varrho + h} - \varrho - h = -\frac{2r}{(1+r)\left[(1-r) + h(1+r)\right]}$$

when $R_0 \leq a - \rho + h$, and

$$\sigma(r) = -(\sqrt{a+h} - \sqrt{h})^2$$

when $R_0 \geq a - \rho + h$.

By considering the graphs of the functions $\eta = \sqrt{h^2 + ha}$ and $\eta = a - \rho + h$ on $0 \leq r < 1$, it is easily established that the inequalities $R_0 \leq a - \rho + h$ and $R_0 \geq a - \rho + h$ are equivalent, respectively, to the inequalities $0 \leq r \leq r(h)$ and $r(h) \leq r < 1$, where $r(h)$ is a root of (9), $r(h) \in (2 - \sqrt{3};\, 1]$.

With regard to the form of the extremal functions $q(z;r)$, here we must proceed in the way indicated in [4, 5], taking formula (4) into account. The following lemma can be proved by this same method.

Lemma 5. *Let* $\operatorname{Re} h = 0$, $\operatorname{Im} h = \lambda \in (-\infty; +\infty)$. *Then, on any circumference* $|z| = r < 1$, *for any function* $q(z) \in P_{[h]}$, *the estimates*

$$-\varrho T(r;\lambda) \leqslant \operatorname{Re}\frac{zq'(z)}{q(z)} \leqslant \varrho T(r;\lambda) \tag{16}$$

*are valid, where*

$$T(r;\lambda) = \sqrt{\frac{1+t}{2}}\left[c\sqrt{\frac{1}{2}(2 - b^2 + b^2 t)} + b\sqrt{1-c^2}\sqrt{\frac{1-t}{2}}\right],$$

$$b = \frac{\varrho}{\sqrt{a^2 + \lambda^2}}, \quad c = \frac{a}{\sqrt{a^2 + \lambda^2}}, \quad a = \frac{1+r^2}{1-r^2}, \quad \varrho = \frac{2r}{1-r^2},$$

*while* $t$ *is the root, unique on* $(0;1]$, *of the equation*

$$b^4 t^3 + b^2(2 - b^2)t^2 + c^2(1 - 2b^2)t - c^2 = 0. \tag{17}$$

*The estimates are exact and are achieved for any* $r \in [0; 1)$ *by the functions*

$$q_0(z) = \frac{1 + \varepsilon e^{-2i\gamma} z}{1 - \varepsilon z}, \tag{18}$$

*where* $|\epsilon| = 1$, $\gamma \in (-\pi/2; \pi/2)$, $\operatorname{tg}\gamma = -\lambda$, *and only by them.*

**Proof.** By proceeding in the same way as in the proof of the preceding lemma and by considering first the problem of establishing the lower estimate in (16), we are led to seeking in the circle

$$R^2 - 2(a\cos\vartheta + \lambda\sin\vartheta)R + 1 + \lambda^2 \leqslant 0, \tag{19}$$

lying in the halfplane $R > 0$, $\vartheta \in (-\pi/2; \pi/2)$, the minimum of the function

$$l(R; \vartheta) = A(R)\cos\vartheta + B(R)\sin\vartheta + C(R), \tag{20}$$

where

$$A(R) = \frac{1}{2}\left(R - \frac{1+\lambda^2}{R} - 2a\right), \quad B(R) = -\lambda,$$

$$C(R) = \frac{1}{2R}(R^2 + 1 + \lambda^2), \quad p(z) + i\lambda = Re^{i\vartheta}.$$

Let us show that the minimum of $l(R;\vartheta)$ in the circle (19) is achieved on the circumference of this circle. To do this we show that when $R > 0$, $-\pi/2 < \vartheta < \pi/2$, the inequality

$$\left(\frac{\partial l}{\partial R}\right)^2 + \left(\frac{\partial l}{\partial \vartheta}\right)^2 \neq 0$$

is valid. If this were not so, then from the equations

$$\frac{\partial l}{\partial R} = 0, \quad \frac{\partial l}{\partial \vartheta} = 0$$

we would get

$$\cos\vartheta = \frac{1 + \lambda^2 - R^2}{1 + \lambda^2 + R^2}, \tag{21}$$

$$\operatorname{tg}\vartheta = \frac{2\lambda R}{1 + \lambda^2 - R^2 + 2aR}. \tag{22}$$

From (21) it would follow that

$$R = \sqrt{1+\lambda^2}\left|\operatorname{tg}\frac{\vartheta}{2}\right|.$$

Substituting this expression into (22), we get

$$\operatorname{tg}\vartheta = \frac{\lambda|\operatorname{tg}\vartheta|}{\sqrt{1+\lambda^2} + a|\operatorname{tg}\vartheta|}. \tag{23}$$

If we assume that $\operatorname{tg}\vartheta \neq 0$, then from (23) we get a contradiction. Therefore $\operatorname{tg}\vartheta = 0$ and consequently $R = 0$ and $\vartheta = 0$.

Thus indeed the function $l(R;\vartheta)$ does not have stationary points in the halfplane $R > 0$, $-\pi/2 < \vartheta < \pi/2$, and therefore it achieves its minimum and maximum in the circle (19) on the circumference, which is what we had to prove.

The equation of the circumference of (19) makes it possible to put expression (20) into the form

$$l(\vartheta) \equiv l(R;\vartheta) = \pm\sqrt{(a\cos\vartheta + \lambda\sin\vartheta)^2 - 1 - \lambda^2}\cos\vartheta.$$

In the case of the minimum we must retain the minus sign. By introducing the angles $\vartheta_0$ and $\vartheta_1$ by the formulas

$$a = \sqrt{a^2+\lambda^2}\cos\vartheta_0, \quad |\vartheta_0| < \frac{\pi}{2},$$

$$\varrho = \sqrt{a^2+\lambda^2}\sin\vartheta_1, \quad \vartheta_1 \in \left[0;\ \frac{\pi}{2}\right],$$

and by setting

$$\sin(\vartheta - \vartheta_0) = -\sin\vartheta_1\cdot\sin\theta, \quad \theta \in \left[-\frac{\pi}{2};\ \frac{\pi}{2}\right],$$

we finally obtain

$$l_1(\theta) \equiv l(\vartheta) = \pm\varrho\,[\cos\vartheta_0\sqrt{1 - b^2\sin^2\theta} + b\sin\vartheta_0\sin\theta]\cos\theta.$$

We need to find the minimum of this function on the segment $[-\pi/2;\pi/2]$. If $\sin\vartheta_0 \geq 0$ this minimum is reached in the second half of the segment, while if $\sin\vartheta_0 \leq 0$ it is reached in the first.

The equation $l_1(\theta) = 0$ can be transformed to the form (17) where $t = \sin\theta$. Thus we obtain the lower estimate in (16).

To obtain the upper estimate we need to act in accordance with the same plan. However, it should be noted that the function (20) is connected with the analogous function $l(R;\vartheta)$ of the upper estimate problem by the relation

$$\tilde{l}(R;\vartheta) = -l\left(\frac{1+\lambda^2}{R};\ \vartheta\right),$$

while the equation for the circumference of the circle (19) is unaltered when $R$ is replaced by $(1+\lambda^2)/R$. Thus we obtain the upper estimate in (16).

Finally, by noting that both estimates are reached at points on the circumference of circle (19), we obtain expression (18) for the external functions. Lemma 5 is proved.

**Remark.** Up to now only the following result was known: If $q(z) \in P_{[h]}$, where $\operatorname{Re} h = 0$, $\operatorname{Im} h = \lambda \in (-\infty; +\infty)$, then on every circumference $|z| = r < 1$ we have the estimates

$$-\varrho \leqslant \operatorname{Re}\frac{zq'(z)}{q(z)} \leqslant \varrho, \quad \varrho = \frac{2r}{1-r^2}, \tag{24}$$

which are exact only when $\lambda = 0$. If we consider that in Lemma 5

$$0 < T(r;\ \lambda) \leqslant 1,$$

and moreover that equality holds if and only if $t = 1$, and consequently if $\lambda = 1$, then (24) follows immediately from (16). We have as yet not succeeded in proving a lemma containing Lemmas 3, 4 and 5.

To complete the cycle of lemmas used in the present paper we prove one more lemma concerning the general class of functions which are regular, univalent and normalized in the circle $|z| < 1$, i.e. the class $S$.

**Lemma 6.** *If $f(z) \in S$, then on any circumference $|z| = r < 1$ the estimate*

$$\operatorname{Re}\frac{zf'(z)}{f(z)} \geqslant \lambda(r)$$

*is valid, where*

$$\lambda(r) = \frac{1-r}{1+r} \tag{25}$$

*when $0 \le r \le \operatorname{th} \frac{1}{2}$ and*

$$\lambda(r) = \exp(-\psi \operatorname{ctg} \psi) \cos \psi \tag{26}$$

*when $\operatorname{th} \frac{1}{2} \le r < 1$; here $r = \operatorname{th}(\psi/2 \sin \psi)$, $\psi \in [0; \pi)$. The estimates are exact. The normalized extremal function in case (25) has the form*

$$f_0(z) = z(1+z)^{-2}, \tag{27}$$

*and in case (26),*

$$f_0(z) = \frac{r}{(1-re^{i\alpha})^2}\left(\frac{u-v}{u+v}\right)^2, \tag{28}$$

*where*

$$u=(\sqrt{r}+\sqrt{z})(1-\sqrt{rz})^{\varepsilon_r},$$
$$v=(\sqrt{r}-\sqrt{z})(1-\sqrt{rz})^{\varepsilon_r},$$

*$\epsilon_r = -e^{i\alpha}$, $\alpha$ is a root, different from $\pi$, of the equation*

$$\alpha+\sin\alpha\ln\frac{1+r}{1-r}=\pi$$

*on the interval $(0;\pi]$. The power functions $(1\pm\zeta)^{\epsilon_r}$ are defined so that they equal 1 when $\zeta=0$.*

Estimate (25) is known [3]. However, we shall prove both estimates because our method of proof covers both cases.

Proof. If $f(z)\in S$, then the domain of the quantity

$$\zeta=\ln\frac{zf'(z)}{f(z)}$$

for a fixed value of $z$, $|z|=r<1$, is the circle [7]:

$$|\zeta|\leqslant\ln\frac{1+r}{1-r}. \tag{29}$$

Setting $\rho=\ln((1+r)/(1-r))$, whence $r=\operatorname{th}(\rho/2)$, we introduce into consideration the function $w=e^{\zeta}$ and we seek $\min\operatorname{Re}w$ in the circle (29). It is enough to consider $\operatorname{Re}w$ on the circumference $|\zeta|=\rho$. Then it is easy to prove that when $0\le\rho\le1$ the minimum of $\operatorname{Re}w$ on the circumference $|\zeta|=\rho$ equals $e^{-\rho}=1-r/1+r$. However, if $\rho>1$, then

$$\lambda(r)=\min_{|\zeta|=\varrho}\operatorname{Re}w=\exp(\varrho\cos\varphi)\cos(\varrho\sin\varphi),$$

where $\phi$ is a root, unique on $(0;\pi)$, of the equation

$$\varphi+\varrho\sin\varphi=\pi.$$

Setting $\psi=\pi-\phi$ we get $\rho=\psi/\sin\psi$. Thus in this case

$$\lambda(r)=\exp(-\psi\operatorname{ctg}\psi)\cos\psi,\quad r=\operatorname{th}\left(\frac{1}{2}\frac{\psi}{\sin\psi}\right),\quad \psi\in[0;\pi],$$

which is what we had to prove.

As regards the extremal functions (27) and (28), the first of them is obtained from (28) when $\alpha=\pi$, so that we should properly speak of only the one extremal function (28) whose form and uniqueness under the normalization adopted in this paper ensue from the results of the investigations in [8].

Remark. It is interesting to note that on the interval $(\operatorname{th}\frac{1}{2};1)$ we have

$$\lambda(r)=C(r)\frac{1+r}{1-r},$$

where $C(r) = \exp(-\psi \operatorname{ctg}(\psi/2)) \cos\psi$ and $|C(r)| < 1$, $C(r) \to -1$ as $r \to 1$, and, moreover, the variation of $C(r)$ is monotonic since $C'(r) < 0$ when $r \in [\operatorname{th} \tfrac{1}{2}; 1)$.

## §3.

We consider certain theorems which are corollaries of Theorem 1 and of the lemmas in the preceding section.

Theorem 2. *Let there be given the classes* $R[\lambda(r); r_0]$ *and* $R_1 = P_{(h)}$, *where* $h \geq 0$. *We introduce the function*

$$\Phi(r) = \frac{2r}{(1-r)^2 \lambda(r)} - \frac{1+r}{1-r},$$

*where* $r \in [0; r_0]$. *When* $r_0 = 1$ *we assume* $\Phi(1) = \infty$. *If* $\Phi(r_0) \geq 0$, *then for* $0 \leq h \leq \Phi(r_0)$ *if* $r_0 < 1$ *or for all* $h \geq 0$ *if* $r_0 = 1$, *the radius* $\tilde{r}$ *of starlikeness for the class* $Q(R, R_1)$ *is equal to the root, unique on* $(0; r_0]$, *of the equation*

$$\Phi(r) = h.$$

*If the number* $\tilde{r}$ *possesses the l-property, then it is equal also to the radius of univalence for this class. The normalized extremal function (not unique, in general) is determined by the formula*

$$f(z; \tilde{r}) = F(z; \tilde{r}) \frac{1-z}{1+(1-2\alpha) z},$$

*where* $h = \alpha/(1-\alpha)$.

This theorem is easily proved on the basis of Theorem 1 and Lemma 3.

We present certain corollaries of this theorem.

Corollary 1. *Let* $R = S^*_\beta$, $\beta \in [0; 1)$, $R_1 = P_{(h)}$, $h \geq 0$. *Then the radius of starlikeness and the radius of univalence for the class* $Q(R, R_1)$ *are both equal to the root, unique on* $(0; 1)$, *of the equation*

$$h(1-r)^2[1+(2\beta-1)r] = (1+r)[(4-2\beta)r - 1 + (2\beta-1)r^2]. \tag{30}$$

This follows from Lemmas 2 and 3 and Theorem 2. In the given case the number $\tilde{r}$ possesses the $l$-property because the normalized extremal function for the class $Q(S^*_\beta, P_{(h)})$ is, as is easily verified, typically-real.

Three simple special cases of equation (30) correspond to the assumptions: 1) $h = 0$, 2) $h = 1$, 3) $\beta = 1/2$. In these cases we obtain the following formulas for the radius $\tilde{r}$ of univalence (and starlikeness):

$$\tilde{r}=(2-\beta+\sqrt{3-2\beta+\beta^2})^{-1}, \tag{31}$$

$$\tilde{r}=2(3-2\beta+\sqrt{9-4\beta+4\beta^2})^{-1}, \tag{32}$$

$$\tilde{r}=(1+2\sqrt{1-\alpha})^{-1}, \tag{33}$$

where $h = \alpha/1 - \alpha$.

Special cases of formulas (31) and (32) when $\beta = 0$ have been obtained by MacGregor [1, 2]. However, he also treated two problems which essentially are equivalent to two particular cases of the problem we have considered, namely when $\beta = 1/2$, $h = 0$ and when $\beta = 1/2$, $h = 1$ (see the remark to Lemma 1). The corresponding values of $\tilde{r}$ are obtained from formula (33) [1, 2].

Corollary 2. *Let $R = S^*_{(\beta)}$ and $R_1 = P_{(h)}$, $h \geq 0$. Then the radius $\tilde{r}$ of univalence (and simultaneously of starlikeness) for the class $Q(R, R_1)$ is equal to the root, unique on the interval* (0;1), *of the equation*

$$h(1-r)^3=2r[1+(1-2\beta)r]-(1-r)^2(1+r). \tag{34}$$

This is obtained at once from Theorem 2 and Lemmas 2 and 3.

Remark. If $h = 1$, then from (34) we get

$$\tilde{r}=2(3+\sqrt{9-8\beta})^{-1}.$$

A second simple case corresponds to the assumption $\alpha = \beta$, where $\alpha = h/(1 + h)$. Here we get

$$\tilde{r}=(2-\alpha+\sqrt{3-4\alpha+\alpha^2})^{-1}.$$

Corollary 3. *Let $R = S$, $R_1 = P_{(h)}$, $h \geq 0$. If $0 \leq h \leq e(e^2-3)/2$, then the radius of univalence (and simultaneously of starlikeness) for the class $Q(R, R_1)$ is equal to the root, unique on $(2-\sqrt{3};\ \mathrm{th}\,\tfrac{1}{2}]$, of the equation*

$$h(1-r)^3=(1+r)(4r-1-r^2).$$

*However, if $h > e(e^2 - 3)/2$, then the radius of starlikeness (but, in general, not of univalence) for the class being considered is determined by the formula*

$$\tilde{r}=\mathrm{th}\left(\frac{\psi}{2\sin\psi}\right), \tag{35}$$

*where* $\psi$ *is the root, unique on* $(0;\pi/2)$, *of the equation*

$$h=\frac{1}{\cos\psi}\exp(\psi\csc\psi)\,[\operatorname{sh}(\psi\csc\psi)\;\exp(\psi\operatorname{ctg}\psi)-\cos\psi]. \qquad (36)$$

To prove this we need to apply Theorem 2 and Lemma 6, taking into account that $\tilde{r}$ possesses the $l$-property when $h\in[0;e(e^2-3)/2]$ but does not possess it, in general, when $h>e(e^2-3)/2$.

**Remark.** Formulas (35) and (36) can be interpreted as the parametric solution of the problem on the radius of starlikeness for the class $Q(S,P_{(h)})$ when $h>e(e^2-3)/2$. The classes $Q(S,P)$ and $Q(S,P_{(1)})$ (in our notation; see the remark to Lemma 1) have been considered by MacGregor [1, 2]. His results in these cases were refined by Krzyż and Reade [3]. The possibility of such refinement had also been pointed out by G. Kuz'mina (RŽ Mat. 1964, no. 3, review 3Б173). We go on to establish certain results of this type, related to Lemma 4.

**Theorem 3.** *Let there be given the classes* $R[\lambda(r);r_0]$ *and* $R_1=P_{[h]}$, $h\geq 0$. *Consider the functions*

$$T(r)=\frac{(1-r)^3}{(1+r)(4r-1-r^2)},\quad U(r)=\frac{2r}{(1+r)^2\lambda(r)}-\frac{1-r}{1+r},$$

$$V(r)=\frac{1}{4\lambda(r)}\left[\frac{1+r^2}{1-r^2}-\lambda(r)\right]^2$$

*and denote by* $r_h$ *the root, unique on* $(2-\sqrt{3};1)$, *of the equation* $T(r)=h$. *If* $r_0\leq r_h$ *and* $U(r_0)\geq h$, *or if* $r_0\geq r_h$ *and* $U(r_h)\geq h$, *then the radius* $\tilde{r}$ *of starlikeness for the class* $Q(R,P_{[h]})$ *is equal to the root, unique on* $(0;r_0]$, *of the equation*

$$U(r)=h. \qquad (37)$$

*If, however,* $r_0\geq r_h$ *and* $V(r_0)\geq h$, *but* $V(r_h)\leq h$, *then this radius is equal to the root, unique on* $(r_h;r_0)$, *of the equation*

$$V(r)=h. \qquad (38)$$

*If in any of these cases the number* $\tilde{r}$ *possesses the l-property, then it is also equal to the radius of univalence for the class* $Q(R,P_{[h]})$. *The extremal function (not unique, in general) is determined by*

$$f(z;\tilde{r})=F(z;\tilde{r})\,q(z;\tilde{r})$$

*(see the remarks to the definitions of* $R$ *and* $R_1$*).*

By $q(z;\bar{r})$ we should understand (13) in the case (37) and (14) in the case (38).

The proof of this theorem is based on the application of Theorem 2 and Lemma 4, and also on the monotonicity of the functions $T(r)$, $U(r)$, $V(r)$ and on their continuity on $(0; r_0)$. We omit the corresponding discussion because it presents no important difficulties.

Corollary 1. *If* $r_0 = 1$ *in Theorem* 1, *then, denoting by* $r_\beta$ *the root, unique on* $[2-\sqrt{3};1)$, *of the equation*

$$\lambda(r) = \frac{4r-1-r^2}{1-r^2} \equiv W(r)$$

*and setting* $h_\beta = T(r_\beta)$, *we get equation* (37) *for* $\bar{r}$ *when* $h \in [0; h_\beta]$ *and equation* (38) *when* $h > h_\beta$. *The remarks on the l-property and on the form of the extremal function remain in force.*

In order to prove this it suffices to note that when $r_0 = 1$ we always have $T(r_0) \le h$ and that the inequalities $U(r_h) \ge h$ *and* $U(r_h) \le h$ are equivalent to the inequalities $\lambda(r_h) \le W(r_h)$ and $\lambda(r_h) \ge W(r_h)$.

Corollary 2. *If* $R = S^*_\beta$ *in Theorem* 3, *then we must make use of the preceding corollary and, moreover,*

$$r_\beta = 2(3-\beta+\sqrt{5-2\beta+\beta^2})^{-1},$$

*while equations* (37) *and* (38) *acquire, respectively, the forms*

$$1+h+2(\beta+\beta h-2)r+(h-1)(2\beta-1)r^2 = 0$$

*and*

$$h = \frac{r^2(\beta r+1-\beta)^2}{(1+r)(1-r)^2[1+(2\beta-1)r]}.$$

Corollary 3. *If* $R = S^*_{(\beta)}$ *in Theorem* 3, *then once again we must make use of Corollary* 1, *and here for* $r_\beta$ *we obtain*

$$1-(2+\beta)r-2(1-2\beta)r^2+(1-\beta)r^3 = 0,$$

*while equations* (37) *and* (38) *take, respectively, the forms*

$$h = \frac{-1+3r+(3-4\beta)r^2-r^3}{(1+r)^2(1-r)}$$

*and*

$$h = \frac{r^2(1-\beta+r-\beta r^2)^2}{(1-r)^2(1+r)[1+(1-2\beta)r]}.$$

## BIBLIOGRAPHY

[1] T. H. MacGregor, *The radius of univalence of certain analytic functions.* I, Proc. Amer. Math. Soc. **14** (1963), 514–520. MR 26 #6388.

[2] ———, *The radius of univalence of certain analytic functions.* II, Proc. Amer. Math. Soc. **14** (1963), 521–524. MR 26 #6389.

[3] J. Krzyż and M. O. Reade, *The radius of univalence of certain analytic functions,* Michigan Math. J. **11** (1964), 157–159. MR 29 #237.

[4] V. A. Zmorovič, *On certain theorems of the theory of extremal estimates in special classes of analytic functions,* Dopovidi Akad. Nauk. Ukraïn. RSR **1965**, 980–984. (Ukrainian) MR **33** #5874.

[5] ———, *On a class of extremal problems associated with regular functions with positive real part in the circle* $|z| < 1$, Ukrain. Mat. Ž. **17** (1965), 12–21; English. transl., Amer. Math. Soc. Transl. (2) 80 (1969), p. MR **33** #5901.

[6] M. S. Robertson, *Extremal problems for analytic functions with positive real part and applications,* Trans. Amer. Math. Soc. **106** (1963), 236–253. MR 26 #325.

[7] G. M. Goluzin, *Geometric theory of functions of a complex variable,* GITTL, Moscow, 1952; English transl., Transl. Math. Monographs, Amer. Math. Soc., Providence, R. I. (to appear). MR **15**, 112.

[8] N. A. Lebedev, *Majorizing region for the expression* $I = \ln z^{\lambda}[f'(z)]^{1-\lambda}/[f(z)]^{\lambda}$ *in the class S,* Vestnik Leningrad. Univ. **10** (1955), 29–41; English transl., Amer. Math. Soc. Transl.(2) **22** (1962), 43–57. MR 17, 248.

Translated by:
N. H. Choksy

# ESTIMATE OF THE CAUCHY INTEGRAL ALONG AN ANALYTIC CURVE

UDC 519.5

M. S. MEL'NIKOV

Let $E$ be a bounded closed set in the complex plane, and let $B(E)$ be the set of all functions $f(z)$, analytic outside $E$, such that $|f(z)| \leq 1$ for $z \in CE$ and $f(\infty) = 0$. The quantity

$$\gamma(E) = \sup_{f \in B(E)} \lim_{z \to \infty} |zf(z)| = \sup_{f \in B(E)} |f'(\infty)| = \sup_{f \in B(E)} \left| \frac{1}{2\pi} \int_{\Gamma} f(\zeta)\, d\zeta \right|,$$

where $\Gamma$ is a contour containing $E$, is called the *analytic content* ([1], [2]) of the set $E$.

The analytic content of an arbitrary bounded set is defined as the upper bound of the analytic contents of its closed subsets.

A. G. Vituškin proposed the following problem ([3], § 4). Let $G$ be any region, $\partial G$ its boundary, $E$ a closed set lying inside $G$, and $f(z)$ a function analytic in the region $G \setminus E$, where $|f(z)| \leq 1$ in $G \setminus E$. For which regions $G$ is the estimate

$$\left| \int_{\partial G} f(\zeta)\, d\zeta \right| \leqslant c(G)\gamma(E)$$

justified, where $c(G)$ is a constant depending only on $G$? In particular, is the estimate true if $G$ is the unit disk?

## § 1. Estimate of the integral

Theorem 1. *Let $E$ be a closed set lying in the disk $|z| \leq 1$, and let $f(z)$ be a continuous function analytic outside $E$ in the disk, where $|f(z)| \leq 1$. Then*

$$\left| \int_{|\zeta|=1} f(\zeta)\, d\zeta \right| \leqslant c\gamma(E),$$

*where $c$ is an absolute constant.**

We first prove several preliminary assertions.

Lemma 1. *Let $E$ be a bounded closed set, and let $f(z)$ be analytic outside $E$, $f(\infty) = 0$, $\operatorname{Im} f(z) \leq \alpha$, $\alpha > 0$. Then $|f'(\infty)| \leq 2\alpha\gamma(E)$.*

---

*We shall designate all absolute constants by the letter $c$.

In fact, if $f_1(z) = f(z)/(f(z) - 2i\alpha)$, then $f_1(z)$ is analytic outside $E$, $|f_1(z)| \le 1$, $f_1(\infty) = 0$, and $|f_1'(\infty)| \le \gamma(E)$ by the definition of $\gamma(E)$. Then

$$|f'(\infty)| \leqslant 2\alpha\, |f_1'(\infty)| \leqslant 2\alpha\gamma(E).$$

**Lemma 2.** *Let $E$ be a bounded closed set, let the point $a$ belong to the component of the complement of $E$ which contains the point $\infty$, let $\rho = \rho(a, E)$ be the distance from $a$ to $E$, and let $f(z)$ be analytic outside $E$, where $|f(z)| \le 1$. Then*

$$|f'(a)|\, \rho\, [\rho - \gamma(E)] \leqslant 2\gamma(E).$$

Without loss of generality $E$ may be assumed to have rectifiable boundary $\partial E$. Then

$$|f'(a)| = \left| \frac{1}{2\pi} \int_{\partial E} \frac{f(\zeta)\, d\zeta}{(\zeta - a)^2} \right| = \left| \frac{1}{2\pi} \int_{\partial E} \left[ \frac{f(\zeta) - f(a)}{(\zeta - a)^2} - \frac{f'(a)}{\zeta - a} \right] d\zeta \right|.$$

But the function $(f(z) - f(a))/(z - a)^2 - f'(a)/(z - a)$ is analytic outside the set $E$, equal to zero at the point $\infty$, and less than $2/\rho^2 + |f'(a)|/\rho$ in absolute value. Then, by the definition of $\gamma(E)$,

$$|f'(a)| \leqslant \left[ \frac{2}{\rho^2} + \frac{|f'(a)|}{\rho} \right] \gamma(E),$$

which was to be proved.

If $f(a) = 0$, then $\rho[\rho - \gamma(E)]\, |f'(a)| \le \gamma(E)$.

**Lemma 3.** *Let the closed set $E$ lie in the annulus $r \le |z| \le 1$, and let $E*$ be the set symmetrical to the set $E$ relative to the unit circle. Then $r^2\gamma(E*) \le 2\gamma(E)$.*

This lemma follows immediately from Lemma 2 if one considers the function $\phi(z) = \overline{f}(1/\overline{z})$, where $f(z)$ is the Ahlfors function [2] of the set $E*$.

**Theorem 2.** *Let $E$ be a closed set in the disk $|z| \le 1$, $0 \notin E$, and let $E*$ be symmetrical to the set $E$ relative to the unit circle. Then*

$$\gamma(E \cup E^*) \leqslant c\,[\gamma(E) + \gamma(E^*)],$$

*where $c$ is an absolute constant.*

**Proof.** Let $E_\delta$ be the $\delta$-neighborhood of the set $E$ and $E_\delta^*$ be the neighborhood of the set $E*$ symmetrical to $E_\delta$. From the definition of analytic content it follows that $\gamma(E_\delta) \to \gamma(E)$, $\gamma(E^*_\delta) \to \gamma(E^*)$ as $\delta \to 0$. Therefore it suffices to show that $\gamma(E \cup E^*) \leq c[\gamma(E_\delta) + \gamma(E^*_\delta)]$ for any $\delta > 0$. For this we prove that

$$|f'(\infty)| \leqslant c[\gamma(E_\delta) + \gamma(E^*_\delta)]$$

for any function $f(z)$ analytic outside $E \cup E^*$, where $|f(z)| \leq 1$, $f(\infty) = 0$. We note that $f(z)$ may be assumed continuous and even smooth in the whole plane and analytic outside $E_\delta \cup E^*_\delta$.

Let $\phi_+(z) = f(z) + \bar{f}(1/\bar{z})$; $\phi_-(z) = f(z) - \bar{f}(1/\bar{z})$. Then $\phi_+(z)$ and $\phi_-(z)$ are smooth functions analytic outside $E_\delta \cup E^*_\delta$; $|\phi_+(z)| \leq 2$ and $|\phi_-(z)| \leq 2$; the function $\phi_+(z)$ takes real values on the unit circle, and the function $\phi_-(z)$ imaginary ones. We represent the function $\phi_+(z)$ as the sum of two functions $\phi_1(z)$ and $\phi_2(z)$ analytic, respectively, outside $E_\delta$ and $E^*_\delta$:

$$\varphi_1(z) = \begin{cases} \varphi_+(z) - \dfrac{1}{2\pi i} \displaystyle\int_{|\zeta|=1} \dfrac{\varphi_+(\zeta)\, d\zeta}{\zeta - z} & \text{for } |z| < 1, \\ -\dfrac{1}{2\pi i} \displaystyle\int_{|\zeta|=1} \dfrac{\varphi_+(\zeta)\, d\zeta}{\zeta - z} & \text{for } |z| > 1, \end{cases}$$

$$\varphi_2(z) = \begin{cases} \dfrac{1}{2\pi i} \displaystyle\int_{|\zeta|=1} \dfrac{\varphi_+(\zeta)\, d\zeta}{\zeta - z} & \text{for } |z| < 1, \\ \varphi_+(z) + \dfrac{1}{2\pi i} \displaystyle\int_{|\zeta|=1} \dfrac{\varphi_+(\zeta)\, d\zeta}{\zeta - z} & \text{for } |z| > 1. \end{cases}$$

(The direction of integrating around the circle is counterclockwise.)

The potential $(1/2\pi i)\int_{|\zeta|=1} \phi_+(\zeta)(\zeta - z)^{-1} d\zeta$ possesses the following properties. Its imaginary part is continuous in the whole plane, and the real part, being a double shell potential [4], does not exceed $\max_{|\zeta|=1}|\phi_+(\zeta)|$ in absolute value and on the unit circle has a discontinuity equal to $\phi_+(z)$, where $|z| = 1$. Therefore the functions $\phi_1(z)$ and $\phi_2(z)$ can be extended by continuation at points of the unit circle not belonging to $E_\delta$, to get functions analytic, respectively, outside $E_\delta$ and $E^*_\delta$, where

$$|\mathrm{Re}\,\varphi_1(z)| \leqslant 2\max|\varphi_+(z)| \leqslant 4, \quad |\mathrm{Re}\,\varphi_2(z)| \leqslant 4, \quad \varphi_1(\infty) = 0, \quad |\varphi_2(\infty)| \leqslant 1,$$

$$\varphi_1(z) + \varphi_2(z) = \varphi_+(z).$$

Applying Lemma 1 to the functions $\phi_1(z)$ and $\phi_2(z)$, we get

$$|\varphi_+'(\infty)| \leqslant |\varphi_1'(\infty)| + |\varphi_2'(\infty)| \leqslant c[\gamma(E_\delta) + \gamma(E_\delta^*)].$$

Analogously, for the function $\phi_-(z)$ we get

$$|\varphi_-'(\infty)| \leqslant c[\gamma(E_\delta) + \gamma(E_\delta^*)].$$

Since $f(z) = ½[\phi_+(z) + \phi_-(z)]$, we have $|f'(\infty)| \leq c[\gamma(E_\delta) + \gamma(E_\delta^*)]$. The theorem is proved.

**Theorem 3.** *Let $E$ be a closed set lying in the annulus $r \leq |z| \leq 1$, and let $f(z)$ be a continuous function analytic in the unit disk outside the set $E$, where $|f(z)| \leq 1$. Then*

$$\left| \int_{|\zeta|=1} f(\zeta)\, d\zeta \right| \leqslant \frac{c}{r^2} \cdot \gamma(E),$$

*where $c$ is an absolute constant.*

**Proof.** As in Theorem 2, without loss of generality $f(z)$ may be assumed to be a smooth function.

We consider the functions

$$\varphi_+(z) = \begin{cases} f(z) & \text{for } |z| < 1, \\ \bar f\left(\frac{1}{\bar z}\right) & \text{for } |z| > 1, \end{cases} \qquad \varphi_-(z) = \begin{cases} f(z) & \text{for } |z| < 1, \\ -\bar f\left(\frac{1}{\bar z}\right) & \text{for } |z| > 1. \end{cases}$$

These functions are analytic outside $E \cup E^*$ and have, respectively, imaginary and real discontinuities on the unit circle. Removing this discontinuity with the aid of the potential (as in Theorem 2), we get the functions

$$\varphi_1(z) = \varphi_+(z) - \frac{1}{2\pi i} \int_{|\zeta|=1} \frac{f(\zeta) - \bar f(\zeta)}{\zeta - z}\, d\zeta,$$

$$\varphi_2(z) = \varphi_-(z) - \frac{1}{2\pi i} \int_{|\zeta|=1} \frac{f(\zeta) + \bar f(\zeta)}{\zeta - z}\, d\zeta,$$

which may be extended by continuation at points of the unit circle not belonging to the set $E$, to get functions analytic outside $E \cup E^*$, where $|\mathrm{Im}\, \phi_1(z)| \leq 3$ and $|\mathrm{Re}\, \phi_2(z)| \leq 3$. Then by Lemma 1

$$|\varphi_1'(\infty)| \leqslant 6\gamma(E \cup E^*), \quad |\varphi_2'(\infty)| \leqslant 6\gamma(E \cup E^*),$$

and by Theorem 2

$$|\varphi_1'(\infty)| \leqslant c[\gamma(E) + \gamma(E^*)], \quad |\varphi_2'(\infty)| \leqslant c[\gamma(E) + \gamma(E^*)];$$

but, since by Lemma 3

$$\gamma(E^*) \leqslant \frac{2}{r^2}\gamma(E),$$

we have

$$|\varphi_1'(\infty)| \leqslant c\cdot\frac{1}{r^2}\gamma(E), \quad |\varphi_2'(\infty)| \leqslant c\cdot\frac{1}{r^2}\gamma(E).$$

Further, we have

$$\left|\int_{|\zeta|=1} f(\zeta)\,d\zeta\right| = \pi|\varphi_1'(\infty)+\varphi_2'(\infty)| \leqslant c\frac{1}{r^2}\gamma(E).$$

The theorem is proved.

It is easy to prove the following well-known result (see [3], § 4).

Lemma 4. *Let the closed set $E$ lie in the region $G$ at a given distance $r$ from the boundary $\partial G$, and let the continuous function $f(z)$ be analytic in $G\setminus E$, where $|f(z)| \leq 1$. Then*

$$\left|\int_{\partial G} f(\zeta)\,d\zeta\right| \leqslant c(G, r)\gamma(E),$$

*where $c(G, r)$ is a constant depending only on $G$ and $r$. For the unit disk $c(G, r) = c \cdot \ln(1 + 1/r)$.*

Thus if a set lies at a given distance from the unit circle the integral is estimated by Lemma 4, while if the set lies at a given distance from the point 0 the integral is estimated by Theorem 3. Theorem 1 will follow from these two assertions.

Proof of Theorem 1. Without loss of generality it may be assumed that the set $E$ is bounded by a finite number of rectifiable closed curves and that $f(z)$ is analytic in the unit disk outside $E$ and everywhere smooth. We designate by $\beta(z)$ real-valued, finitely differentiable function on the unit disk with the following properties: $\beta(z) = 1$ for $1 \geq |z| \geq 2/3$; $\beta(z) = 0$ for $|z| \leq 1/3$; $0 \leq \beta(z) \leq 1$; $|\partial\beta/\partial\bar{z}| \leq 4$. Then $\beta f + (1-\beta)f = f$.

By the Cauchy-Green formula we have $\beta f = f_1 + \phi_1$ and $(1-\beta)f = f_2 + \phi_2$, where $f_1(z)$ is analytic in the unit disk outside the part of the set $E$ which lies in the annulus $1/3 \leq |z| \leq 1$, and $f_2(z)$ is analytic in the unit disk outside the part of $E$ which lies in the disk $|z| \leq 2/3$; here $|f_1(z)| \leq c$, $|f_2(z)| \leq c$ (since $|\partial\beta/\partial\bar{z}| \leq 4$); $\phi_1(z) + \phi_2(z) = 0$ and $f_1(z) + f_2(z) = f(z)$. Then

$$\left|\int_{|\zeta|=1} f(\zeta)\,d\zeta\right| \leqslant \left|\int_{|\zeta|=1} f_1(\zeta)\,d\zeta\right| + \left|\int_{|\zeta|=1} f_2(\zeta)\,d\zeta\right|,$$

and Theorem 1 immediately follows from Theorem 3 with $r = 1/3$ and Lemma 4 with $r = 1/3$, since the analytic content of a part of a set is not greater than the content of the set itself.

Corollary 1. *If sets $E_i$ lie, respectively, in non-overlapping disks $K_i$, then $\gamma(\bigcup_i E_i) \leq c\,\Sigma_i\,\gamma(E_i)$.*

Corollary 2. *Let $E$ be a closed set in the annulus $\frac{1}{2} \leq |z| \leq 1$, and let $f(z)$ be a continuous function analytic in this annulus outside the set $E$, where $|f(z)| \leq 1$. Then*

$$\left|\int_{|\zeta|=1} f(\zeta)\,d\zeta - \int_{|\zeta|=\frac{1}{2}} f(\zeta)\,d\zeta\right| \leqslant c\gamma(E).$$

**Proof.** Without loss of generality it may be assumed that the set $E$ has rectifiable boundary $\partial E$. Then

$$\int_{|\zeta|=1} f(\zeta)\,d\zeta - \int_{|\zeta|=\frac{1}{2}} f(\zeta)\,d\zeta = \int_{\partial E} f(\zeta)\,d\zeta.$$

As in the proof of Theorem 1, let $\beta(z)$ be a real-valued, finitely differentiable function in the annulus $\frac{1}{2} \leq |z| \leq 1$ with the following properties: $\beta(z) = 1$ for $|z| = 1$; $\beta(z) = 0$ for $|z| = \frac{1}{2}$; $|\partial\beta/\partial\bar{z}| \leq 3$; $0 \leq \beta(z) \leq 1$. Then $\beta f + (1-\beta) f = f$.

By the Cauchy-Green formula we may write $\beta f = f_1 + \phi_1$ and $(1-\beta) f = f_2 + \phi_2$, where $f_1(z)$ is analytic outside the set $E$ in the unit disk, $f_2(z)$ is analytic outside the set $E$ outside the disk $|z| \leq \frac{1}{2}$, $|f_1(z)| \leq c$, $|f_2(z)| \leq c$, $\phi_1 + \phi_2 = 0$. Then Corollary 2 is obtained from Theorem 1 and from the assertion that the estimate $|\int_{\partial E} f(\zeta)\,d\zeta| \leq c\,\gamma(E)$ holds if the function $f(z)$ is analytic outside the set $E$ outside the disk $|z| \leq \frac{1}{2}$, where $|f(z)| \leq 1$. (This assertion is proved completely analogously to Theorem 1.)

We now estimate the integral for regions with analytic boundary.

Theorem 4. *Let the region $G$ be bounded by a simple closed analytic curve $\Gamma$, let the closed set $E \subset \overline{G}$, and let $f(z)$ be a continuous function analytic in the region $G \setminus E$, where $|f(z)| \leq 1$. Then*

$$\left|\int_\Gamma f(\zeta)\, d\zeta\right| \leqslant c(G)\cdot\gamma(E),$$

*where $c(G)$ is a constant depending only on $G$.*

**Proof.** Let $I(G, E) = \sup_{f(z)} |\int_\Gamma f(\zeta)\, d\zeta|$, where the sup is taken over all continuous functions $f(z)$ analytic in the region $G\setminus E$ for which $|f(z)| \leq 1$. We first prove the following assertion.

**Lemma 5.** *If the region $G_1$ maps conformally onto the region $G_2$ by the function $w(z)$, where $0 < m < |w'(z)| < M$, $z \in G_1$, and $E_2$ is the image under this map of the set $E_1 \subset G_1$, then*

$$\frac{1}{m} I(G_2, E_2) \geqslant I(G_1, E_1) \geqslant \frac{1}{M} I(G_2, E_2).$$

In fact, let $\Gamma_1$, $\Gamma_2$ be the boundaries of regions $G_1$, $G_2$. We have

$$\int_{\Gamma_2} f(w)\, dw = \int_{\Gamma_1} f(w(z))\, w'(z)\, dz.$$

Therefore, if $f_2(w) = f_2(z(w))$, where $z(w)$ is the map inverse to $w(z)$, if $f_2(z) = m \cdot f_1(z)/w'(z)$, where $f_1(z)$ is such that $|\int_{\Gamma_1} f_1(z)\, dz| = I(G_1, E_1)$, and if the function $f_1(z)$ is analytic in the region $G_1\setminus E_1$, $|f_1(z)| \leq 1$, then $|f_2(w)| \leq 1$ and

$$I(G_2, E_2) \geqslant \left|\int_{\Gamma_2} f_2(w)\, dw\right| = \left|\int_{\Gamma_1} \frac{mf_1(z)}{w'(z)} w'(z)\, dz\right| = mI(G_1, E_1).$$

The second inequality is proved in a completely analogous way.

Let $G$ be bounded by a simple analytic curve, and let $K$ be the unit disk; then there exists a conformal map $w(z)$ of a given neighborhood $UG$ of $G$ onto the neighborhood $UK$ of $K$, which takes $G$ into $K$, and $0 < m < |w'(z)| < M$, $z \in UG$. Let $E_1$ be the image of $E$ under this map. Then, by Lemma 5,

$$I(G, E) \leqslant \frac{1}{m} I(K, E_1) \tag{1}$$

and

$$I(UK, E_1) \leqslant MI(UG, E). \tag{2}$$

By Theorem 3, inequality (1) means that $I(G, E) \leq c\gamma(E_1)/m$, and by Lemma 4 (since the set $E$ lies at a given distance from the boundary of $UG$) inequality (2) means that

$$\gamma(E_1) \leqslant I(UK, E_1) \leqslant MI(UG, E) \leqslant Mc(G)\gamma(E),$$

i.e.

$$I(G, E) \leqslant \frac{M}{m} c(G)\gamma(E).$$

Theorem 4 is proved.

Corollary 3. *If two sets* $E_1$ *and* $E_2$ *are separated by an analytic curve* $\Gamma$ (*in particular, if they lie on opposite sides of a straight line*), *then* $\gamma(E_1 \cup E_2) \leq c(\Gamma)[\gamma(E_1) + \gamma(E_2)]$.

Remark 1. The requirement of continuity of the function $f(z)$ in the theorem on the estimate of the integral may be replaced by the requirement of continuity only in a neighborhood of the boundary of the region or may be removed completely under the condition that the set $E$ does not pass outside the boundary of $G$, and that the integral is taken along a contour inside the region which contains $E$.

Remark 2. The quantity

$$\alpha(E) = \sup \lim_{z\to\infty} |z \cdot f(z)|,$$

where sup is taken over all functions $f(z)$ *continuous* on the plane and analytic outside $E$ for which $|f(z)| \leq 1$, $f(\infty) = 0$, is called the analytic $C$-content of the set $E$. Just as we did for the estimate of the integral using analytic content, we can obtain an estimate of the Cauchy integral of a continuous function using analytic $C$-content under the condition that the set $E$ does not pass outside the boundary of $G$.

We shall indicate two applications of our results.

## § 2. Necessary and sufficient conditions for a peak point

Let $E$ be a bounded closed set in the complex plane, and let $A(E)$ be the Banach algebra of limits uniform on $E$ of functions analytic on $E$, i.e. analytic on some neighborhood of the set $E$ which depends on the function.

The point $\zeta \in E$ is called a peak point relative to $A(E)$ if there exists a function $f(z) \in A(E)$ such that $|f(\zeta)| > |f(z)|$ for all $z \in E$, $z \neq \zeta$ (see [5] or [6]).

In [5] sufficient conditions are deduced for the point $\zeta \in E$ to be a peak point relative to $A(E)$, in terms of the diameter of the complement of $E$, which lies in the disk of radius $r$ with center at the point $\zeta$.

Below we deduce necessary and sufficient conditions for the point $\zeta$ to be a peak point, in terms of the analytic content of the complement of $E$. Let $K(n)$ be the annulus $1/2^n \geq |\zeta - z| \geq 1/2^{n+1}$ and $\gamma_n = \gamma[CE \cap K(n)]$.

Theorem 5. *For the point $\zeta \in E$ to be a peak point relative to $A(E)$, it is necessary and sufficient that*

$$\sum_{n=0}^{\infty} 2^n \gamma_n = \infty. \tag{3}$$

Proof. Necessity. We suppose that $\zeta \in E$ is a peak point relative to $A(E)$ and the series (3) converges. Let the function $g(z) \in A(E)$, $|g(\zeta)| > |g(z)|$ for $z \in E$, $z \neq \zeta$, and $g(\zeta) = 1$. Then the function $g^k(z) \in A(E)$, since $A(E)$ is an algebra.

Let $N$ be such that $\sum_{n=N}^{\infty} 2^n \gamma_n < \delta$ $(\delta > 0)$, and let the number $k$ be such that

$$|g^k(z)| \leqslant \frac{1}{4} \quad \text{for } z \in E,\ |z - \zeta| \geqslant 2^{-N}.$$

Since $g^k(z) \in A(E)$, then for any $\epsilon > 0$ there is a function $f(z)$ analytic in a neighborhood of $E$, such that $|f(z) - g^k(z)| \leq \epsilon$ $(z \in E)$. Then there exists a set $G \subset CE$ with rectifiable boundary, such that

$$f(\zeta) = \frac{1}{2\pi i} \int\limits_{|z-\zeta|=2^{-N}} \frac{f(z)\,dz}{z-\zeta} - \frac{1}{2\pi i} \int\limits_{\Gamma} \frac{f(z)\,dz}{z-\zeta},$$

where $\Gamma$ is the boundary of the intersection of the set $G$ with the disk of radius $2^{-N}$ (we are assuming that the function $f(z)$ has been extended continuously onto the whole plane without increasing the absolute value). But

$$\left| \frac{1}{2\pi i} \int\limits_{|z-\zeta|=2^{-N}} \frac{f(z)\,dz}{z-\zeta} \right| \leqslant \frac{1}{2\pi} \int\limits_{|z-\zeta|=2^{-N}} \frac{|f(z)|\,|dz|}{|z-\zeta|} \leqslant \frac{1}{4} + \varepsilon;$$

$$\left| \frac{1}{2\pi i} \int\limits_{\Gamma} \frac{f(z)\,dz}{z-\zeta} \right| = \left| \sum_{n=N}^{\infty} \left[ \frac{1}{2\pi i} \int\limits_{|z-\zeta|=2^{-n}} \frac{f(z)\,dz}{z-\zeta} - \frac{1}{2\pi i} \int\limits_{|z-\zeta|=2^{-n-1}} \frac{f(z)\,dz}{z-\zeta} \right] \right|,$$

and, in view of the estimate of the integral for an annulus (Corollary 2), since $|f(z)/(z-\zeta)| \leq (1+\epsilon)\, 2^n$ for $|z-\zeta| = 2^{-n}$, we get

$$\left| \frac{1}{2\pi i} \int\limits_{\Gamma} \frac{f(z)\,dz}{z-\zeta} \right| \leqslant c \sum_{n=N}^{\infty} 2^n \gamma_n \leqslant c\delta(1+\varepsilon).$$

This means that

$$|f(\zeta)| \leqslant \frac{1}{4} + \varepsilon + c\delta(1+\varepsilon);$$

but since $\delta$ and $\epsilon$ are arbitrary this contradicts the fact that

$$|f(\zeta)| \geqslant |g(\zeta)| - \varepsilon = 1 - \varepsilon.$$

Sufficiency. Let $\Sigma_{n=0}^{\infty} 2^n \gamma_n = \infty$. According to Lemma 1 of [5], to prove that $\zeta \in E$ is a peak point it is sufficient to construct a "nonstrict" peak, i.e. for given numbers $0 < \beta < \alpha < 1$ and arbitrary $\delta > 0$ to construct a function $f(z) \in A(E)$ having the properties

$$|f(z)| \leqslant 1, \quad |f(\zeta)| > \alpha, \quad |f(z)| < \beta \quad \text{for } z: \quad |z - \zeta| \geqslant \delta.$$

From the semiadditivity of the analytic content of sets separated by a circle or a straight line (Corollary 3) it immediately follows that in each annulus $K(n)$ there exists a closed set $l_n \subset CE$ such that its diameter $d(l_n) \leq 2^{-n-1}/6$ and $\Sigma_{n=0}^{\infty} 2^n \gamma(l_n) = \infty$. We designate by $\phi_n(z)$ the Ahlfors function [2] of the set $l_n$. This function has the following properties: a) $\phi_n(z)$ is analytic outside $l_n$; b) $|\phi_n(z)| \leq 1$; c) $|\phi_n(z)| \leq \gamma(l_n)/\rho(z, l_n)$, where $\rho(z, l_n)$ is the distance from the point $z$ to the set $l_n$; d) $\phi_n(z) = \Sigma_{k=1}^{\infty} a_n^k/(z - z_n)^k$, where

$$z_n \in l_n, \ |z - z_n| \geqslant 2d(l_n), \ a_n^1 = \gamma(l_n), \quad |a_n^k| \leqslant c\gamma(l_n)\,[d(l_n)]^{k-1},$$

where $c$ is an absolute constant (see [2]).

From the last condition it follows that for $|z - z_n| \geq 6d(l_n)$ we have

$$|\varphi_n(z)| \geqslant c_1 \cdot \frac{\gamma(l_n)}{|z - z_n|}, \tag{4}$$

where $c_1$ is an absolute constant.

Now, using these conditions, we construct from the functions $\phi_n(z)$ a "nonstrict" peak. Let $\delta > 0$ be given. Choose numbers $N_1$, $N_2$ so that

$$2^{-N_1} \leqslant \frac{9}{20} c_1 (\delta - 2^{-N_1}), \tag{5}$$

where $c_1$ is the constant of inequality (4), and

$$3 \geqslant \sum_{n=N_1}^{N_2} 2^n \gamma(l_n) \geqslant 1. \tag{6}$$

Let $f(z) = (1/9)\Sigma_{n=N_1}^{N_2} \phi_n(z) \cdot \beta_n$, where $|\beta_n| = 1$ and $\arg \beta_n = -\arg \phi_n(\zeta)$. Then, if $|z - \zeta| \geq \delta$, then $\rho(z, l_{N_1}) \geq \rho(z, k(N_1)) \geq \delta - 2^{-N_1}$ and, since

$\gamma(l_n) \le 2^{-n}$, by virtue of c) and (5) we obtain

$$|f(z)| \leqslant \frac{1}{9} 2 \frac{2^{-N_1}}{\delta - 2^{-N_1}} \leqslant \frac{c_1}{10} \quad \text{for } z\colon |z - \zeta| \geqslant \delta. \tag{7}$$

We now prove that $|f(z)| \le 1$. To do this we bound by one the absolute values of the functions $\phi_{j-1}(z)$, $\phi_j(z)$ and $\phi_{j+1}(z)$ [where $j$ is such that $z \in k(j)$] by condition b), and for the remaining $n$ we note that $\rho(z, k(n)) \ge 2^{-n-1}$; therefore by c) and (6) we obtain

$$|f(z)| \leqslant \frac{1}{9}\left[3 + 2 \sum_{n=N_1}^{N_2} 2^n \gamma(l_n)\right] \leqslant 1.$$

We bound $|f(\zeta)|$ from below by using inequalities (4) and (6):

$$|f(\zeta)| = \frac{1}{9} \sum_{n=N_1}^{N_2} |\varphi_n(\zeta)| \geqslant \frac{c_1}{9} \sum_{n=N_1}^{N_2} 2^n \gamma(l_n) \geqslant \frac{c_1}{9}. \tag{8}$$

Conditions (7), (8) and the inequality $|f(z)| \le 1$ for $z \in E$ means that we have constructed a "nonstrict" peak with $\beta = c_1/10$, $\alpha = c_1/9$. The theorem is proved.

## § 3. A theorem from the theory of approximation

Using results of A. G. Vituškin [7] and Theorem 4, we shall determine a geometric condition on the set $E$ sufficient for any function continuous on $E$ and analytic at its interior points to permit approximation by rational fractions uniformly on $E$.

The set of those boundary points of a bounded closed set $E$ which do not belong to the boundaries of components of the complement of $E$ is called the inner boundary of $E$. In [7] it is proved that any function continuous on a bounded closed set $E$ and analytic at its interior points can be approximated by rational fractions (uniformly on $E$), if the inner boundary of $E$ consists of a finite number of points.

Theorem 6. *If the inner boundary of a bounded closed set $E$ lies on a finite number of analytic curves, then any function continuous on $E$ and analytic at its interior points permits approximation by rational fractions uniformly on $E$.*

Proof. We dwell on the case when the inner boundary lies on one closed analytic curve $\Gamma$, since the theorem reduces to it in the same way as the case of a finite number of points of the inner boundary is considered in [7].

By Theorem 3 of [7] it is sufficient to prove that for any disk $K$

$$\alpha(\overline{CE} \cap K) \leqslant c(E)\,\alpha(CE \cap K), \tag{9}$$

where $c(E)$ is a constant depending only on $E$ and $\alpha(l)$ is the analytic $C$-content of the set $l$. If $K$ does not intersect $\Gamma$ inequality (9) follows from [7].

Let $K$ intersect $\Gamma$. To prove inequality (9) it is sufficient to establish that

$$\Big|\int_{\partial K} \varphi(\zeta)\,d\zeta\Big| \leqslant c(E)\,\alpha(CE \cap K), \tag{10}$$

where $\phi(\zeta)$ is the continuous Ahlfors function [2] of the set $\overline{CE} \cap K$, and $\partial K$ is the boundary of the disk $K$. We designate by $E_1$ the closure of the interior of $\Gamma$, and by $E_2$ the closure of the exterior of $\Gamma$. Let

$$\Gamma_1 = [E_1 \cap \partial K] \cup [\Gamma \cap K], \quad \Gamma_2 = [E_2 \cap \partial K] \cup [\Gamma \cap K].$$

Then

$$\int_{\partial K} \varphi(\zeta)\,d\zeta = \int_{\Gamma_1} \varphi(\zeta)\,d\zeta + \int_{\Gamma_2} \varphi(\zeta)\,d\zeta.$$

By Theorem 7 of [7] the function $\phi(z)$ can be approximated with arbitrary accuracy on $E_1$ by a function $\phi_1(z)$ analytic in $E_1 \setminus G_1$, where $G_1$ is a set lying strictly inside $E_1$, and $G_1 \subset [CE \cap K]$; therefore, according to Theorem 4 and Remark 2,

$$\Big|\int_{\Gamma_1} \varphi_1(\zeta)\,d\zeta\Big| = \Big|\int_{\Gamma} \varphi_1(\zeta)\,d\zeta\Big| \leqslant c(\Gamma)\,\alpha(CE \cap K),$$

and since $\phi_1(z)$ approximates $\phi(z)$, we also have

$$\Big|\int_{\Gamma_1} \varphi(\zeta)\,d\zeta\Big| \leqslant c(\Gamma)\,\alpha(CE \cap K).$$

In a completely analogous way we get that

$$\Big|\int_{\Gamma_2} \varphi(\zeta)\,d\zeta\Big| \leqslant c(\Gamma)\,\alpha(CE \cap K).$$

Thus inequality (10) is valid. The theorem is proved.

In conclusion, the author bears deep gratitude to A. G. Vituškin for attention given to the paper and for valuable advice.

## BIBLIOGRAPHY

[1] L. V. Ahlfors, *Bounded analytic functions,* Duke Math. J. 14 (1947), 1–11. MR 9, 24.

[2] A. G. Vituškin, *Analytic capacity of sets and some of its properties,* Dokl. Akad. Nauk SSSR 123 (1958), 778–781. (Russian) MR 21 #2056.

[3] ———, *A problem of Denjoy,* Izv. Akad. Nauk SSSR Ser. Mat. 28 (1964), 745–756. (Russian) MR 29 #6024.

[4] I. G. Petrovskiĭ, *Lectures on partial differential equations,* Fizmatgiz, Moscow, 1961; English transl. of 1st ed., Interscience, New York, 1964. MR 25 #2308.

[5] A. A. Gončar, *The minimal boundary of the algebra* $A(E)$, Izv. Akad. Nauk SSSR Ser. Mat. 27 (1963), 949–955. (Russian) MR 27 #3812.

[6] E. Bishop, *A minimal boundary for function algebras,* Pacific J. Math. 9 (1959), 629–642. MR 22 #191.

[7] A. G. Vituškin, *Necessary and sufficient conditions on a set in order that any continuous function analytic at the interior points of the set may admit uniform approximation by rational fractions,* Dokl. Akad. Nauk SSSR 171 (1966), 1255–1258 = Soviet Math. Dokl. 7 (1966), 1622–1625. MR 35 #391.

Translated by:
N. Koblitz

# ESTIMATES OF THE CAUCHY INTEGRAL

UDC 519.5

A. G. VITUŠKIN

In this article an estimate of the Cauchy integral of a function along a sufficiently smooth contour will be given using the analytic content of the set of singular points of the function enclosed by this contour. This estimate will prove useful in the theory of approximating functions by rational fractions.

## §1. Formulation of the result and some corollaries

**Definition.** A curvilinear polygonal path $\gamma$, which does not intersect itself and which consists of a finite number of links joined to one another (Jordan arcs $\gamma_1, \cdots, \gamma_n$), is called a Ljapunov curve if each link $\gamma_i$ has a tangent at every point and if the argument $\arg \tau(z, \gamma_i)$ of the tangent vector $\tau(z, \gamma_i)$ to the arc $\gamma_i$ at the point $z$, as a function of the point $z \in \gamma_i$, satisfies the Hölder condition with constant $L$ and exponent $\alpha$, i.e. if $|\arg \tau(z_1, \gamma_i) - \arg \tau(z_2, \gamma_i)| \leq L|z_1 - z_2|^\alpha$ for every $z_1$ and $z_2$ on $\gamma_i$. A region whose boundary consists of a finite number of Ljapunov curves will be called a Ljapunov region.

**Notation.** $A(e, m)$ is the set of all functions each of which is analytic outside some closed subset of the set $e$, is bounded in absolute value by the constant $m$, and is equal to zero at the point at infinity; $C(e, m)$ is the subset of functions of $A(e, m)$ continuous on the whole plane; $\gamma(e, f) = \lim_{z\to\infty} zf(z)$ is the residue of the function; $\gamma(e) = \sup|\gamma(e, f)|$, $f \in A(e, 1)$, is the analytic content of the set $e$; $\alpha(e) = \sup|\gamma(e, f)|$, $f \in C(e, 1)$ is the analytic $C$-content of the set $e$; $\partial g$ is the boundary of the region $g$; $C(g, e, m)$ is the set of functions continuous on $\bar{g} = g \cup \partial g$, bounded on $\bar{g}$ by the constant $m$, and analytic in $g \setminus e$; $h_1(e)$ is the Hausdorff length of the set $e$, $\sigma(z, \delta)$ is the open disk of radius $\delta$ with center at the point $z$; $\kappa(e) = \sup h_1(e')/\gamma(e')$ is the least upper bound taken over all closed sets $e' \subset \partial e$ for which $\gamma(e') > 0$.

For Ljapunov curves $\kappa(e) < \infty$, since on these curves content is commensurable with length (cf. [1]).

**Theorem 1.** *Let $g$ be a Ljapunov region, $e$ a closed subset of $\bar{g}$, $f(z)$ a function continuous on $\bar{g}$ with modulus of continuity $\omega(\delta)$ and analytic in $g\setminus e$. Then $|\int_{\partial g} f(z)dz| \leq C(g)\omega[\alpha(e)]\alpha(e)$, where $C(g)$ is a constant depending only on the region $g$.*

From the proof of Theorem 1 given in §8 it is evident that $C(g)$ depends only on $n$ (the common number of links of the boundary), on $L$ and $\alpha$ (the Hölder constant and exponent for the links) and on $\kappa(g)$. This means that for regions bounded by a finite number of straight line segments, $C(g)$ depends only on the number of these segments. It seems likely that $C(g)$ is completely determined by the number $\kappa(g)$.

As $h_1(\partial g)$ grows the quantity $C(g)$ becomes arbitrarily large (cf. [2]). In [2], for $k$ arbitrary, we constructed a region $g_k$ bounded by a smooth curve, a set $e \subset g_k$, and a function $f(z) \in C(g, e, 1)$ such that $|\int_{\partial g_k} f(z)dz| \geq k\gamma(e)$.

Based on this example, it is possible to construct a region with rectifiable boundary for which Theorem 1 is not true.

**Theorem 2.** *Let* $g$ *be a Ljapunov region,* $e$ *a closed subset of* $\bar{g}$, $f(z)$ *a function bounded on* $\bar{g}$ *by the constant* $m$, *analytic in* $g \setminus e$ *and continuous almost everywhere on* $\partial g$. *Then* $|\int_{\partial g} f(z)dz| \leq C(g) \cdot m \cdot \gamma(e)$.

This theorem follows from Theorem 1. For the case of a region bounded by an analytic curve Theorem 2 was proved by M. S. Mel'nikov (cf. [3], [7][1]). However, for some of the simplest types of regions, for example the square, this estimate of the integral has not been proved.

We shall state two corollaries of Theorem 1.

**Theorem 3.** *Let* $e$ *be a closed bounded set, and let* $f(z)$ *be a function continuous on the whole plane with modulus of continuity* $\omega(\delta)$. *For* $f(z)$ *to be uniformly approximable on* $e$ *to any precision by rational fractions, it is necessary and sufficient that every square* $\xi(z, \delta)$ *with center at point* $z$ *and with side* $\delta$ *satisfy the inequality*

$$\left| \int_{\partial \xi(z,\delta)} f(z)\, dz \right| \leqslant C \cdot \omega(\delta) \cdot \gamma\,[\xi(z, \delta) \setminus e],$$

*where* $C$ *is a constant common for all* $z$ *and* $\delta$.

The sufficiency of the conditions stated is proved in [4], and necessity follows easily from Theorem 2.

**Theorem 4.** *If the inner boundary of a closed bounded set* $e$ *lies on a countable number of Ljapunov curves, then every function continuous on* $e$ *and analytic at interior points of* $e$ *can be approximated to any precision*

---

1) Article [7] appears on pp. 503–514 of the present issue of Matematičeskiĭ Sbornik (and is translated in the present issue of Amer. Math. Soc. Transl., series 2).

*uniformly on e by rational fractions.*

Theorem 4 follows from Theorem 1 and from Theorems 3 and 7 of [5].

## §2. Construction of a special system of partitions of unity

We shall show that there exist a family of real functions $g_k(z, \delta)$ (here $k$ runs through all natural numbers, and the parameter $\delta$ through all positive real values) and a constant $\lambda > 0$ satisfying the following conditions.

1) For every $\delta$ we have the identity

$$\sum_{k=1}^{\infty} g_k(z, \delta) \equiv 1.$$

2) For every $k$ and $\delta$ the function $g_k(z, \delta)$ is nonnegative and finitely differentiable, and

$$\max_z |\operatorname{grad} g_k(z, \delta)| \leqslant \frac{\lambda}{\delta}.$$

3) For every $k$ and $\delta$ the function $g_k(z, \delta)$ is finite, the diameter of the set $\mu_k^\delta$ arising as the closure of the support of this function is less than $\delta$, and the set $\mu_k^\delta$ is convex.

4) For every $\delta$ and every disk $\sigma$ of radius $\delta$ the number of values of the index $k$ for which $\sigma \cap \mu_k^\delta \neq \emptyset$ does not exceed $\lambda$.

We emphasize that $\lambda$ does not depend on either $k$ or $\delta$.

To construct the indicated family of functions we fix any nonnegative, finitely differentiable function having the properties $g(z) = 0$ for $|z| \geq 1$, and $\int g(z)\, dx\, dy = 1$. We divide the plane into equal squares $\xi_k$ $(k = 1, 2, \cdots)$ with pairwise disjoint interiors and with side of unit length, and we set $g_k(z) = \int_{\xi_k} g(\zeta - z) d\xi\, d\eta$. The functions $\{g_k(z)\}$ are finitely differentiable and finite; since $\psi(z) = \int g(\zeta - z) d\xi\, d\eta \equiv 1$ we also have $\Sigma_{k=1}^{\infty} g_k(z) \equiv 1$. Therefore the function $g_k(6z/\delta)$ may be taken as $g_k(z, \delta)$.

Verification of conditions 1)–4) is left to the reader.

## §3. Partitioning of the set of singularities of a function

Below we shall often use the expressions

$$\int g(\zeta) \frac{\partial f}{\partial \bar{\zeta}}\, d\bar{\zeta}\, d\zeta \quad \text{and} \quad \int \frac{\partial f}{\partial \bar{\zeta}} \frac{g(\zeta)}{\zeta - z}\, d\bar{\zeta}\, d\zeta,$$

where $g(\zeta)$ is a continuously differentiable function and $f(\zeta)$ is a continuous, but in general not a differentiable, function. If $f(\zeta)$ is nondifferentiable the indicated expressions require clarification. First of all, we recall that $\partial f/\partial\bar{\zeta}$ is conditional notation for the expression

$$-\frac{1}{2i}\left[\frac{\partial u}{\partial y}+\frac{\partial v}{\partial x}-i\frac{\partial u}{\partial x}+i\frac{\partial v}{\partial y}\right],$$

where $u$ and $v$ are the real and imaginary parts of the function $f$. It is well known that a natural meaning can be ascribed to the expressions indicated above, for example, with the aid of Stokes' formula

$$\int_g \frac{\partial\varphi}{\partial\bar{\zeta}}\,d\bar{\zeta}\,d\zeta=\int_{\partial g}\varphi(\zeta)\,d\zeta$$

(here $g$ is a region with smooth boundary, and $\phi(z)$ is a continuously differentiable function). We shall assume that

$$\int_g \frac{g(\zeta)}{\zeta-z}\frac{\partial f}{\partial\bar{\zeta}}\,d\bar{\zeta}\,d\zeta=2\pi i f(z)\,g(z)+\int_{\partial g} f(\zeta)\frac{g(\zeta)}{\zeta-z}\,d\zeta-\int_g \frac{f(\zeta)}{\zeta-z}\frac{\partial g}{\partial\bar{\zeta}}\,d\bar{\zeta}\,d\zeta.$$

The expressions on the right of this equation already do not contain derivatives of $f(z)$.

Let $f(z)$ be a continuous finite function. We fix $\delta>0$ and represent this function in the form

$$f(z)=\sum_{k=1}^{\infty} f_k^{\delta}(z),\ \text{where}\ f_k^{\delta}(z)=\frac{1}{2\pi i}\int\frac{g_k(\zeta,\delta)}{\zeta-z}\frac{\partial f}{\partial\bar{\zeta}}\,d\bar{\zeta}\,d\zeta.$$

The equation $f(z)=\Sigma_{k=1}^{\infty} f_k^{\delta}(z)$ is valid, since

$$f(z)=\frac{1}{2\pi i}\int\frac{1}{\zeta-z}\frac{\partial f}{\partial\bar{\zeta}}\,d\bar{\zeta}\,d\zeta=\frac{1}{2\pi i}\int\frac{1}{\zeta-z}\frac{\partial f}{\partial\bar{\zeta}}\sum_{k=1}^{\infty} g_k(\zeta,\delta)\,d\bar{\zeta}\,d\zeta$$

$$=\sum_{k=1}^{\infty}\frac{1}{2\pi i}\int\frac{g_k(\zeta,\delta)}{\zeta-z}\frac{\partial f}{\partial\bar{\zeta}}\,d\bar{\zeta}\,d\zeta=\sum_{k=1}^{\infty} f_k^{\delta}(z).$$

**Lemma 3.1.** *If $f(z)$ is a function continuous in the whole plane with modulus of continuity $\omega(\delta)$, and $g(z)$ is a continuously differentiable finite function whose support has diameter no greater than $\delta$, then the function*

$$f_g(z)=\frac{1}{2\pi i}\int\frac{g(\zeta)}{\zeta-z}\frac{\partial f}{\partial\bar{\zeta}}\,d\bar{\zeta}\,d\zeta$$

*has the following properties*:

a) *It is continuous on the whole plane and* $f_g(\infty) = 0$.

b) *It is analytic in the region* $e_f \cup e_g$, *where* $e_f$ *is the region of analyticity of the function* $f(z)$, *and* $e_g$ *is the interior of the set* $\{g(z) = 0\}$.

c) $\max_z |f_g(z)| \leq 2\delta\,\omega(\delta) \max_z |\partial f/\partial \bar z|$.

Applying this lemma to the function $f_k^\delta(z)$, we get the function $f(z)$ represented as the sum of functions $\{f_k^\delta(z)\}$, each of which is bounded in absolute value by the quantity $2\lambda\omega(\delta)$, i.e. is small for small $\delta$, and the set of singularities of the function $f_k^\delta(z)$ is contained in the intersection of the set of singularities of the original function and the support of the function $f_k^\delta(z)$.

**Proof of Lemma 3.1.** Since the function $f(\zeta)\,\partial g/\partial \bar\zeta$ is bounded, the function

$$f_g(z) = f(z)\,g(z) - \frac{1}{2\pi i}\int f(\zeta)\,\frac{1}{\zeta - z}\,\frac{\partial g}{\partial \bar\zeta}\,d\bar\zeta\,d\zeta$$

is continuous and $f(\infty) = 0$. Let $\mu - \{g(z) = 0\}$, $z_0 \in c_f \cup e_g$, and $\sigma$ be the disk with center at point $z_0$ wholly lying in $e_f \cup e_g$. We shall show that $\partial f_g(z_0)/\partial \bar z = 0$.

Since $\int f(\zeta)\,(\zeta - z)^{-1}(\partial g/\partial \bar\zeta)\,d\bar\zeta\,d\zeta$ is analytic at the point $z_0$, we have

$$\begin{aligned}
\frac{\partial f_g(z_0)}{\partial \bar z} &= \frac{\partial}{\partial \bar z}\left(f(z)\,g(z) - \frac{1}{2\pi i}\int_\mu \frac{f(\zeta)}{\zeta - z}\,\frac{\partial g}{\partial \bar\zeta}\,d\bar\zeta\,d\zeta\right)\\
&= \frac{\partial}{\partial \bar z}\left(f(z)\;g(z) - \frac{1}{2\pi i}\int_\sigma \frac{f(\zeta)}{\zeta - z}\,\frac{\partial g}{\partial \bar\zeta}\,d\bar\zeta\,d\zeta\right)\\
&= \frac{\partial}{\partial \bar z}\left[f(z)\,g(z) - \frac{1}{2\pi i}\left(2\pi i f(z)\,g(z) + \int_{\partial\sigma} \frac{f(\zeta)\,g(\zeta)}{\zeta - z}\,d\zeta - 0\right)\right]\\
&= \frac{\partial}{\partial \bar z}\left(\frac{1}{2\pi i}\int_{\partial\sigma} \frac{f(\zeta)\,g(\zeta)}{\zeta - z}\,d\zeta\right) = 0,
\end{aligned}$$

i.e. $f_g(z)$ is analytic in $e_f \cup e_g$. From properties a) and b) of the function $f_g(z)$, which have already been proved, it follows that there exists a point $z_1 \in \mu$ such that $f_g(z_1) = \max |f_g(z)|$. Since

$$f_g(z_1) = \frac{1}{2\pi i}\int \frac{g(\zeta)}{\zeta - z}\frac{\partial f}{\partial \bar\zeta}\, d\bar\zeta\, d\zeta = \frac{1}{2\pi i}\int_\mu \frac{g(\zeta)}{\zeta - z}\frac{\partial}{\partial \bar\zeta}[f(\zeta) - f(z_1)]\, d\bar\zeta\, d\zeta,$$

we have

$$|f_g(z_1)| \leqslant \frac{1}{2\pi}\,\omega(\delta) \max_z \left|\frac{\partial g}{\partial \bar z}\right| \int_\mu \frac{|d\bar\zeta\, d\zeta|}{|\zeta - z|},$$

and since

$$\int_\mu \frac{|d\bar\zeta\, d\zeta|}{|\zeta - z|} = \int_\mu \frac{2\, d\xi\, d\eta}{|\zeta - z|} \leqslant 2 \int_{|\zeta| \leqslant \delta} \frac{d\xi\, d\eta}{|\zeta|} = 4\pi\delta,$$

we have

$$|f_g(z)| \leqslant 2\delta\omega(\delta) \max_z \left|\frac{\partial g}{\partial \bar z}\right|,$$

i.e. property c) is also proved.

## §4. Concerning the additivity of analytic content

Let $e$ be an arbitrary set. We represent it in the form $e = \bigcup_k e_k^\alpha$, where $e_k^\alpha = e \cap \mu_k^\alpha$ ($\alpha = \alpha(e)$; for $\mu_k^\alpha$ see §1) and prove that

$$\lambda_1 \alpha(e) \leqslant \sum_k \alpha(e_k^\alpha) \leqslant \lambda_2 \alpha(e),$$

where $\lambda_1$ and $\lambda_2$ are positive constants not depending on $e$.

The first of these inequalities is simple to prove. In fact, let $f(z) \in C(e, 1)$. We represent this function in the form

$$f(z) = \sum_k f_k^\alpha(z) = \sum_k \frac{1}{2\pi i}\int \frac{g_k^\alpha(\zeta)}{\zeta - z}\frac{\partial f}{\partial \bar\zeta}\, d\bar\zeta\, d\zeta.$$

From Lemma 3.1 it follows that $f_k^\alpha(z) \in C(e_k^\alpha, m_\alpha)$ and $m_\alpha \le 2\lambda\omega(\alpha) \le 4\lambda$. Choosing $f(z)$ so that $|\gamma(e, f)| \ge \frac{1}{2}\alpha(e)$, we obtain

$$\alpha(e) \leqslant 2\left|\frac{1}{2\pi i}\int_{\partial e} f(z)\, dz\right| \leqslant 2\sum_k \left|\frac{1}{2\pi i}\int_{\partial e_k^\alpha} f_k^\alpha(z)\, dz\right|$$

$$\leqslant 2\sum_k m_\alpha \cdot \alpha(e_k^\alpha) \leqslant 8\lambda \sum_k \alpha(e_k^\alpha),$$

i.e. $\lambda_1 \alpha(e) \le \Sigma_k \alpha(e_k^\alpha)$, where $\lambda_1 = 1/8\lambda$. The first inequality is proved.

The second inequality will be obtained from the lemmas following below.

Lemma 4.1. *Let* $p$ *be a natural number and* $\{C_{k,n} \leq C\}$ *nonnegative numbers* ($n = 1, 2, \cdots, pk$; $k = 1, 2, \cdots$; $C > 0$). *Then*

$$\sum_{k=1}^{\infty}\sum_{n=1}^{pk} C_{k,n} \geqslant \frac{C}{4p}\left(\sum_{k=1}^{\infty}\sum_{n=1}^{pk}\frac{C_{k,n}}{kC}\right)^2.$$

Proof. We set $m = m(\{C_{k,n}\}) = \Sigma_{k=1}^{\infty}\Sigma_{n=1}^{pk} C_{k,n}/kC$. We shall alter the numbers $\{C_{k,n}\}$, increasing those with small index $k$ and decreasing those whose index $k$ has large value, while preserving the value of $m(\{C_{k,n}\})$. Then $\mu(\{C_{k,n}\}) = \Sigma_{k=1}^{\infty}\Sigma_{n=1}^{pk} C_{k,n}$ will be decreased. Therefore it is sufficient to prove the inequality only for the case when the numbers $\{C_{k,n}\}$ take the maximum possible values (for a fixed value of $m$) and occupy places with the least values of the index $k$, i.e. for the case when $C_{k,n} = C$ for $n=1, \cdots, pk$; $k = 1, \cdots, q$; $C_{k,n} = 0$ for $k \geq q + 2$. From the condition $\Sigma_{k=1}^{\infty}\Sigma_{n=1}^{pk} C_{k,n}/kC = m$ it follows that $m \leq \Sigma_{k=1}^{q+1}\Sigma_{n=1}^{pk} C/kC = p(q+1)$, i.e. $q + 1 \geq m/p$. If $q \geq 1$, then

$$\mu(\{C_{k,n}\}) \geqslant \sum_{k=1}^{q}\sum_{n=1}^{pk} C = Cp\frac{q(q+1)}{2} \geqslant \frac{1}{4}Cp(q+1)^2 \geqslant \frac{1}{4}Cp\left(\frac{m}{p}\right)^2,$$

i.e. $\mu(\{C_{n,k}\}) = Cm^2/4p$. For $q = 0$ we have $\mu(\{C_{k,n}\}) = C\mu^{-1}(\{C_{k,n}\}) \cdot m^2 \geq Cm^2/p$. Thus

$$\mu(\{C_{k,n}\}) = \min\left(\frac{1}{p}, \frac{1}{4p}\right) \cdot Cm^2 = \frac{1}{4p}Cm^2.$$

The lemma is proved.

Lemma 4.2. *Let* $\{e_i\}$ *be a system of subsets of the set* $e$ *such that for any disk of radius* $\alpha(e)$ *the number of sets of this system having common points with this disk does not exceed the constant* $q$. *Then for functions* $\{f_i(z) \in C(e_i, 1)\}$ *the inequality* $\max_z \Sigma_i |f_i(z)| \leq M(q)$ *is fulfilled, where* $M(q) > 0$ *is a constant depending only on* $q$.

Proof. We hold $\{f_i(z) \in C(e_i, 1)\}$ fixed and set $m = \max_z \Sigma_i |f_i(z)|$. We fix the point $\zeta$ and index the set $\{e_i\}$ by indexes $k$ and $n$ so that $k = [\rho(\zeta, e_{k,n})/\alpha(e)]$, where $\rho(\zeta, e_{k,n})$ is the distance from the point $\zeta$ to the set $e_{k,n} \in \{e_i\}$. Here it turns out that the index $n$ takes no more than $pk = p(q)k$ values. We set $\mu(\zeta) = \Sigma_{k,n}\alpha(e_{k,n})/k\alpha(e)$, $\mu = \max \mu(\zeta)$ and $\phi(z) = \Sigma_i \phi_i(z)$, where $\phi_i(z) \in C(e_i, 1)$ and $\gamma(e_i, f) = \frac{1}{2}\alpha(e_i)$. Since

$|\phi_i(z)| \leq \min(1, \alpha(e)/\rho(\zeta, e_i))$, we have

$$|\varphi(\zeta)| \leqslant \sum_i |\varphi_i(\zeta)| \leqslant M_1(q)\left(1 + \sum_{k,n} \frac{\alpha(e_{k,n})}{k\,\alpha(e)}\right) \leqslant M_1(q)(1+\mu) = \mu_1,$$

i.e. $\phi(z) \in C(e, \mu_1)$. Therefore $\alpha(e) \geq \mu_1^{-1}|\gamma(e, f)| = \mu_1^{-1}\Sigma_i \frac{1}{2}\alpha(e_i)$, i.e. $2\mu_1\alpha(e) \geq \Sigma_{k,n}\alpha(e_{k,n})$. We recall that the indexing of the sets $\{e_{k,n}\}$ depends on the point $\zeta$. However, the last equation is valid for the renumbering at any $\zeta$, in particular for the point $\zeta$ for which $\mu(\zeta) = \mu$. Since $\Sigma_{k,n}\alpha(e_{k,n}) \geq \alpha(e)\mu^2/4p$ by virtue of Lemma 4.1, we have $\mu^2/4p \leq 2\mu_1 = 2M_1(q)(1+\mu)$, i.e. $\mu$ can not be too large a number, as $\mu^2$ is on the left of the inequality and a quantity of order $\mu$ on the right. From this inequality it follows that $\mu \leq M_2(q)$, where $M_2(q)$ is a constant depending only on $q$. By analogy with how this was done for the function $\phi(z)$, it is proved that $m \leq M_1(q)(1+\mu) \leq M_1(q)(1 + M_2(q)) = M(q)$. The lemma is proved.

Lemma 4.3. *If the sets $\{e_i \subset e\}$ satisfy the conditions of Lemma 4.2, then $\Sigma_i \alpha(e_i) \leq M(q) \cdot \alpha(e)$.*

The proof follows from Lemma 4.2, if it is applied to the functions $\phi_i(z) \in C(e_i, 1)$ $(\gamma(e_i, \phi_i) = (1-\epsilon)\alpha(e_i))$, where $\epsilon$ is a constant close to zero. From Lemma 4.3 and property 4) of the functions $g_k(z, \delta)$ (see §2) we obtain the second part of the inequality introduced at the beginning of this section.

## §5. Change in content under conformal deformations of a set

Lemma 5.1. *Let $g$ be a Ljapunov region, and let $z' = \phi(z)$ be a conformal map of the region $g$ onto the region $g' = \phi(g)$ which is continuously extended onto $\bar{g}$ and one-to-one on $\bar{g}$, and such that on $\bar{g}$ $\partial\phi/\partial z$ satisfies the Hölder condition, with exponent $\alpha_1 > 0$ and constant $L_1$. Then for any closed sets $e \subset \bar{g}$ and $e' = \phi(e)$ we have the inequality $\alpha(e') \leq M(g, \phi)\alpha(e)$, where $M(g, \phi)$ is a constant not depending on $e$.*

Proof. By the definition of $C$-content there exists a function $f(z') \in C(e', 1)$ such that $\gamma(e', f) = \frac{1}{2}\alpha(e')$. We set $f_1(z) = f[\phi(z)] = f_2(z) + f_3(z)$, where

$$f_2(z) = \frac{1}{2\pi i}\int_{\partial e} \frac{f_1(\zeta)}{\zeta - z}\, d\zeta \text{ and } f_3(z) = \frac{1}{2\pi i}\int_{\partial g} \frac{f_1(\zeta)}{\zeta - z}\, d\zeta.$$

The function $f_3(z)$ is analytic in $g$. We shall show that it is bounded.

Since $f(z')$ is analytic outside $g'$ and $f(\infty) = 0$, we have

$$\int_{\partial g} \frac{f_1(\zeta)}{\varphi(\zeta)-\varphi(z)} \frac{d\varphi}{d\zeta} d\zeta = \int_{\partial g'} \frac{f(\zeta')}{\zeta'-z'} d\zeta' = 0 \quad \text{for} \quad z \in g$$

and hence

$$f_3(z) = f_3(z-0) = \frac{1}{2\pi i} \int_{\partial g} f_1(\zeta) \varphi(\zeta, z) d\zeta,$$

where

$$\varphi(\zeta, z) = \frac{1}{\zeta - z} - \frac{1}{\varphi(\zeta)-\varphi(z)} \frac{d\varphi}{d\zeta} = \frac{\varphi(\zeta)-\varphi(z)-(\zeta-z)\dfrac{d\varphi}{d\zeta}}{(\zeta-z)[\varphi(\zeta)-\varphi(z)]}.$$

Since the map is one-to-one and $d\phi/d\zeta$ satisfies the Hölder condition, we have

$$|\varphi(\zeta)-\varphi(z)| \geqslant C_1 |\zeta - z|$$

and

$$\left| \varphi(\zeta)-\varphi(z)-(\zeta-z)\frac{d\varphi}{d\zeta} \right| \leqslant C_2 |\zeta - z|^{1+\alpha_1}.$$

Therefore

$$|f_3(z)| \leqslant \frac{1}{2\pi} \int_{\partial g} |f_1(\zeta)\varphi(\zeta, z) d\zeta| \leqslant \frac{1}{2\pi} \int_{\partial g} \frac{C_2}{C_1} \cdot \frac{|\zeta - z|^{1+\alpha_1}}{|\zeta - z|^2} |d\zeta| \leqslant C_3.$$

Thus the function $f_3(z)$ is bounded. Consequently, $|f_2(z)| = |f_1(z) - f_3(z)| \leq 1 + C_3$. We set

$$f_4(z) = \int_{\partial e} \frac{f_2(\zeta)}{\zeta - z} \frac{d\varphi}{d\zeta} d\zeta.$$

Since

$$f_4(z) = 2\pi i f_2(z) - \int_{\partial g} \frac{f_2(\zeta)}{\zeta - z} \frac{d\varphi}{d\zeta} d\zeta$$

and for $z \in g$

$$\left| \int_{\partial g} \frac{f_2(\zeta)}{\zeta - z} \frac{d\varphi}{d\zeta} d\zeta \right| = \left| \int_{\partial g} \frac{f_2(\zeta)}{\zeta - z} \left( \frac{d\varphi(\zeta)}{d\zeta} - \frac{d\varphi(z)}{d\zeta} \right) d\zeta \right|$$

$$\leqslant (1 + C_3) \int_{\partial g} \frac{L_1 |\zeta - z|^{\alpha_1}}{|\zeta - z|} |d\zeta| \leqslant C_4,$$

we have $f_4(z) \in C(e, C_4)$. The residue of this function equals

$$\frac{1}{2\pi i}\int_{\partial g} f_4(z)\,dz = \int_{\partial g} f_2(\zeta)\frac{d\varphi}{d\zeta}\,d\zeta = \int_{\partial g} f_1(\zeta)\frac{d\varphi}{d\zeta}\,d\zeta = \int_{\partial g'} f(z')\,dz' = 2\pi i\,\frac{\alpha(e')}{2}.$$

Consequently

$$\alpha(e') = \frac{1}{\pi i}\cdot\frac{1}{2\pi i}\int_{\partial g} f_4(z)\,dz \leqslant \frac{C_4}{\pi}\,\alpha(e).$$

The lemma is proved.

**Lemma 5.2.** *Let $g$ be a simply-connected Ljapunov region bounded by a smooth curve (a curve without points of discontinuity of the argument of the tangent). Then for any closed set $e \subset \bar{g}$ and any function $f(z) \in C(e, m)$ one has*

$$\left|\int_{\partial g} f(z)\,dz\right| \leqslant C_1(g)\cdot m\alpha(e),$$

*where $C_1(g)$ is a constant depending only on $g$.*

**Proof.** By virtue of a theorem of Kellogg (see [6], 1952 edition, p. 468), there exists a conformal map $z' = \phi(z)$ of a closed region onto the closed disk $\sigma$, whose derivative satisfies the Hölder condition. Therefore from Lemma 5.1 we get that $\alpha(e') \leq M(g, \phi)\,\alpha(e)$ for every $e \subset \bar{g}$ and $e' = \phi(e)$.

The function $1/(d\phi/dz)$ is analytic in $g$ and bounded. Since for every closed set lying strictly inside the disk the estimate of the integral is proved, i.e. for every function $\psi(z) \in C(\sigma, e', m)$ and set $e'$ not containing boundary points of the disk we have $|\int_{\partial\sigma}\psi(z')dz'| \leq C(g)\cdot m\alpha(e)$ (see [3], [7]), it follows that

$$\left|\int_{\partial g} f(z)\,dz\right| = \left|\int_{\partial g}\frac{f(z)}{\frac{d\varphi}{dz}}\frac{d\varphi}{dz}\,dz\right|$$

$$= \left|\int_{\partial\sigma}\frac{f[\varphi^{-1}(z')]}{\frac{d}{dz}\{\varphi[\varphi^{-1}(z')]\}}\,dz'\right| \leqslant C(g)\max_{z'\in\sigma}\left|\frac{f[\varphi^{-1}(z')]}{\frac{d}{dz}\{\varphi[\varphi^{-1}(z')]\}}\right|\alpha(e')$$

$$\leqslant C_5\cdot m\alpha(e') \leqslant C_5\cdot M(g,\varphi)\cdot m\alpha(e) = C_1(g)\cdot m\alpha(e).$$

Thus, if a closed set $e$ lies strictly inside the region $g$ and $f(z) \in C(g, e, m)$, then $|\int_{\partial g} f(z)\,dz| \leq C_1(g)\cdot m\alpha(e)$. If $e$ is an arbitrary closed set of $\bar{g}$ and $f(z) \in C(g, e, m)$, then by Theorem 7 of [5] the function $f(z)$ can be uniformly approximated to any precision by a function $f^*(z) \in C(g, e, m)$ which is analytic in a neighborhood of the boundary of the region $g$. Since for $f^*(z)$ the estimate of the integral has already been obtained, i.e.

$$\left| \int_{\partial g} f^*(z)\, dz \right| \leqslant C_1(g)(m+\varepsilon)\,\alpha(e),$$

for the function $f(z)$ it follows that $|\int_{\partial g} f(z)\,dz| \leq C_1(g)\cdot m\alpha(e)$.

The lemma is proved.

## §6. Lemmas concerning smooth functions

Lemma 6.1. *Let $y = f(x)$ be a real function of the real variable $x$, defined on the interval $r = \{-c \leq x \leq c\}$; $y' = df/dx$ satisfies the Hölder condition with constant $L$ and exponent $\alpha$; $f(0) = \epsilon$; everywhere on the interval $r$ the function $f(x) > 0$. Then*

$$\left| \frac{df}{dx} \right| \leqslant C(L, \alpha)\left( \frac{\varepsilon}{c} + \varepsilon^{\frac{\alpha}{1+\alpha}} \right).$$

Proof. For definiteness we assume that $p = df(0)/dx < 0$. We consider the function $g(x) = \epsilon - px + x^{1+\alpha}L/(1+\alpha)$ and the point $x_0 = (p/L)^{1/\alpha}$. On the interval $[0, x_0]$ the function $g(x)$ decreases monotonically and $g(x) \geq f(x)$ everywhere on the interval $0 \leq x \leq \min(c, x_0)$.

We now examine two cases: $x_0 \geq c$ and $x_0 \leq c$. We first assume that $x_0 \geq c$. Since on the interval $[0, c]$ we have $g(x) \geq f(x) > 0$ and $g(0) = f(0)$, it follows that $g(0) - g(c) \leq \epsilon$, i.e. $pc - c^{1+\alpha}L/(1+\alpha) \leq \epsilon$, or in other words $p \leq \epsilon/c + L\cdot c^{\alpha}/(1+\alpha)$. In the case under consideration $dg(c)/dx \leq 0$, i.e. $-p + c^{\alpha}L \leq 0$; this means that $p - p/(1+\alpha) \leq \epsilon/c$. Thus in the first case $p \leq (1+\alpha)\epsilon/\alpha c$.

Now let $x_0 \leq c$. Since $g(0) - g(x_0) \leq \epsilon$, we have $px_0 - Lx_0^{1+\alpha}/(1+\alpha) \leq \epsilon$. Since $dg(x_0)/dx = 0$, we have $-p + x_0^{\alpha}L = 0$. Consequently

$$p \leqslant \left( \frac{1+\alpha}{\alpha} \right)^{\frac{\alpha}{1+\alpha}} L^{\frac{1}{1+\alpha}} \varepsilon^{\frac{\alpha}{1+\alpha}}.$$

Therefore, in any case

$$p \leqslant \frac{1+\alpha}{\alpha}\frac{\varepsilon}{c} + \left( \frac{1+\alpha}{\alpha} \right)^{\frac{\alpha}{1+\alpha}} L^{\frac{1}{1+\alpha}} \varepsilon^{\frac{\alpha}{1+\alpha}} \leqslant C(L, \alpha)\left( \frac{\varepsilon}{c} + \varepsilon^{\frac{\alpha}{1+\alpha}} \right).$$

The lemma is proved.

**Lemma 6.2.** *Let $f_1(x)$ and $f_2(x)$ satisfy the Hölder condition with constant $L$ and exponent $\alpha$ on the interval $r = \{-c \le x \le c\}$: $f(x) = |f_1(x) - f_2(x)| > 0$ on the interval $r$, and $f(0) = \epsilon \le c$. Then*

$$\int_{-c}^{\varepsilon} f(x)\, x^{-2}\, dx + \int_{\varepsilon}^{c} f(x)\, x^{-2}\, dx \leqslant C_1(L, \alpha) = \text{const.}$$

**Proof.** Integrating by parts, we obtain

$$\int_{\varepsilon}^{c} f(x)\, x^{-\cdot}\, dx = \frac{f(\varepsilon)}{\varepsilon} - \frac{f(c)}{c} + \int_{\varepsilon}^{c} \frac{1}{x} \frac{df}{dx}\, dx$$

$$= \frac{f(\varepsilon)}{\varepsilon} - \frac{f(c)}{c} + \frac{df(0)}{dx} \ln \frac{c}{\varepsilon} + \int_{\varepsilon}^{c} \frac{1}{x} \left( \frac{df(x)}{dx} - \frac{df(0)}{dx} \right) dx$$

$$\leqslant \frac{f(\varepsilon)}{\varepsilon} + C_2(L, \alpha) \left( \frac{\varepsilon}{c} + \varepsilon^{\frac{\alpha}{1+\alpha}} \right) \ln \frac{\varepsilon}{c} + \int_{\varepsilon}^{c} 2Lx^{\alpha} \cdot \frac{1}{x}\, dx \leqslant \frac{f(\varepsilon)}{\varepsilon} + C_2(L, \alpha)$$

(cf. Lemma 6.1). But

$$f(\varepsilon) \leqslant \varepsilon + \varepsilon \max_{x \leqslant \varepsilon} \left| \frac{df}{dx} \right| \leqslant \varepsilon + \varepsilon \left\{ C(2L, \alpha) \left( \frac{\varepsilon}{c} + \varepsilon^{\frac{\alpha}{1+\alpha}} \right) + 2L\varepsilon^{\alpha} \right\} \leqslant \epsilon C_3(L, \alpha).$$

Consequently

$$\int_{\varepsilon}^{c} f(x)\, x^{-2}\, dx \leqslant C_3(L, \alpha) + C_2(L, \alpha) = \frac{1}{2} C_1(L, \alpha).$$

The second integral is estimated analogously.

**Lemma 6.3.** *Let $\gamma$ be an arc with endpoints $z'$ and $z''$ the argument of whose tangent vector satisfies the Hölder condition with constant $L$ and exponent $\alpha$; $z_0 = x_0 + iy_0$ is a point of this arc distant from each endpoint by not less than $2c$, $0 \le y_0 \le c$; the arc $\gamma$ does not intersect the line $y = 0$. Then there exists a constant $c_0$ depending only on $L$ and $\alpha$ such that if $c \le c_0$, then there exists a function $f(x)$ defined on the interval $\{-c \le x \le c\}$ whose graph lies on the arc $\gamma$; the derivative of this function satisfies the Hölder condition with constant $2L$ and exponent $\alpha$.*

The proof of this lemma is easily obtained from Lemma 6.1.

## §7. Estimate of the content of a set bounded by Ljapunov curves

**Lemma 7.1.** *Let* $f_1(x)$ *and* $f_2(x)$ *be real functions defined on the interval* $r' = \{-4c \le x \le 4c\}$; *let* $f(x) = |f_1(x) - f_2(x)| > 0$; *let* $df_1/dx$ *and* $df_2/dx$ *satisfy the Hölder condition with constant* $L$ *and exponent* $\alpha$; $|df_1/dx| \le 1/8$, $|df_2/dx| \le 1/8$; *let* $x_1, \cdots, x_p$ *be points of the x-axis such that* $x_1 = -2c$, $x_{p-1} \le 2c$, $x_p \ge 2c$, *and* $x_j = x_{j-1} + f(x_{j-1})$ $(j = 2, 3, \cdots, p)$; *let* $e_j$ *be a set lying in the closed region* $\xi_j$ *and bounded by the graphs of the functions* $f_1(x)$, $f_2(x)$ *and the straight lines* $x = x_j$ *and* $x = x_{j+1}$; *the projection of the set* $e = \mathbf{U}_j e_j$ *is contained in the interval* $r = \{-c \le x \le c\}$. *Then* $\Sigma_j \alpha(e_j) \le C_4(L, \alpha)\, \alpha(e)$. *We emphasize that* $C_4(L, \alpha)$ *depends only on* $L$ *and* $\alpha$.

**Proof.** For definiteness let $f_2(x) > f_1(x)$. We designate by $\gamma$ the graph of the function $f_3(x) = f_2(x) + f(x)$ and by $\gamma_i$ the part of $\gamma$ contained between the straight lines $x = x_j$ and $x = x_{j+1}$. Since $|df_1/dx| \le 1/8$ and $|df_2/dx| \le 1/8$, we have $h_1(\partial\xi_j) \le 6(x_{j+1} - x_j)$ and therefore

$$\alpha(e_j) \leqslant \frac{1}{2\pi} h_1(\partial\xi_j) \leqslant \frac{6}{2\pi}(x_{j+1} - x_j) < (x_{j+1} - x_j).$$

As $h_1(\gamma_j) \ge x_{j+1} - x_j$, on the arc $\gamma_j$ it is possible to lay out an arc $\beta_j$ of length $h_1(\beta_j) = \alpha(e_j)$ which does not pass over the straight lines $x = x_{j+1}$, $x = x_j$.

We set $\beta = \mathbf{U}_j \beta_j$. Since for a subset of Ljapunov curves the analytic content and the length are commensurable (see [3]), we have

$$\gamma(\beta) \geqslant C_5(L, \alpha)\, h_1(\beta) \geqslant C_5(L, \alpha) \sum_j h_1(\beta_j) = C_5(L, \alpha) \sum_j \alpha(e_j).$$

Consequently there exists a function $\phi(z) \in A(\beta, 1)$ such that $\gamma(\beta, \phi(z)) = \gamma(\beta) \ge C_5(L, \alpha)\, \Sigma_j \alpha(e_j)$.

Let $\phi_j(z) = (2\pi i)^{-1} \int_{\beta_j} \phi(\zeta)(\zeta - z)^{-1} d\zeta$ (here the integration proceeds along a contour close to $\beta_j$ which encloses the arc $\beta_j$ and does not enclose points of the other arcs $\{\beta_s\}$); $\psi_j(z) \in C(e_j, 1)$ is a function for which $\gamma(e_j, \psi_j) = \frac{1}{2}\alpha(e_j)$; $\delta_j = (2/\alpha(e_j)) \cdot \gamma(\beta_j, \phi_j)$; $\psi(z) = \Sigma_j \delta_j \psi_j(z)$. The number $\delta_j$ is selected so that the equation $\gamma(e_j, \delta_j \psi_j) = \gamma(\beta_j, \phi_j)$ holds. Consequently

$$\gamma(e, \psi) = \gamma(\beta, \varphi) \geqslant C_5(L, \alpha) \sum_j \alpha(e_j).$$

If we assume that at any point of the interval $r$ the inequality $f(x) \geq c$ is fulfilled, then $f(x) \geq c/4$ everywhere on $r$, and therefore the number of values of the index $j$ will not exceed the number $p \leq 17$. Consequently $\Sigma_j \alpha(e_j) \leq 17 \max_j \alpha(e_j) \leq 17\, \alpha(e)$, and the assertion of the lemma is proved in this case.

We next assume that $f(x) \leq c$ on the interval $r$. This permits us to use Lemma 6.2.

We estimate $m = \max_z |\psi(z)|$. Since $\psi(z)$ is analytic outside $e$, its maximum absolute value is attained at some point $z = x + iy \in e$. Let $z \in e_s$ and $a_j \in \beta_j$. We estimate the expression

$$\psi(z) - \varphi(z) \leqslant \sum_j |\delta_j \psi_j(z) - \varphi_j(z)|.$$

For any $j \neq s$ we have

$$\psi_j(z) = \frac{1}{2\pi i} \int_{\partial \xi_j} \frac{\psi_j(\zeta)}{\zeta - z} d\zeta = \frac{1}{a_j - z} \frac{1}{2\pi i} \int_{\partial \xi_j} \psi_j(\zeta)\, d\zeta + \frac{1}{2\pi i} \int_{\partial \xi_j} \psi_j(\zeta)\, \lambda_j(\zeta, z)\, d\zeta,$$

where $\lambda_j(\zeta, z) = 1/(\zeta - z) - 1/(a_j - z)$, i.e.

$$\psi_j(z) = \frac{|\gamma(e_j, \psi_j)|}{a_j - z} + \frac{1}{2\pi i} \int_{\partial \xi_j} \psi_j(\zeta)\, \lambda_j(\zeta, z)\, d\zeta.$$

Analogously we get that

$$\varphi_j(z) = \frac{1}{a_j - z} \cdot \frac{1}{2\pi i} \int_{\partial \beta_j} \varphi(\zeta)\, d\zeta + \frac{1}{2\pi i} \int_{\partial \beta_j} \varphi(\zeta)\, \lambda_j(\zeta, z)\, d\zeta$$

$$= \frac{\gamma(\beta_j, \varphi_j)}{a_j - z} + \frac{1}{2\pi i} \int_{\partial \beta_j} \varphi(\zeta)\, \lambda_j(\zeta, z)\, d\zeta.$$

Thus for $j \neq s$

$$|\delta_j \psi_j(z) - \varphi_j(z)| \leqslant \left| \frac{1}{a_j - z} [\delta_j \gamma(e_j, \psi_j) - \gamma(\beta_j, \phi_j)] \right| + \left| \frac{\delta_j}{2\pi i} \int_{\partial \xi_j} \psi_j(\zeta) \lambda_j(\zeta, z)\, d\zeta \right|$$

$$+ \left| \frac{1}{2\pi i} \int_{\partial \beta_j} \varphi(\zeta)\, \lambda_j(\zeta, z)\, d\zeta \right| = \left| \frac{\delta_j}{2\pi i} \int_{\partial \xi_j} \psi_j(\zeta)\, \lambda_j(\zeta, z)\, d\zeta \right|$$

$$+ \left| \frac{1}{2\pi i} \int_{\partial \beta_j} \varphi(\zeta)\, \lambda_j(\zeta, z)\, d\zeta \right|,$$

insofar as

$$\delta_j = \frac{2}{\alpha(e_j)}\gamma(\beta_j, \phi_j) \text{ and } \gamma(e_j, \psi_j) = \frac{1}{2}\alpha(e_j)$$

For $j > s$ and $\zeta \in \beta_j \cup \partial\xi_j$ we have

$$|\lambda_j(\zeta, z)| = \frac{|a_j - \zeta|}{|a_j - z||\zeta - z|} \leqslant \frac{d(\beta_j \cup \partial\xi_j)}{|x_j - x|^2},$$

where $d(\beta_j \cup \partial\xi_j)$ is the diameter of the set $\beta_j \cup \partial\xi_j$, i.e.

$$|\lambda_j(\zeta, z)| \leqslant \frac{M_1(x_{j+1} - x_j)}{|x_j - x|^2} = \frac{M_1 f(x_j)}{|x_j - x|^2}$$

(here $M_1$ is an absolute constant). Therefore

$$\left|\int_{\partial\beta_j} \varphi(\zeta)\lambda_j(\zeta, z)\,d\zeta\right| \leqslant \int_{\partial\beta_j} |\lambda_j(\zeta, z)\,d\zeta|$$

$$\leqslant \frac{M_1 f(x_j)}{|x - x_j|^2}\cdot 2h_1(\beta_j) \leqslant \frac{4M_1 f(x_j)}{(x - x_j)^2}(x_{j+1} - x_j) \leqslant 8M_1 \int_{x_j}^{x_{j+1}} \frac{f(t)}{(x - x_j)^2}\,dt.$$

If $j - s \geq 4$, then $|x - x_j| \geq M_2|x - x_{j+1}|$ and therefore

$$\left|\int_{\partial\beta_j} \varphi(\zeta)\lambda_j(\zeta, z)\,d\zeta\right| \leqslant 8M_1 \int_{x_j}^{x_{j+1}} \frac{f(t)}{(x - x_j)^2}\,dt \leqslant \frac{8M_1}{M_2}\int_{x_j}^{x_{j+1}} \frac{f(t)}{(x - x_{j+1})^2}\,dt$$

$$\leqslant \frac{8M_1}{M_2}\int_{x_j}^{x_{j+1}} \frac{f(t)}{(x - t)^2}\,dt = M_3 \int_{x_j}^{x_{j+1}} \frac{f(t)}{(x - t)^2}\,dt.$$

It is proved analogously that for $s - j \geq 4$

$$\left|\int_{\partial\beta_j} \varphi(\zeta)\lambda_j(\zeta, z)\,d\zeta\right| \leqslant M_3 \int_{x_j}^{x_{j+1}} \frac{f(t)}{(x - t)^2}\,dt.$$

Thus for $|s - j| \geq 4$

$$\left|\int_{\partial\beta_j} \varphi(\zeta)\lambda_j(\zeta, z)\,d\zeta\right| \leqslant M_3 \int_{x_j}^{x_{j+1}} \frac{f(t)}{(x - t)^2}\,dt.$$

It is similarly proved that for $|s-j| \geq 4$

$$\left|\frac{1}{2\pi i}\int_{\partial\xi_j}\psi_j(\zeta)\,\lambda_j(\zeta, z)\,d\zeta\right| \leqslant M_4\int_{x_j}^{x_{j+1}}\frac{f(t)}{(x+t)^2}\,dt\,.$$

Since

$$|\gamma(\beta_j, \varphi_j)| = \frac{1}{2\pi}\left|\int_{\partial\beta_j}\varphi(\zeta)\,d\zeta\right| \leqslant \frac{1}{2\pi}2h_1(\beta_j) = \frac{\alpha(e_j)}{\pi},$$

we have $\delta_j \leq 2/\pi < 1$, and for $|s-j| \geq 4$ we obtain

$$|\delta_j\psi_j(z)-\varphi_j(z)| \leqslant \left|\frac{\delta_j}{2\pi i}\int_{\partial\xi_j}\psi_j(\zeta)\lambda_j(\zeta, z)\,d\zeta\right| + \left|\frac{1}{2\pi i}\int_{\beta_j}\varphi(\zeta)\lambda_j(\zeta, z)\,d\zeta\right|$$

$$\leqslant \frac{|\delta_j|}{2\pi}M_4\int_{x_j}^{x_{j+1}}\frac{f(t)}{(x-t)^2}\,dt + \frac{1}{2\pi}M_3\int_{x_j}^{x_{j+1}}\frac{f(t)}{(x-t)^2}\,dt \leqslant M_5\int_{x_j}^{x_{j+1}}\frac{f(t)}{(x-t)^2}\,dt.$$

Thus

$$|\psi(z)-\varphi(z)| \leqslant \sum_j|\delta_j\psi_j(z)-\varphi_j(z)|$$

$$\leqslant \sum_{|j-s|\leqslant 3}|\delta_j\psi_j(z)-\varphi_j(z)| + M_5\sum_{|j-s|\geqslant 4}\left|\int_{x_j}^{x_{j+1}}\frac{f(t)}{(x-t)^2}\,dt\right|.$$

From the fact that $|df/dx| \leq 1/4$ it follows that $x_{s+3} \geq x + f(x)$ and $x_{s-2} \leq x - f(x)$, and therefore, by virtue of Lemma 6.2,

$$\sum_{|j-s|\geqslant 4}\int_{x_j}^{x_{j+1}}\frac{f(t)}{(x-t)^2}\,dt \leqslant \int_{x-2c}^{x-f(x)}\frac{f(t)}{(x-t)^2}\,dt + \int_{x+f(x)}^{x+2c}\frac{f(t)}{(x-t)^2}\,dt \leqslant C_1(L, \alpha).$$

Consequently

$$|\psi(z)-\varphi(z)| \leqslant \sum_{|j-s|\leqslant 3}|\delta_j\psi_j(z)-\varphi_j(z)| + M_5C_1(L, \alpha) \leqslant \sum_{|j-s|\leqslant 3}|\delta_j\psi_j(z)|$$

$$+ \sum_{|j-s|\leqslant 3}|\varphi_j(z)| + M_5C_1(L, \alpha) \leqslant 7 + \sum_{|j-s|\leqslant 3}|\varphi_j(z)| + M_5C_1(L, \alpha),$$

insofar as $|\delta_j| \leq 1$ and $|\psi_j(z)| \leq 1$. Since the distance $\rho(z, \beta_j) \geq \min_{x_s \leq x \leq x_{s+1}} f(x) \geq \frac{1}{2}f(x_s)$ and for $|s-j| \leq 3$ the length

$$h_1(\beta_j) = \alpha(e_j) \leqslant x_{j+1} - x_j = f(x_j) \leqslant \frac{4}{3} f(x_{j-1})$$

$$\leqslant \left(\frac{4}{3}\right)^2 f(x_{j-2}) \leqslant \left(\frac{4}{3}\right)^3 f(x_{j-3}),$$

i.e. $h_1(\beta_j) \leq (4/3)^3 f(x_s)$, we have

$$\sum_{|j-s|\leqslant 3} |\varphi_j(z)| = \sum_{|j-s|\leqslant 3} \left| \frac{1}{2\pi i} \int_{\partial\beta_j} \frac{\varphi(\zeta)}{\zeta - z} d\zeta \right| \leqslant' \sum_{|j-s|\leqslant 3} \frac{1}{2\pi} \cdot \frac{2}{f(x_s)} 2h_1(\beta_j)$$

$$\leqslant \sum_{|j-s|\leqslant 3} \frac{1}{\pi} \left(\frac{4}{3}\right)^3 = 7 \cdot \frac{2}{\pi} \left(\frac{4}{3}\right)^3 = M_6.$$

Thus $|\psi(z) - \phi(z)| \leq 7 + M_6 + M_5 C_1(L, \alpha) = C_6(L, \alpha)$. Consequently $|\psi(z)| \leq C_6(L, \alpha) + |\phi(z)| \leq C_6(L, \alpha) + 1$. As $\gamma(e, \psi) \geq C_5 \Sigma_j \alpha(e_j)$, from the definition of analytic content we obtain

$$\alpha(e) \geqslant \gamma(e, \psi)(\max_z |\psi(z)|)^{-1} \geqslant C_5 \sum_j \alpha(e_j)(C_6 + 1)^{-1},$$

i.e.

$$\sum_j \alpha(e_j) \leqslant \frac{C_6(L, \alpha) + 1}{C_5} \alpha(e) = C_4(L, \alpha) \cdot \alpha(e).$$

The lemma is proved.

## §8. Proof of the fundamental theorem

We proceed to the proof of Theorem 1. We divide the boundary of the region $g$ into smooth arcs $\gamma_1, \cdots, \gamma_n$ so that the fluctuation of the argument of the tangent vector on each of these arcs is sufficiently small, for example, smaller than $\pi/4$, and so that $\bigcup_j \gamma_j = \partial g$. The first condition on the arcs is, in essence, superfluous and complicates the natural estimation of the constant $C(g)$, but it simplifies the proof for a rough estimation of the constant $C(g)$. We extend the arc $\gamma_j$ to the closed curve $d_j$, i.e. we construct a closed (without joining itself) smooth Ljapunov curve $d_j \supset \gamma_j$. We designate the region bounded by the curve $d_j$ by $g_j$. The curve $d_j$ is constructed so that the region $g_j$ adjoins to $\gamma_j$ from the same side as the region $g$. If the region $g$ borders on $\gamma_j$ from both sides, then we shall consider that two copies of $\gamma_j$ have been taken, and that $g$ borders on each copy from only one side. Let $z'_j$ and $z''_j$ be the endpoints of the arc $\gamma_j$ and

$g^* = \mathbf{U}_j(z'_j \cup z''_j)$. We set $\alpha = \alpha(e)$ and examine the functions

$$f_k^\alpha(z) = \frac{1}{2\pi i}\int_g \frac{\partial f}{\partial \bar\zeta}\,\frac{g_k(\zeta, \alpha)}{\zeta - z}\,d\bar\zeta\,d\zeta$$

(cf. §3). Here it is assumed that $f(z)$ is extended onto the whole plane with module of continuity $\omega(\delta)$ and is finite. By definition, $f(z) = \Sigma_k f_k^\alpha(z)$ and therefore $\int_{\partial g} f(z)dz = \Sigma_k \int_{\partial g} f_k^\alpha(z)\,dz$. Since $f_k^\alpha(z)$ is analytic outside the set $\mu_k^\alpha$ (see Lemma 3.1; for $\mu_k^\alpha$ see §2), we have

$$\int_{\partial g} f_k^\alpha(z)\,dz = \int_{\partial g_{k,\alpha}} f_k^\alpha(z)\,dz, \text{where } g_{k,\alpha} = g \cap \mu_k^\alpha.$$

We estimate $\Sigma_k \left|\int_{\partial g_{k,\alpha}} f_k^\alpha(z)\,dz\right|$. We fix the number $q > 0$ and agree to designate by $\Sigma_{1,k}$ the summation over all values of $k$ for which the distance $\rho(g^*, \mu_k^\alpha) \le q\alpha$; by $\Sigma_{2,k}$ the summation over $k$ for which $\mu_k^\alpha \cap \partial g = \varnothing$ and $\mu_k^\alpha \subset g$; by $\Sigma_{3,k}$ the summation over $k$ for which $\mu_k^\alpha \cap \gamma_j \ne \varnothing$ only for one value of $j$ and $\rho(g^*, \mu_k^\alpha) \ge q\alpha$; and by $\Sigma_{4,k}$ the summation over all $k$ which did not enter into $\Sigma_{1,k}$, $\Sigma_{2,k}$, or $\Sigma_{3,k}$.

Several times in the proof we shall bound the number $q$ from below and the number $\alpha$ from above. These restrictions are determined only by the properties of the region $g$. For "large" values of the number $\alpha$ an estimate of the integral may be given by means of the length of the boundary of the region $g$.

1. We estimate $\Sigma_{1,k}\left|\int_{\partial g_{k,\alpha}} f(z)dz\right|$. For any point $z \in g^*$ the number of $k$ for which $\rho(\mu_k^\alpha, z) \le q\alpha$, does not exceed $q_1 \cdot \lambda \cdot q^2$ (see property 4) in §2), where $q_1$ is an absolute constant. Therefore the number of terms in $\Sigma_{1,k}$ does not exceed $n \cdot q_1 \cdot \lambda \cdot q^2$.

We now estimate each of those terms. By Lemma 3.1

$$|f_k^\alpha(z)| \le 2\lambda\omega(\alpha)$$

and therefore

$$\left|\int_{\partial g_{k,\alpha}} f_k^\alpha(z)\,dz\right| \leqslant 2\lambda\omega(\alpha)\,h_1(\partial g_{k,\alpha}) \leqslant 2\lambda\omega(\alpha)\,[h_1(\mu_k^\alpha \cap \partial g) + h_1(g \cap \partial\mu_k^\alpha)].$$

Since $\mu_k^1$ is convex, $h_1(g \cap \partial\mu_k^\alpha) \le h_1(\partial\mu_k^\alpha) \le 2\pi\alpha$. We recall that for sets lying on a Ljapunov curve the analytic content and length are commensurable. Consequently

$$h_1(\mu_k^\alpha \cap \partial g) \leqslant C_4(g)\,\alpha(\partial g_{k,\alpha}) \leqslant C_4(g)\,\alpha(\mu_k^\alpha) \leqslant C_4(g)\,\alpha,$$

insofar as the diameter of the set $\mu_k^\alpha$ does not exceed $\alpha$ (see property 3) in §2). Thus

$$\left|\int_{\partial g_{k,\alpha}} f_k^\alpha(z)\,dz\right| \leqslant 2\lambda\omega(\alpha)\,[C_4(g)\,\alpha + 2\pi\alpha]$$

and therefore

$$\sum_{1,k}\left|\int_{\partial g_{k,\alpha}} f_k^\alpha(z)\,dz\right| \leqslant nq_1\lambda q^2\cdot 2\lambda\omega(\alpha)\,[C_4(g)\,\alpha + 2\pi\alpha] \leqslant C_5(g)\,\omega(\alpha)\cdot\alpha.$$

2. We estimate $\Sigma_{2,k}|\int_{\partial g_{k,\alpha}} f_k^\alpha(z)dz|$. Since $\mu_k^\alpha \cap \partial g = \varnothing$ and $f_k^\alpha(z) \in C(e \cap \mu_k^\alpha, 2\lambda\omega(\alpha))$ (see Lemma 3.1), we have

$$\int_{\partial g_{k,\alpha}} f_k^\alpha(z)\,dz = \int_{\partial\mu_k^\alpha} f_k^\alpha(z)\,dz = 2\pi i\gamma\,(e \cap \mu_k^\alpha, f_k^\alpha(z)),$$

and therefore from Lemma 3.1 we obtain

$$\left|\int_{\partial g_{k,\alpha}} f_k^\alpha(z)\,dz\right| \leqslant \alpha(e \cap \mu_k^\alpha)\,[\max_z |f_k^\alpha(z)|]^{-1} \leqslant 2\lambda\omega(\alpha)\,\alpha\,(e \cap \mu_k^\alpha).$$

Since $\Sigma_k\,\alpha(e \cap \mu_k^\alpha) \le \lambda_2\,\alpha(e)$ (see §4), we have

$$\sum_{2,k}\left|\int_{\partial g_{k,\alpha}} f_k^\alpha(z)\,dz\right| \leqslant \sum_k 2\lambda\omega(\alpha)\,\alpha\,(e \cap \mu_k^\alpha) \leqslant 2\lambda\lambda_2\omega(\alpha)\cdot\alpha = q_2\omega(\alpha)\cdot\alpha.$$

3. We estimate $\Sigma_{3,k}|\int_{\partial g_{k,\alpha}} f_k^\alpha(z)dz|$. Let $\mu_k^\alpha \cap \gamma_j \neq \varnothing$. Since $d_j$ has a continuously changing tangent and does not intersect itself, and since $\rho(\mu_k^\alpha, z_j' \cup z_j'') > q\alpha$ $(q \geq 1)$, we have $\mu_k^\alpha \cap d_j = \mu_k^\alpha \cap \gamma_j$ for sufficiently small $\alpha$, and consequently $\mu_k^\alpha \cap g = \mu_k^\alpha \cap g_j$. We emphasize that the smallness of $\alpha$ required here is determined only by the smoothness of the curves $\{d_j\}$, i.e. the smoothness of the arcs $\{\gamma_j\}$, and does not depend on the lengths of the arcs $\{d_j\}$.

From the condition $\mu_k^\alpha \cap g = \mu_k^\alpha \cap g_j$ it follows that

$$\int_{\partial g_{k,\alpha}} f_k^\alpha(z)\,dz = \int_{\partial(g_j \cap \mu_k^\alpha)} f_k^\alpha(z)\,dz.$$

Since $f_k^\alpha(z)$ is analytic outside $\mu_k^\alpha$ (Lemma 3.1), i.e. is analytic in the region $g_j\setminus\mu_k^\alpha$, we have $\int_{\partial(g_j\cap\mu_k^\alpha)} f_k^\alpha(z)dz = \int_{\partial g_j} f_k^\alpha(z)dz$. Therefore from Lemma 5.2 we get

$$\left|\int_{\partial g_j} f_k^\alpha(z)\,dz\right| \leqslant C_1(g_j)\max_z|f_k^\alpha(z)|\cdot\alpha(e\cap\mu_k^\alpha)\leqslant C_1(g_j)\cdot 2\lambda\omega(\alpha)\cdot\alpha(e\cap\mu_k^\alpha).$$

Consequently

$$\sum_{3,k}\left|\int_{\partial g_{k,\alpha}} f_k^\alpha(z)\,dz\right| \leqslant \max C_1(g_j)\,2\lambda\omega(\alpha)\sum_k\alpha(e\cap\mu_k^\alpha)\leqslant C_6(g)\,\omega(\alpha)\cdot\alpha,$$

as $\Sigma_k\,\alpha(e\cap\mu_k^\alpha)\le\lambda_2\alpha(e)$ (see §4).

The estimation of $\Sigma_{4,k}|\int_{\partial g_{k,\alpha}} f_k^\alpha(z)\,dz|$ will require a further lemma.

Lemma 8.1. *Let $g$ be a simply-connected region whose boundary consists of four smooth Ljapunov curves joined to one another; at corner points the argument of the tangent vector to the boundary of the region undergoes a jump of no more than $\pi/4$, and on each link of the boundary the fluctuation of the argument of the tangent does not exceed $\pi/4$. The diameter of each link of the boundary is no less than $\mu h_1(\partial g)$.*

*Then for every $e\subset\bar g$ and $f(z)\in C(g, e, m)$ we have $|\int_{\partial g} f(z)dz|\le C_7(g)m\alpha(e)$, where $C_7(g)$ is a constant depending only on $\mu$, the Hölder constant $L$ and exponent for the links of the boundary of the region $g$.*

The proof of this lemma follows from parts 1–3 of this section, as for small $\alpha(e)$ the sum $\Sigma_{4,k}$ does not contain a single term.

4. We estimate, finally, $\Sigma_{4,k}|\int_{\partial g_{k,\alpha}} f_k^\alpha(z)dz|$. We designate by $\sigma_i$ the union of all components of the set $\bar g\cap\mu_k^\alpha$ ($k$ fixed) which contain points of $\gamma_i$. If $\sigma_i$ does not contain points of $\partial g\setminus\gamma_i$, then, repeating the arguments set forth in part 3, it can be shown that $|\int_{\partial\sigma_i} f_k^\alpha(z)\,dz|\le C_6(g)\omega(\alpha)\cdot\alpha(e\cap\mu_k^\alpha)$.

Let $a_i\in\gamma_i\cap\sigma_i$ and $a_j\in\gamma_i\cap\sigma_j$ $(i\ne j)$. For convenience we shall assume that the straight line $y=0$ is tangent to $\gamma_i$ at the point $a_i$. From Lemma 6.3 it follows that if $q$ is sufficiently large and $\alpha$ small (the smallness of $\alpha$ is determined only by the smoothness of the curves $\{\gamma_s\}$), then it is possible to construct real functions $\phi_i(x)$ and $\phi_j(x)$ whose derivatives satisfy the Hölder condition on the interval $r'=\{4\alpha+a_i\ge x\ge a_i-4\alpha\}$, while their graphs lie on $\gamma_i$ and $\gamma_j$ and contain the points $a_i$ and $a_j$, respectively, i.e. $\gamma_i$ and $\gamma_j$ behave like single-valued functions in a neighborhood of $\mu_k^\alpha$. On the interval $r'$ the function $\phi(x)=|\phi_i(x)-\phi_j(x)|>0$. By virtue of Lemma 6.1 we may require that $|d\phi_j/dx|\le 1/8$ and $|d\phi_i/dx|\le 1/8$ on the interval $r'$. We designate by $\xi$ the region bounded by the curves

$\phi_i(x)$ and $\phi_j(x)$ and the straight lines $x = a_i \pm 4\alpha$. From Lemma 6.3 it follows that for small $\alpha$ there are no points of $\partial g$ inside $\xi$. This means that $\partial \gamma_i$ consists only of points of $\partial \mu_k^\alpha$ and points of the graphs of $\phi_i(x)$ and $\phi_j(x)$.

Since $f_k^\alpha(z)$ is analytic outside $e \cap \mu_k^\alpha$ (Lemma 3.1), we have $\int_{\partial\sigma_i} f_k^\alpha(z)\,dz = \int_{\partial\xi} f_k^\alpha(z)dz$. We fix points $x_1, \cdots, x_p$ on the axis $y = 0$ such that $x_1 = a_i - 4\alpha$, $x_p \geq a_i + 4\alpha$, $x_{p-1} \leq a_i + 4\alpha$, and $x_s = x_{s-1} + \phi(x_{s-1})$. We designate by $\xi_s$ the region bounded by the graphs of $\phi_i(x)$, $\phi_j(x)$ and the straight lines $x = x_s$ and $x = x_{s+1}$. By virtue of Lemma 8.1 we have

$$\left|\int_{\partial\xi_s} f_k^\alpha(z)\,dz\right| \leqslant C_7(\xi_s) \max_z |f_k^\alpha(z)| \cdot \alpha[e \cap \mu_k^\alpha \cap \xi_s]$$

$$\leqslant C_8(g)\,\omega(\alpha) \cdot \alpha[e \cap \mu_k^\alpha \cap \xi_s].$$

From Lemma 7.1 we have $\Sigma\alpha[e \cap \mu_k^\alpha \cap \xi_s] \leq C_9(g)\alpha(e \cap \sigma_i)$; consequently

$$\left|\int_{\sigma_i} f_k^\alpha(z)\,dz\right| \leqslant \sum_s \left|\int_{\partial\xi_s} f_k^\alpha(z)\,dz\right| \leqslant C_8(g)\,\omega(\alpha) \sum_s \alpha[e \cap \mu_k^\alpha \cap \xi_s]$$

$$\leqslant C_8(g)\,C_9(g)\,\omega(\alpha) \cdot \alpha(e \cap \sigma_i) \leqslant C_8(g)\,C_9(g)\,\omega(\alpha)\,\alpha(e \cap \mu_k^\alpha).$$

Thus for any $i$

$$\left|\int_{\partial\sigma_i} f_k^\alpha(z)\,dz\right| \leqslant [C_6(g) + C_8(g)\,C_9(g)]\,\omega(\alpha)\,\alpha(e \cap \mu_k^\alpha),$$

and since the index $i$ takes no more than $n$ values,

$$\left|\int_{\partial g_{k,\alpha}} f_k^\alpha(z)\,dz\right| \leqslant \sum_i \left|\int_{\partial\sigma_i} f_k^\alpha(z)\,dz\right| \leqslant n\,[C_6(g) + C_8(g)\,C_9(g)]\,\omega(\alpha)\,\alpha(e \cap \mu_k^\alpha)$$

$$\leqslant C_{10}(g)\,\omega(\alpha)\,\alpha(e \cap \mu_k^\alpha).$$

Since $\Sigma_k\, \alpha(e \cap \mu_k^\alpha) \leq \lambda_2\, \alpha(e)$ (see §4), we have

$$\sum_{4,k} \left|\int_{\partial g_{k,\alpha}} f_k^\alpha(z)\,dz\right| \leqslant \lambda_2 C_{10}(g)\,\omega(\alpha)\,\alpha(e).$$

Combining the results of parts 1–4, we obtain

$$\left|\int_{\partial g} f(z)\,dz\right| \leqslant \sum_{i=1}^{4} \sum_{i,k} \left|\int_{\partial g_{k,\alpha}} f_k^\alpha(z)\,dz\right| \leqslant C(g)\,\omega(\alpha)\,\alpha(e).$$

The theorem is proved.

## §9. Estimate of the Cauchy integral along a closed polygon

**Theorem 5.** *Let $g$ be a region whose boundary consists of $n$ straight line segments, $e$ a closed subset of $\bar{g}$, $f(z)$ a function in $C(g, e, m)$ with modulus of continuity $\omega(\delta)$. Then $|\int_{\partial g} f(z)dz| \leq Cn\omega(\alpha)\alpha(e)$, where $C$ is an absolute constant.*

Recalling the proof of Theorem 1, it is easy to see that in the case under consideration $C(g) = Cn$. Using the results of §8.1, it is possible to obtain the more precise estimate

$$\left| \int_{\partial g_{k,\alpha}} f_k^{\alpha}(z)\,dz \right| \leqslant 2\lambda\omega(\alpha)(n_k\alpha + 2\pi\alpha) + n\lambda_4\omega(\alpha)\,\alpha(e \cap \mu_k^{\alpha})$$

($\lambda_4$ is an absolute constant; $n_k$ is the number of points of the set $g^*$ no more than $q\alpha$ distant from $\mu_k^{\alpha}$), and therefore

$$\sum_{1,k} \left| \int_{\partial g_{k,\alpha}} f_k^{\alpha}(z)\,dz \right| \leqslant \lambda_5 n\omega(\alpha)\cdot\alpha,$$

where $\lambda_5$ is an absolute constant.

## BIBLIOGRAPHY

[1] L. D. Ivanov, *On Denjoy's conjecture*, Uspehi Mat. Nauk 18 (1963), no. 4 (112), 147–149. (Russian) MR 28 #236.

[2] A. G. Vituškin, *A problem of Denjoy*, Izv. Akad. Nauk SSSR Ser. Mat. 28 (1964), 745–756. (Russian) MR 29 #6024.

[3] M. S. Mel'nikov, *Analytic capacity and the Cauchy integral*, Dokl. Akad. Nauk SSSR 172 (1967), 26–29 = Soviet Math. Dokl. 8 (1967), 20–23. MR 35 #4382.

[4] A. G. Vituškin, *Approximation of a function by rational fractions*, Dokl. Akad. Nauk SSSR 171 (1966), 1023–1026 = Soviet Math. Dokl. 7 (1966), 1582–1585. MR 35 #390.

[5] ———, *Necessary and sufficient conditions on a set in order that any continuous function analytic at the interior points of the set may admit uniform approximation by rational fractions*, Dokl. Akad. Nauk SSSR 171 (1966), 1255–1258 = Soviet Math. Dokl. 7 (1966), 1622–1625. MR 35 #391.

[6] G. M. Goluzin, *Geometric theory of functions of a complex variable*, GITTL, Moscow, 1952; English transl., Transl. Math. Monographs, Amer. Math. Soc., Providence, R. I., (to appear). MR 15, 112.

[7] M. S. Mel'nikov, *Estimate of the Cauchy integral along an analytic curve*, Mat. Sb. 71 (113) (1966), 503–514; English transl., Amer. Math. Soc. Transl. (2) 80 (1968), 243–255. MR 34 #6120.

Translated by N. Koblitz